LONDON MATHEMATICAL SOCIETY LECTURE NOTE

Managing Editor: Professor N. J. Hitchin, Mathematical Institute,
University of Oxford, 24–29 St Giles, Oxford OX1 3LB, United Kingdom

CW00702573

The titles below are available from booksellers, or, from Cambridge University Pr

London Mathematical Society Lecture Note Series. 347

Surveys in Contemporary Mathematics

Edited by

NICHOLAS YOUNG
University of Leeds

and

YEMON CHOI
University of Manitoba

CAMBRIDGE
UNIVERSITY PRESS

CAMBRIDGE UNIVERSITY PRESS

Cambridge, New York, Melbourne, Madrid, Cape Town, Singapore, São Paulo, Delhi

Cambridge University Press
The Edinburgh Building, Cambridge CB2 8RU, UK

Published in the United States of America by Cambridge University Press, New York

www.cambridge.org

© Cambridge University Press 2008

First published 2008

Printed in the United Kingdom at the University Press, Cambridge

A catalogue record for this publication is available from the British Library

Library of Congress Cataloguing in Publication data

ISBN-13   978-0-521-70564-6 paperback

Contents

Preface

This is the second of two volumes that showcase young scientists who are continuing the outstanding tradition of Russian mathematics in their home country. There remain numerous strong research groups, particularly in Moscow and St. Petersburg, despite the familiar difficulties: academic salaries in Russia remain low, many leading figures have departed and there are plentiful opportunities for employment in university positions abroad or in sectors in Russia that offer a living wage. It is hoped that the articles in this book give a picture of the interests and achievements of mathematicians that participate in some of the active seminars in the country. Seven have something of the character of a survey, but also contain many original results and give extensive bibliographies; the eighth is a revised and expanded version of a 2002 research article.

The first of the two volumes (LMS Lecture Notes 338) was entitled *Surveys in Geometry and Number Theory*; this one is mainly on combinatorial and algebraic geometry and topology. Both volumes contain papers based on courses of lectures given at British universities by the authors under the 'Young Russian Mathematicians' scheme, which the London Mathematical Society set up to help such mathematicians visit the UK and to provide them with financial support.

In the nineties sheer subsistence was difficult for Russian academics. Over the last five years things have improved, and the salaries of university employees, though not generous, are closer to sufficing for the necessities of life. It still remains difficult for young scientists to get established in a career, and two of the contributors to this volume have since chosen other paths, one in the mathematical diaspora and one in industry.

Nicholas Young
Department of Pure Mathematics
Leeds University

Yemon Choi
Department of Mathematics
University of Manitoba.

Rank and determinant functions for matrices over semirings

Alexander E. Guterman

Introduction

The difference between semirings and rings is the lack of additive inverses in semirings. The most common examples of semirings which are not rings are the non-negative integers $\mathbb{Z}^+$, the non-negative rationals $\mathbb{Q}^+$ and the non-negative reals $\mathbb{R}^+$ with usual addition and multiplication. There are classical examples of non-numerical semirings as well. One of the first examples appeared in the work of Dedekind [29] in connection with the algebra of ideals of a commutative ring (one can add and multiply ideals but it is not possible to subtract them). Later Vandiver [62] proposed the class of semirings as the best class of algebraic structures which includes both rings and bounded distributive lattices. Boolean algebras, max-algebras, tropical semirings and fuzzy scalars are other important examples of semirings. See the monographs [37, 38, 44] for more details.

During the last few decades, matrices with entries from various semirings have attracted the attention of many researchers working both in theoretical and applied mathematics. It should be emphasized that the majority of examples of semirings arise in various applications of algebra. To illustrate an application of semirings, we present a situation that arises constantly in parallel computations. To make this situation more apparent, we give its 'real life' analogue. Let us consider k seaports. A ship with certain goods arrives at the i-th port at the moment x_i, $i = 1, \ldots, k$. Also there are l airports with a airplane in each airport. Each airplane has to depart at the time b_j, $j = 1, \ldots, l$. All goods are delivered by trains from the sea-ports to the airports. Let $t_{i,j}$ denote the travel time from the i-th port to the j-th airport. The problem is to find x_i, $i = 1, \ldots, k$, (or certain relations between them) in such a way that all goods coming on ships will be delivered further by airplanes on the same day and in optimal time, if b_j, $j = 1, \ldots, l$, (or certain relations between them) are known.

This problem is not linear in its original formulation since maximum is not a linear function. However, it can be formulated as a linear problem using the semiring $(\mathbb{R}, \max, +)$. Namely, given a $l \times k$ matrix $T = [t_{i,j}]$ and a vector $\mathbf{b} = [b_1, \ldots, b_l]^t$, we need to find a vector $\mathbf{x} = [x_1, \ldots, x_k]^t$ such that $T\mathbf{x} = \mathbf{b}$ (or in some cases $T\mathbf{x} \leq \mathbf{b}$).

Different analogues of classical matrix invariants in the semiring context have been investigated and applied actively. In the present paper we survey several possible definitions of ranks and determinants for matrices over semirings. The relationships between these notions are discussed, and the arithmetic behaviour of determinant and rank functions for matrices over semirings is investigated. In particular, we provide sharp estimates for ranks of sums and products of matrices in the context of semirings. Semiring analogues of the Dieudonné and Frobenius theorems on linear transformations preserving matrix invariants are given.

Other applications of semirings include: automata theory [25, 33]; optimization theory [6, 22, 26, 48, 60]; optimal control [3, 23]; discrete event systems [4]; operational research [3]; ergodic control [1, 2]; mathematical economics [1]; the assignment problem [21, 22]; graph theory [22, 40]; and algebraic geometry [31, 53, 63]. For a detailed index of applications see [38, pp. 355-356].

1 Preliminaries

Definition 1.1. A *semiring* $(\mathcal{S}, +, \cdot)$ is a set $\mathcal{S}$ with two binary operations, addition $(+)$ and multiplication $(\cdot)$, such that:

- $\mathcal{S}$ is an Abelian monoid under addition (the identity element is denoted by 0);
- $\mathcal{S}$ is a semigroup under multiplication (the identity element, if it exists, is denoted by 1);
- multiplication is distributive over addition on both sides;
- $s0 = 0s = 0$ for all $s \in \mathcal{S}$.

In this paper we assume that there exists a multiplicative identity 1 in $\mathcal{S}$ which is different from 0.

Definition 1.2. A semiring $\mathcal{S}$ is called *commutative* if the multiplication in $\mathcal{S}$ is commutative.

Definition 1.3. A semiring S is called *antinegative* (or *zero-sum-free*) if $a + b = 0$ implies that $a = b = 0$.

In other words the zero element is the only element with an additive inverse.

Definition 1.4. We say that a semiring S *does not have zero divisors* if from $ab = 0$ in S it follows that either $a = 0$ or $b = 0$.

Definition 1.5. A semiring S is called *Boolean* if S is isomorphic to a certain set of subsets of a given set M where the sum of two subsets is their union and the product is their intersection. The zero element is the empty set and the identity element is the whole set M.

It can be easily seen that a Boolean semiring is commutative and antinegative. If a Boolean semiring S consists of only two subsets of M, namely the empty subset and M, then it is called a *binary Boolean semiring* (or $\{0, 1\}$-*semiring*) and is denoted by **B**.

Definition 1.6. A semiring S is called a *chain* if S is a totally ordered set with universal lower and upper bounds, where addition and multiplication are defined by $a + b = \max\{a, b\}$ and $a \cdot b = \min\{a, b\}$.

Note that that any chain semiring S can be represented as a Boolean algebra of subsets of S by sending each element $x \in S$ to the associated negative cone $\{y \in S \mid y \leq x\}$.

2 Semimodules, bases and dimension

Semimodules over semirings are analogues of vector spaces over fields and modules over rings. The precise definition is as follows.

Definition 2.1. Given a semiring S, we define a *left semimodule* U over S to be an Abelian monoid with neutral element $\mathbf{0}$, equipped with a function

$$S \times U \to U \quad , \quad (s, \mathbf{u}) \to s\mathbf{u}$$

called *scalar multiplication* such that for all $\mathbf{u}$ and $\mathbf{v}$ in U and $r, s \in S$

    (i) $(sr)\mathbf{u} = s(r\mathbf{u})$,
    (ii) $(s + r)\mathbf{u} = s\mathbf{u} + r\mathbf{u}$,
    (iii) $s(\mathbf{u} + \mathbf{v}) = s\mathbf{u} + s\mathbf{v}$,
    (iv) $1\mathbf{u} = \mathbf{u}$,
    (v) $s\mathbf{0} = \mathbf{0} = 0\mathbf{u}$.

Right semimodules and bi-semimodules can be defined in a similar way.

Definition 2.2. Let U be a semimodule and let $P \subseteq U$ be a nonempty subset. An element $\mathbf{u}$ in U is called a *left* (respectively, *right*) *linear combination of elements from* P if there exists $k \in \mathbb{N}$, $s_1, \ldots, s_k \in \mathcal{S}$, $\mathbf{u}_1, \ldots, \mathbf{u}_k \in P$ such that $\mathbf{u} = \sum_{i=1}^{k} s_i \mathbf{u}_i$ (respectively, $\mathbf{u} = \sum_{i=1}^{k} \mathbf{u}_i s_i$).

Definition 2.3. The *left* (respectively, *right*) *linear span* $\langle P \rangle_{\mathcal{S}}$ of the set P is the set of all left (respectively, right) linear combinations of elements from P with coefficients from $\mathcal{S}$. We say that the set P *generates* a subset $V \subseteq U$ if $V \subseteq \langle P \rangle_{\mathcal{S}}$.

Note that all linear combinations that we consider are finite and nonempty by Definition 2.2.

Definition 2.4 ([19, 27, 55]). A subset P of elements in a semimodule U over a semiring $\mathcal{S}$ is called *left* (respectively, *right*) *linearly independent* if there is no element in P that can be expressed as a left (respectively, right) linear combination of other elements of P with coefficients in $\mathcal{S}$. The subset P is called *linearly dependent* if it is not linearly independent.

Note that in contrast with vector spaces over fields, there are several ways to define the notion of independence for semimodules. For example, in [1, 36] the following definition is used.

Definition 2.5. A system of elements, P, in a semimodule U is *left* (respectively, *right*) *linearly dependent* if there are two left (respectively, right) linear combinations of elements of P which are equal to each other. A system is called *linearly independent* if it is not linearly dependent.

Definition 2.5 is stronger than Definition 2.4. That is, any system of elements which is independent in the sense of Definition 2.5 is also independent in the sense of Definition 2.4. The converse is not always true, as is shown by the following example.

Example 2.6. The system $\{3, 5\}$ of 1-vectors over $\mathbb{Z}$ is linearly independent in the sense of Definition 2.4 (for any $\alpha, \beta \in \mathbb{Z}$ neither $3\alpha = 5$ nor $5\beta = 3$). However, it is linearly dependent in the sense of Definition 2.5 since $3 \cdot 5 = 5 \cdot 3$.

Definition 2.7. Let U be a semimodule and let B be a collection of elements of U which are left (respectively, right) linearly independent in the sense of Definition 2.4. The set B is called a *left* (respectively, *right*) *basis* of the semimodule U if the left (respectively, right) linear span of B is equal to U. The *left* (respectively, *right*) *dimension* of U is the minimal number of elements in any left (respectively, right) basis of U.

From now on we will consider only left linear independence (in the sense of Definition 2.4) and left dimension. However, all our considerations will be valid for right linear independence and right dimension as well.

3 Rank functions

The theory of matrices over semirings has been an object of intense study over the last few decades: see for example the monographs [37, 47] and the references therein. We first introduce some notation.

Let $\mathcal{M}_{m,n}(\mathcal{S})$ denote the set of $m \times n$ matrices with entries in the semiring $\mathcal{S}$: we write $\mathcal{S}^n$ for $\mathcal{M}_{n,1}(\mathcal{S})$ and $\mathcal{M}_n(\mathcal{S})$ for $\mathcal{M}_{n,n}(\mathcal{S})$. The set $\mathcal{M}_n(\mathcal{S})$ is a semiring under the standard matrix addition and multiplication.

Let I_n denote the $n \times n$ identity matrix, let $J_{m,n}$ denote the $m \times n$ matrix with all entries equal to 1, and let $O_{m,n}$ be the $m \times n$ zero matrix. We omit the subscripts if the size is clear from the context, writing I, J and O respectively.

The matrix $E_{i,j}$ denotes the matrix with (i,j) entry equal to 1 and the other entries equal to zero. $C_i = \sum_{j=1}^{m} E_{i,j}$ is the ith column matrix, $R_j = \sum_{i=1}^{n} E_{i,j}$ is the jth row matrix. We denote by $\text{diag}(a_1, \ldots, a_n)$ the diagonal matrix $\sum_{i=1}^{n} a_i E_{i,i} \in \mathcal{M}_n(\mathcal{S})$ with $a_1, \ldots, a_n \in \mathcal{S}$ on the main diagonal.

Finally. for $A \in \mathcal{M}_{m,n}(\mathcal{S})$ let $A^t \in \mathcal{M}_{n,m}(\mathcal{S})$ be the transposed matrix of A.

Definition 3.1. For $A, B \in \mathcal{M}_{m,n}(\mathcal{S})$, we say that A *dominates* B, $A \geq B$, if and only if $b_{i,j} \neq 0$ implies that $a_{i,j} \neq 0$.

Definition 3.2. Let $A = [a_{i,j}]$, $B = [b_{i,j}] \in \mathcal{M}_{m,n}(\mathcal{S})$ be such that $A \geq B$. Then we denote by $A \backslash B$ the matrix $C = [c_{i,j}] \in \mathcal{M}_{m,n}(\mathcal{S})$ such that

$$c_{i,j} = \begin{cases} 0 & \text{if } b_{i,j} \neq 0 \\ a_{i,j} & \text{otherwise.} \end{cases}$$

Many authors have investigated various rank functions for matrices over semirings and their properties, see [15, 16, 17, 42, 59, 64]. It is well known that the concept of matrix rank over a field has a geometric interpretation as the dimension of the image space of the corresponding linear transformation. The situation is more complicated over semirings; this geometric approach leads to some surprising properties of the corresponding rank function. For example, it may occur that the rank of a submatrix is greater than the rank of a matrix. The reason is that the dimension of a subsemimodule may exceed the dimension of the whole semimodule.

Example 3.3 ([10]). Let S be an arbitrary antinegative semiring without zero divisors. Consider the semimodule U_4 over S generated by the vectors $\mathbf{u}_1 = [0, 1, 0]^t$, $\mathbf{u}_2 = [0, 0, 1]^t$, $\mathbf{u}_3 = [1, 0, 1]^t$, $\mathbf{u}_4 = [1, 1, 0]^t$. Then $\dim U_4 = 4$, but U_4 is a proper subsemimodule of the 3-dimensional S-semimodule S^3.

Indeed, it is easy to see that none of u_1, u_2, u_3, u_4 is a linear combination of the others. The reason why U_4 has no bases of fewer than 4 elements is that $[1, 0, 0]^t$ cannot be expressed as a linear combination of vectors from U_4 due to antinegativity.

We may easily construct the following matrix example based on Example 3.3.

Example 3.4. Let S be any antinegative semiring without zero divisors. Consider

$$Y = \begin{bmatrix} 1 & 0 & 0 & 1 & 1 \\ 0 & 1 & 0 & 0 & 1 \\ 0 & 0 & 1 & 1 & 0 \end{bmatrix}$$

and its proper submatrix

$$X = \begin{bmatrix} 0 & 0 & 1 & 1 \\ 1 & 0 & 0 & 1 \\ 0 & 1 & 1 & 0 \end{bmatrix}.$$

The image of the linear operator corresponding to Y is the whole space S^3, so its dimension is 3. However, the dimension of the image of the linear operator corresponding to X is 4, by Example 3.3.

3.1 Definitions

It turns out that there are many essentially different rank functions for matrices over semirings. The geometric approach leads to the following definitions based on the notion of linear independence introduced in Definition 2.4.

Definition 3.5. Let $A \in \mathcal{M}_{m,n}(\mathcal{S})$. The *row rank* $r(A)$ is defined to be the dimension of the linear span of the rows of A. The *column rank* $c(A)$ is defined to be the dimension of the linear span of the columns of A.

It is proved in [17] that $r(B) \leq r(A)$ if B is obtained by deleting some columns of A, and correspondingly $c(B) \leq c(A)$ if B is obtained by deleting some rows of A. Example 3.4 shows, however, that in general the rank of a submatrix can be greater than the rank of a matrix.

Let us consider one of the most important semiring rank functions.

Definition 3.6. A matrix $A \in \mathcal{M}_{m,n}(\mathcal{S})$ is of *factor rank* k if there exist matrices $B \in \mathcal{M}_{m,k}(\mathcal{S})$ and $C \in \mathcal{M}_{k,n}(\mathcal{S})$ with $A = BC$ and k is the smallest positive integer for which such a factorization exists. By definition the only matrix with factor rank zero is the zero matrix O.

The factor rank of A is denoted by $\mathrm{rank}(A)$.

Note that the factor rank of A is equal to the minimum number of matrices of factor rank 1 with sum equal to A (see the proof in [9, 24]). Also, for any submatrix B of A we have $\mathrm{rank}(B) \leq \mathrm{rank}(A)$, see [17].

The notion of factor rank is important in various applications. It was used in [45] for demographic investigations, in [65] for combinatorial optimization problems, and in statistics (see [24] for details).

Definition 3.7. Let $\mathcal{S}$ be a subsemiring of a certain field or division ring. For a matrix $A \in \mathcal{M}_{m,n}(\mathcal{S})$ we define $\rho(A)$ to be the dimension of the linear span of rows of A over this field or division ring.

It turns out that the values of all aforementioned rank functions may be different for a matrix over a semiring.

Example 3.8 ([55]). Let

$$A = \begin{bmatrix} 2 & 3 & 0 & 0 & 0 & 1 \\ 0 & 0 & 2 & 2 & 2 & 2 \\ 0 & 0 & 2 & 4 & 0 & 2 \\ 0 & 0 & 2 & 0 & 4 & 2 \\ 2 & 3 & 1 & 0 & 2 & 2 \end{bmatrix} \in \mathcal{M}_{5,6}(\mathbb{Z}_+).$$

Then $\rho(A) = 3, \mathrm{rank}(A) = 4, r(A) = 5, c(A) = 6$.

Note that the basis of a linear span of a set of elements need not lie in this set, see Examples 3.19 and 3.20 below. Therefore we consider also the following definitions of row and column ranks.

Definition 3.9. Let $A \in \mathcal{M}_{m,n}(\mathcal{S})$. The *spanning row rank* $sr(A)$ is the minimum number of rows that span all rows of A, and the *spanning column rank* $sc(A)$ is the minimum number of columns that span all columns of A.

In addition, it appears that over semirings there is no correspondence between minimal spanning and maximal linearly independent subsets. Thus we obtain another pair of row and column ranks.

Definition 3.10. A matrix $A \in \mathcal{M}_{m,n}(\mathcal{S})$ is said to be of *maximal row rank* k (written as $mr(A) = k$) if it has k linearly independent rows and any $(k + 1)$ rows are linearly dependent.

Similarly, A is said to be of *maximal column rank* k (written as $mc(A) = k$) if it has k linearly independent columns and any $(k + 1)$ columns are linearly dependent.

Clearly, all the rank functions introduced above coincide for matrices over fields. However, they are essentially different for matrices over semirings. In the following sections we provide some examples to illustrate this difference and investigate the relationship between these functions.

Now we consider the so-called *combinatorial ranks*, which are useful in graph theory, transversal theory and communication networks (see [20, 38, 58] and references therein). Note that these ranks do not coincide with the usual rank function even if $\mathcal{S}$ is a field.

Definition 3.11. A *line* of a matrix is a row or a column of this matrix.

Definition 3.12. Let $A \in \mathcal{M}_{m,n}(\mathcal{S})$. The *term rank* $t(A)$ is defined to be the minimum number of lines needed to include all nonzero elements of A.

We shall denote by $t_c(A)$ the minimum number of columns needed to include all nonzero elements of A and by $t_r(A)$ the minimum number of rows needed to include all nonzero elements of A.

Definition 3.13. A *generalized diagonal* of a matrix $A \in \mathcal{M}_{m,n}(\mathcal{S})$ is a set of $\min\{m, n\}$ positions in A such that no row or column contains two of these positions.

Proposition 3.14 (König Theorem, cf. [20, Theorem 1.2.1]). *Let $\mathcal{S}$ be* a semiring, $A \in \mathcal{M}_{m,n}(\mathcal{S})$. Then the term rank of A is the maximum number of nonzero entries in some generalized diagonal of A.

The term rank can also be characterized in terms of zero submatrices of maximal size, as follows.

Proposition 3.15 ([54]). *Let $\mathcal{S}$ be a semiring, $A \in \mathcal{M}_{m,n}(\mathcal{S})$. Then $t(A) = k$ if and only if there exist positive integers s, r, $s + r = n + m - k$, such that a certain permutation of rows and columns of A contains a submatrix $O_{r,s}$ and no permutation of rows and columns of A contains $O_{p,q}$ if $p + q > n + m - k$.*

The following 'dual' matrix invariant is also useful.

Definition 3.16. Let $A \in \mathcal{M}_{m,n}(\mathcal{S})$. The *zero-term rank* $z(A)$ is the minimum number of lines needed to include all zero elements of A.

Note that some of the rank functions discussed above, especially the factor rank, have been rediscovered several times by many researchers and are known in the literature under different names. It should be also pointed out that the list of rank functions introduced here is not complete. We have presented here only the most common notions. There are other natural ways to define rank functions over semirings, see for example [5, 30, 38, 37], but we plan to discuss these functions and interrelations between them elsewhere.

3.2 Relationships between different rank functions

As was already mentioned, if the semiring $\mathcal{S}$ is a field then

$$\rho(A) = \text{rank}(A) = r(A) = c(A) = sr(A) = sc(A) = mr(A) = mc(A)$$

for any matrix $A \in \mathcal{M}_{m,n}(\mathcal{S})$. The situation is quite different for matrices over general semirings. The only relations that hold for an arbitrary semiring are the following inequalities:

Proposition 3.17 ([9]). *Let $A \in \mathcal{M}_{m,n}(\mathcal{S})$. Then:*

   (i) $\text{rank}(A) \leq \min\{r(A), c(A)\}$;
   (ii) $r(A) \leq sr(A) \leq mr(A) \leq m$ *and* $c(A) \leq sc(A) \leq mc(A) \leq n$;
   (iii) $\text{rank}(A) \leq t(A)$;
   (iv) *if $\mathcal{S}$ is a subsemiring of a field then* $\rho(A) \leq \text{rank}(A)$.

In [15] the authors investigate how the column rank and the factor rank of matrices over a certain algebraic system depend on this system. In [17] the authors consider the sets of numbers r such that *every* A in $\mathcal{M}_{m,n}(\mathcal{S})$ with $c(A) = r$ has rank$(A) = r$. They find upper bounds of such sets for different classes of semirings and establish that these upper bounds depend considerably on the semiring $\mathcal{S}$.

In contrast to the situation with rank functions over fields, row and column ranks may not coincide even over commutative semirings.

Example 3.18. Let $\mathcal{S}$ be any antinegative semiring without zero divisors. We consider the matrix X defined in Example 3.4. It follows from Example 3.3 and Proposition 3.17(ii) that $c(X) = sc(X) = mc(X) = 4$ whereas $r(X) = sr(X) = mr(X) = 3$.

Actually, the maximal column and row ranks may exceed the respective spanning ranks, and each of these in turn may exceed the column and row rank, as the following examples show.

Example 3.19 ([59]). Consider $A = [3 - \sqrt{7}, \sqrt{7} - 2] \in \mathcal{M}_{1,2}(\mathbb{Z}[\sqrt{7}]^+)$. We have $sc(A) = 2$ since $3 - \sqrt{7} \neq \alpha(\sqrt{7} - 2)$ and $\alpha(3 - \sqrt{7}) \neq \sqrt{7} - 2$ in $\mathbb{Z}[\sqrt{7}]^+$. However, $c(A) = 1$ since $1 = (3 - \sqrt{7}) + (\sqrt{7} - 2)$ generates the column space of A.

Example 3.20 ([9]). Consider $A = [4 - \sqrt{7}, \sqrt{7} - 2, 1] \in \mathcal{M}_{1,3}(\mathbb{Z}[\sqrt{7}]^+)$. We have $sc(A) = 1$, since 1 spans all columns of A. However, in similar fashion to the previous example, one can see that $mc(A) = 2$.

All above examples are given for matrices over antinegative semirings. Below we shall present examples showing that row and column ranks may be different for matrices over commutative rings as well.

The following example shows that spanning column rank may be greater than column rank and can even be greater than maximal row rank for matrices over several commutative rings.

Example 3.21. Let

$$A = \begin{bmatrix} 2 & 3 \\ 2 & 3 \end{bmatrix}$$

be considered as a matrix with the entries either from the ring $\mathbb{Z}$ of integers, or from the ring $\mathbb{Z}_6$ of integers modulo 6. In both cases it is easy to see that $sc(A) = 2$ but $c(A) = mr(A) = 1$.

The following example shows that maximal column rank may be

greater than spanning column rank and can be greater than maximal row rank for matrices over several commutative rings.

Example 3.22. Let

$$A = \begin{bmatrix} 1 & 2 & 3 \\ 1 & 2 & 3 \end{bmatrix} \in M_{2,3}(\mathbb{Z}).$$

Then $mc(A) = 2$ but $sc(A) = mr(A) = 1$.

The next example shows that in general $\max\{r(A), c(A)\} \not\leq t(A)$.

Example 3.23. Let X be the matrix from Example 3.4: then $c(X) = 4$ and $r(X) = t(X) = 3$.

The inequality $\min\{r(A), c(A)\} \leq t(A)$ does not hold over $\mathbb{Z}^+$.

Example 3.24 ([9]). For

$$A = \begin{bmatrix} 3 & 5 & 7 \\ 5 & 0 & 0 \\ 7 & 0 & 0 \end{bmatrix} \in M_{3,3}(\mathbb{Z}^+)$$

one has $r(A) = c(A) = 3$, $t(A) = 2$.

Remark 3.25. It is straightforward to see that if $S = \mathbf{B}$ is a binary Boolean semiring then $z(A) = t(J \setminus A)$ for any $A \in M_{m,n}(S)$.

3.3 Arithmetic behaviour of rank

The behaviour of the usual rank function ρ over fields with respect to matrix multiplication and addition is illustrated by the following classical inequalities.

- *Bounds for the rank of a sum:*

$$| \rho(A) - \rho(B) | \leq \rho(A + B) \leq \rho(A) + \rho(B). \tag{3.1}$$

- *Sylvester's laws:*

$$\rho(A) + \rho(B) - n \leq \rho(AB) \leq \min\{\rho(A), \rho(B)\}. \tag{3.2}$$

- *The Frobenius inequality:*

$$\rho(AB) + \rho(BC) \leq \rho(ABC) + \rho(B). \tag{3.3}$$

Here A, B, C are compatible matrices with entries from a field.

Following [9], we investigate the behaviour of different rank functions over semirings with respect to matrix addition and multiplication. It turns out that arithmetic properties of rank functions over a semiring depend considerably on the structure of the semiring.

Definition 3.26. We say that an inequality is *exact* if there is a substitution of variables such that equality holds. We say that an inequality involving rank functions and matrix variables is *best possible* if, for any given values of these rank functions, one can substitute matrices with these values of ranks so that equality holds.

The notion of a 'best possible' inequality for given rank functions will become clearer in the examples to follow: see e.g. Theorem 3.32.

Inequalities for the rank of a sum

Let us show that the lower bound in the inequality (3.1) is not valid for the factor rank over an arbitrary semiring.

Example 3.27 ([9]). Let S be a Boolean semiring. Then

$$\text{rank}(A + B) \not\geq |\text{rank}(A) - \text{rank}(B)|.$$

To be specific, consider

$$A = K_7 = \begin{bmatrix} 0 & 1 & 1 & 1 & 1 & 1 & 1 \\ 1 & 0 & 1 & 1 & 1 & 1 & 1 \\ 1 & 1 & 0 & 1 & 1 & 1 & 1 \\ 1 & 1 & 1 & 0 & 1 & 1 & 1 \\ 1 & 1 & 1 & 1 & 0 & 1 & 1 \\ 1 & 1 & 1 & 1 & 1 & 0 & 1 \\ 1 & 1 & 1 & 1 & 1 & 1 & 0 \end{bmatrix}$$

and $B = I_7$. Then $1 = \text{rank}(J_7) = \text{rank}(A + B)$, but over any Boolean semiring $\text{rank}(K_7) \leq 5$ due to the factorization

$$K_7 = \begin{bmatrix} 1 & 0 & 1 & 0 & 1 \\ 1 & 1 & 0 & 0 & 0 \\ 0 & 1 & 0 & 1 & 1 \\ 0 & 1 & 1 & 1 & 0 \\ 0 & 1 & 1 & 0 & 1 \\ 0 & 0 & 1 & 1 & 1 \\ 1 & 0 & 1 & 1 & 0 \end{bmatrix} \begin{bmatrix} 0 & 0 & 1 & 1 & 1 & 1 & 0 \\ 1 & 0 & 0 & 0 & 0 & 1 & 1 \\ 0 & 1 & 1 & 0 & 0 & 0 & 0 \\ 1 & 1 & 0 & 0 & 1 & 0 & 0 \\ 0 & 0 & 0 & 1 & 0 & 0 & 1 \end{bmatrix}$$

(see [28] for more details). Thus

$$|\operatorname{rank}(A) - \operatorname{rank}(B)| = \operatorname{rank}(B) - \operatorname{rank}(A) \geq 7 - 5 = 2 > 1 = \operatorname{rank}(A + B).$$

However, the following bounds are true.

Proposition 3.28 ([9]). *Let $\mathcal{S}$ be an antinegative semiring and let $A, B \in \mathcal{M}_{m,n}(\mathcal{S})$. Then:*

(i) $\operatorname{rank}(A + B) \leq \min\{\operatorname{rank}(A) + \operatorname{rank}(B), m, n\};$

(ii) $\operatorname{rank}(A + B) \geq \begin{cases} \operatorname{rank}(A) & \text{if} \quad B = O \\ \operatorname{rank}(B) & \text{if} \quad A = O \\ 1 & \text{if} \quad A \neq O \text{ and } B \neq O. \end{cases}$

These bounds are exact: the upper bound is best possible and the lower bound is the best possible over Boolean semirings.

Even in the case when $\mathcal{S}$ is a subsemiring of $\mathbb{R}^+$, the lower bound in (3.1) is not valid for the factor rank as the following example shows.

Example 3.29 ([9]). Let $r, s \geq 4$ and $s < n - 4$. Let us consider

$$A' = \begin{bmatrix} 1 & 2 & 3 & 4 \\ 1 & 1 & 1 & 1 \\ 1 & 0 & 1 & 0 \\ 0 & 0 & 2 & 2 \end{bmatrix} \quad \text{and} \quad B' = \begin{bmatrix} 0 & 0 & 0 & 0 \\ 0 & 0 & 0 & 0 \\ 0 & 1 & 0 & 1 \\ 0 & 1 & 0 & 1 \end{bmatrix}.$$

Note that $\operatorname{rank}(A') = 4$, $\rho(A') = 3$, $\operatorname{rank}(B') = \rho(B') = 1$ and

$$\operatorname{rank}(A' + B') = \rho(A' + B') = 2.$$

Let

$$A = \begin{bmatrix} A' & O_{4,r-4} & O_{4,n-r} \\ O_{r-4,4} & L_{r-4} & O_{r-4,n-r} \\ O_{m-r,4} & O_{m-r,r-4} & O_{m-r,n-r} \end{bmatrix}$$

and

$$B = \begin{bmatrix} B' & O_{4,1} & O_{4,s-1} & O_{4,n-s-4} \\ O_{s-1,4} & O_{s-1,1} & U_{s-1} & O_{s-1,n-s-4} \\ O_{m-s-3,4} & O_{m-s-3,1} & O_{m-s-3,s-1} & O_{m-s-3,n-s-4} \end{bmatrix},$$

where

$$U_k = \sum_{1 \leq i \leq j \leq k} E_{i,j} \quad \text{and} \quad L_k = \sum_{1 \leq j < i \leq k} E_{i,j} \in \mathcal{M}_k(\mathcal{S}).$$

Then $\mathrm{rank}(A) = r$, $\rho(A) = r - 1$, $\mathrm{rank}(B) = \rho(B) = s$ and

$$\mathrm{rank}(A + B) = |\, r - s \,| -1 \quad < \quad |\,\mathrm{rank}(A) - \mathrm{rank}(B)\,|$$

if S is a subsemiring of $\mathbb{R}^+$. In order to achieve the same result in the case $r = s + 3$, it is necessary to exchange the blocks A' and B' in the matrices A and B.

However, the following is true.

Proposition 3.30 ([9]). *Let $S \subseteq \mathbb{R}^+$, $A, B \in \mathcal{M}_{m,n}(S)$. Then*

$$\mathrm{rank}(A + B) \geq |\rho(A) - \rho(B)|.$$

This bound is exact and best possible.

The following example shows that an analogue of the upper bound in (3.1) is not valid for row and column ranks.

Example 3.31. Consider

$$A = \begin{bmatrix} 1 & 0 & 0 \\ 0 & 1 & 0 \\ 1 & 1 & 0 \\ 1 & 0 & 0 \\ 0 & 1 & 0 \end{bmatrix}, \quad B = \begin{bmatrix} 0 & 2 & 2 \\ 0 & 2 & 0 \\ 0 & 0 & 2 \\ 0 & 6 & 2 \\ 0 & 4 & 6 \end{bmatrix}.$$

It is easy to see that

$$r(A) = r(B) = sr(A) = sr(B) = mr(A) = mr(B) = 2.$$

However, over $\mathbb{Z}_+$ we have

$$r(A + B) = r \begin{bmatrix} 1 & 2 & 2 \\ 0 & 3 & 0 \\ 1 & 1 & 2 \\ 1 & 6 & 2 \\ 0 & 5 & 6 \end{bmatrix} = 5 = sr(A + B) = mr(A + B)$$

Proposition 3.32 ([9]). *Let S be an antinegative semiring. Then for $O \neq A, B \in \mathcal{M}_{m,n}(S)$ we have*

$$1 \leq c(A + B), r(A + B), sr(A + B), sc(A + B), mr(A + B), mc(A + B).$$

These bounds are exact over any antinegative semiring and best possible over Boolean semirings.

Proof. We are going to show that the lower bound is the best possible. Let S be a Boolean semiring. For each pair (r, s), $0 \leq r, s \leq m$ we consider the matrices A_r, B_s given by

$$A_r = J \setminus \left(\sum_{i=1}^{r} E_{i,i} \right)$$

and

$$B_s = \begin{cases} J \setminus \left(\sum_{i=1}^{s} E_{i,i+1} \right) & \text{if } s < m \\ J \setminus \left(\sum_{i=1}^{s-1} E_{i,i+1} \right) + E_{s,1} & \text{if } s = m. \end{cases}$$

Then

$$c(A_r) = r(A_r) = sr(A_r) = sc(A_r) = mr(A_r) = mc(A_r) = r,$$

$$c(B_s) = r(B_s) = sr(B_s) = sc(B_s) = mr(B_s) = mc(B_s) = s,$$

and $A_r + B_s = J$ has row and column ranks equal to 1. Thus, these bounds are the best possible for matrices with Boolean entries.

The rest of the proof follows directly from the definitions. $\qquad\square$

Proposition 3.33 ([9]). *Let S be a subsemiring in $\mathbb{R}^+$. Then for all $A, B \in \mathcal{M}_{m,n}(S)$ we have*

$$\left. \begin{array}{ccc} c(A+B), & r(A+B), & sr(A+B), \\ sc(A+B), & mr(A+B), & mc(A+B) \end{array} \right\} \geq |\rho(A) - \rho(B)|.$$

These bounds are exact and best possible.

Now we turn to the term rank and the zero term rank. The following inequalities hold:

Proposition 3.34 ([9]). *Let S be an arbitrary semiring. For all matrices $A, B \in \mathcal{M}_{m,n}(S)$ we have:*

$$t(A+B) \leq \min\{t(A) + t(B), m, n\}.$$

This bound is exact and best possible.

The following example shows that no analogue of the lower bound in (3.1) can hold for the term rank over arbitrary semirings.

Example 3.35 ([9]). Let $A = B = J_{m,n}$ over a field of characteristic 2. Then $t(A + B) = t(O) = 0$.

However, for antinegative semirings there is a lower bound which is even better than the lower bound presented in (3.1). Namely, the following is true.

Proposition 3.36 ([9]). *Let S be an antinegative semiring. For any matrices $A, B \in \mathcal{M}_{m,n}(S)$ the following inequality holds:*

$$t(A + B) \geq \max\{t(A), t(B)\}.$$

This bound is exact and the best possible.

Proposition 3.37 ([9]). *Let S be an antinegative semiring. For all $A, B \in \mathcal{M}_{m,n}(S)$, we have*

$$0 \leq z(A + B) \leq \min\{z(A), z(B)\}$$

These bounds are exact and best possible.

Proof. Both bounds follow directly from the definition of the zero-term rank function. In order to check that the lower bound is exact and best possible, for each pair (r, s), $0 \leq r, s \leq \min\{m, n\}$ we consider the matrices A_r, B_s given by

$$A_r = J \setminus \left(\sum_{i=1}^{r} E_{i,i} \right)$$

and

$$B_s = \begin{cases} J \setminus \left(\displaystyle\sum_{i=1}^{s} E_{i,i+1} \right) & \text{if } s < \min\{m, n\} \\ J \setminus \left(\displaystyle\sum_{i=1}^{s-1} E_{i,i+1} \right) + E_{s,1} & \text{if } s = \min m, n. \end{cases}$$

Then $z(A_r) = r$, $z(B_s) = s$ by Definition 3.16 and $z(A_r + B_s) = 0$ by antinegativity. Similarly, it can be checked that the upper bound is exact and best possible. $\qquad\square$

Sylvester's inequalities

Firstly, we demonstrate that the analogue of Sylvester's lower bound (3.2) does not hold for the factor rank.

Example 3.38. Let S be a Boolean semiring, let

$$A = \begin{bmatrix} 1 & 0 & 0 & \cdots & 0 \\ 1 & 1 & 0 & \cdots & 0 \\ 1 & 0 & 1 & \cdots & \\ \vdots & \vdots & \vdots & \ddots & \vdots \\ 1 & 0 & 0 & \cdots & 1 \end{bmatrix}$$

and let $B = A^t$. Then $\operatorname{rank}(A) = \operatorname{rank}(B) = n$ and

$$\operatorname{rank}(AB) = 1 \not\geq \operatorname{rank}(A) + \operatorname{rank}(B) - n = n$$

since $AB = J$.

However, we can prove the following result.

Proposition 3.39 ([9]). *Let S be an antinegative semiring and let $A \in \mathcal{M}_{m,n}(S)$, $B \in \mathcal{M}_{n,k}(S)$. Then:*

(i) $\operatorname{rank}(AB) \leq \min\{\operatorname{rank}(A), \operatorname{rank}(B)\}$;
(ii) *provided that S has no zero divisors,*

$$\operatorname{rank}(AB) \geq \begin{cases} 0 & \text{if} \quad \operatorname{rank}(A) + \operatorname{rank}(B) \leq n, \\ 1 & \text{if} \quad \operatorname{rank}(A) + \operatorname{rank}(B) > n. \end{cases}$$

These bounds are exact, the upper bound is best possible and the lower bound is best possible over Boolean semirings.

The next example demonstrates that lower bounds for Sylvester's and Frobenius' inequalities (3.2) and (3.3) for factor rank are not valid in the case $S \subseteq \mathbb{R}^+$.

Example 3.40 ([9]). Let

$$A = \begin{bmatrix} 0 & 1 & 1 & 1 \\ 1 & 0 & 1 & 1 \\ 1 & 1 & 0 & 4 \\ 1 & 1 & 4 & 0 \end{bmatrix} \quad \text{and } B = \begin{bmatrix} 1 & 1 & 0 & 0 \\ 1 & 1 & 0 & 0 \\ 0 & 0 & 1 & 1 \\ 0 & 0 & 1 & 1 \end{bmatrix}.$$

Then $\rho(A) = 3$, $\operatorname{rank}(A) = 4$, $\rho(B) = \operatorname{rank}(B) = 2$ and

$$AB = \begin{bmatrix} 1 & 1 & 2 & 2 \\ 1 & 1 & 2 & 2 \\ 2 & 2 & 4 & 4 \\ 2 & 2 & 4 & 4 \end{bmatrix},$$

so rank$(AB) = 1$. Thus,

$$1 = \text{rank}(AB) \ngeq \text{rank}(A) + \text{rank}(B) - n = 4 + 2 - 4 = 2$$

and

$$6 = 4 + 2 = \text{rank}(AI) + \text{rank}(IB)$$
$$\nleq \text{rank}(AIB) + \text{rank}(I) = 1 + 4 = 5.$$

However, the following generalization of the lower bound holds.

Proposition 3.41 ([9]). *Let $\mathcal{S} \subseteq \mathbb{R}^+$, $A \in M_{m,n}(\mathcal{S})$, $B \in M_{n,k}(\mathcal{S})$. Then*

$$\text{rank}(AB) \geq \begin{cases} 0 \text{ if } \rho(A) + \rho(B) \leq n, \\ \rho(A) + \rho(B) - n \text{ if } \rho(A) + \rho(B) > n. \end{cases}$$

This bound is exact and best possible.

Proposition 3.42 ([9]). *Let $\mathcal{S}$ be an antinegative semiring without zero divisors. For $O \neq A \in M_{m,n}(\mathcal{S})$ and $O \neq B \in M_{n,k}(\mathcal{S})$ such that $c(A) + r(B) > n$, we have*

$$1 \leq c(AB), r(AB), sr(AB), sc(AB), mr(AB), mc(AB) \ . \qquad (3.4)$$

These bounds are exact over any antinegative semiring without zero divisors and best possible over Boolean semirings.

Proof. For $O \neq A \in M_{m,n}(\mathcal{S})$, $O \neq B \in M_{n,k}(\mathcal{S})$ the matrix A has at least $c(A)$ (respectively, $sc(A)$, $mc(A)$) nonzero columns while B has at least $r(B)$ (respectively, $sr(B)$, $mr(B)$) nonzero rows.

Thus, if $c(A) + r(B) > n$, then $AB \neq O$. Hence (3.4) is established. In order to check the exactness we may take $A = B = E_{1,1}$.

Let $\mathcal{S}$ be a Boolean semiring. We consider the matrices

$$A_r = \sum_{i=1}^{r} E_{i,i} + \sum_{i=1}^{m} E_{i,1}, \quad B_s = \sum_{i=1}^{s} E_{i,i} + \sum_{i=1}^{n} E_{1,i} \ ,$$

for each pair (r, s) such that $1 \leq r \leq \min\{m, n\}$, $1 \leq s \leq \min\{k, n\}$. Then

$$c(A_r) = r(A_r) = sr(A_r) = sc(A_r) = mr(A_r) = mc(A_r) = r,$$

$$c(B_s) = r(B_s) = sr(B_s) = sc(B_s) = mr(B_s) = mc(B_s) = s,$$

by definition; and since $A_r B_s = J$,

$$r(A_r B_s) = sr(A_r B_s) = mr(A_r B_s) = 1,$$
$$c(A_r B_s) = sc(A_r B_s) = mc(A_r B_s) = 1,$$

as required.   □

Note that the condition $c(A) + r(B) > n$ is necessary because for $A = I_k \oplus O$ and $B = O \oplus I_j$ we have $AB = O$ while $k + j \le n$.

Proposition 3.43 ([9]). *Let S be a subsemiring in $\mathbb{R}^+$. Then for $A \in \mathcal{M}_{m,n}(S)$, $B \in \mathcal{M}_{n,k}(S)$ we have*

$$\begin{matrix} c(AB), sc(AB), mc(AB), \\ r(AB), sr(AB), mr(AB) \end{matrix} \ge \max \{0, \rho(A) + \rho(B) - n\} .$$

These bounds are exact and best possible.

The following example, given in [59] for the spanning column rank, shows that the analogue for the upper bound in (3.2) does not work for row and column ranks.

Example 3.44 ([59]). Let $A = [3, 7, 7] \in \mathcal{M}_{1,3}(\mathbb{Z}^+)$ and let

$$B = \begin{bmatrix} 1 & 1 & 1 \\ 0 & 1 & 1 \\ 0 & 0 & 1 \end{bmatrix}.$$

Then $c(A) = sc(A) = mc(A) = 2$, $c(B) = sc(B) = mc(B) = 3$ and $c(AB) = c(3, 10, 17) = sc(AB) = mc(AB) = 3$.

However, the following upper bounds are proven in [59].

Proposition 3.45 ([59]). *Let S be an antinegative semiring. For $A \in \mathcal{M}_{m,n}(S)$, $B \in \mathcal{M}_{n,k}(S)$ we have*

$$c(AB) \le c(B), \quad sc(AB) \le sc(B), \quad mc(AB) \le mc(B),$$
$$r(AB) \le r(A), \quad sr(AB) \le sr(A), \quad mr(AB) \le mr(A).$$

These bounds are exact and best possible.

Now we pass to the term rank.

Example 3.46. The inequality $t(AB) \le \min(t(A), t(B))$ does not hold. For instance, take $A = C_1$, $B = R_1$: then $t(AB) = t(J_n) = n > 1$.

The analogue of the lower bound in (3.2) does not hold for matrices over arbitrary semirings.

Example 3.47. Let us consider $A = B = J_n$ over a field whose characteristic divides n. Then $t(AB) = t(nJ_n) = 0$.

However, the following inequalities hold.

Proposition 3.48 ([9]). *Let $\mathcal{S}$ be an antinegative semiring. Then for any $A \in \mathcal{M}_{m,n}(\mathcal{S})$, $B \in \mathcal{M}_{n,k}(\mathcal{S})$, the inequalities*

$$t(AB) \le \min(t_r(A), t_c(B))$$

and

$$t(AB) \ge \begin{cases} 0 & if \quad t(A) + t(B) \le n, \\ t(A) + t(B) - n & if \quad t(A) + t(B) > n \end{cases}$$

hold. These bounds are exact and best possible.

Proposition 3.49 ([9]). *Let $\mathcal{S}$ be an antinegative semiring. For $A \in \mathcal{M}_{m,n}(\mathcal{S})$, $B \in \mathcal{M}_{n,k}(\mathcal{S})$ we have*

$$0 \le z(AB) \le \min\{z(A) + z(B), k, m\} \ .$$

These bounds are exact and best possible for $n > 2$.

Frobenius' inequality

The triple (A, I, A^t) provides a counterexample in the Boolean case to Frobenius' inequality (3.3) for the factor rank, where A is the same matrix as in Example 3.38. Also, Example 3.40 shows that a direct generalization of (3.3) does not work for the non-negative reals. However, we have the following result.

Proposition 3.50 ([9]). *Let $\mathcal{S} \subseteq \mathbb{R}^+$, $A \in \mathcal{M}_{m,n}(\mathcal{S})$, $B \in \mathcal{M}_{n,k}(\mathcal{S})$, and $C \in \mathcal{M}_{k,l}(\mathcal{S})$. Then*

$$\rho(AB) + \rho(BC) \le \mathrm{rank}(ABC) + \mathrm{rank}(B).$$

This bound is exact and best possible.

Example 3.51. Let A and B be the same matrices as in the second part of the proof of Proposition 3.42, where r and s are such that $r + s > n + 1$. Then the triple (A, I, B) is a counterexample to the Frobenius inequalities for the column (respectively, row) ranks.

Example 3.52 ([9]). For an arbitrary semiring, the triple (C_1, R_1, O) is a counterexample to the term rank version of the Frobenius inequality (3.3), since

$$t(C_1 R_1) + t(R_1 O) = n > t(C_1 R_1 O) + t(R_1) = 1.$$

However, if $\mathcal{S}$ is a subsemiring of $\mathbb{R}^+$ then the following trivial version holds:

$$\rho(AB) + \rho(BC) \leq t(ABC) + t(B)$$

Example 3.53. The triple (C_1, I, R_1) is a counterexample to the zero-term rank version of the Frobenius inequality, since

$$z(C_1) + z(R_1) = 2n > z(C_1 R_1) + z(I) = n.$$

4 The bideterminant function

The development of linear algebra over semirings certainly requires such an important matrix invariant as the determinant function, see [38]. It turns out that the determinant cannot be defined in a classical way even for matrices over commutative semirings without zero divisors. The main problem is connected with the fact that if a semiring is not a ring, then there are elements which do not possess additive inverses. The determinant function for matrices over commutative semirings may be naturally replaced by the so-called *bideterminant* (see Definition 4.1 below) which has been known since 1972; see also [49] and [36, 38, 41, 57].

It should be pointed out that the bideterminant is useful for solving various pure algebraic problems (see [57]) as well as in applications. For example, it can be used to solve systems of linear equations, see [36, 39], and can also be applied to certain problems in graph theory, see [49]. Another application is the problem of verifying whether or not a given matrix is sign-nonsingular, see [35]. The authors of [35] showed that this problem is polynomially equivalent to the problem of deciding if the digraph of the matrix contains a cycle of an even length. The proof is based on the properties of max-algebraic determinants. See the monographs [38, 39, 44] for a more detailed exposition of applications of the bideterminant function over semirings.

Definition 4.1 ([38, Chapter 19]). Let $\mathcal{S}$ be a commutative semiring. The *bideterminant* of a matrix $A = [a_{i,j}] \in \mathcal{M}_n(\mathcal{S})$ is the pair $(\|A\|^+, \|A\|^-)$, defined by

$$\|A\|^+ = \sum_{\sigma \in A_n} a_{1,\sigma(1)} \cdots a_{n,\sigma(n)}, \quad \|A\|^- = \sum_{\sigma \in S_n \setminus A_n} a_{1,\sigma(1)} \cdots a_{n,\sigma(n)},$$

where S_n is the symmetric group on the set $\{1, \ldots, n\}$ and A_n denotes the subgroup of even permutations.

The bideterminant function possesses some natural properties: it is invariant under matrix transpose, and for any scalar $\alpha \in \mathcal{S}$

$$(\|\alpha A\|^+, \|\alpha A\|^-) = (\alpha^n \|A\|^+, \alpha^n \|A\|^-).$$

However, some basic properties of the determinant are no longer valid for the bideterminant. For example, if A is invertible then $\|A\|^+ \neq \|A\|^-$, but the converse is not always true.

Example 4.2 ([36]). Let us consider $A = E_{1,1} + 2E_{1,2} + 3E_{2,1} + 4E_{2,2} \in M_2(\mathcal{S})$, where $\mathcal{S} = (\mathbb{Q}^+, \max, \cdot)$ is the set of non-negative rationals with addition defined by $a + b = \max\{a, b\}$ and with the standard multiplication. Then $(\|A\|^+, \|A\|^-) = (4, 6)$ but A is not invertible.

The bideterminant function is not multiplicative in general. However, the following weaker version of multiplicativity still holds.

Proposition 4.3 ([56]). *For all* $A, B \in \mathcal{M}_n(\mathcal{S})$

$$\|AB\|^+ + \|A\|^+\|B\|^- + \|A\|^-\|B\|^+ = \|AB\|^- + \|A\|^+\|B\|^+ + \|A\|^-\|B\|^-.$$

More generally, for $A_1, \ldots, A_s \in \mathcal{M}_n(\mathcal{S})$, *we have*

$$\|A_1 A_2 \cdots A_s\|^+ \quad + \sum_{t_1, \ldots, t_s = \pm \, : \, t_1 \cdots t_s = -} \|A_1\|^{t_1} \|A_2\|^{t_2} \cdots \|A_s\|^{t_s}$$

$$= \|A_1 A_2 \cdots A_s\|^- \quad + \sum_{t_1, \ldots, t_s = \pm \, : \, t_1 \cdots t_s = +} \|A_1\|^{t_1} \|A_2\|^{t_2} \cdots \|A_s\|^{t_s}$$

Now we will concentrate on recognizing various matrix properties in terms of the bideterminant.

Let $\mathcal{R}$ be an associative ring. A matrix $A \in \mathcal{M}_{m,n}(\mathcal{R})$ is called *right singular* if there exists a nonzero element $\mathbf{x} \in \mathcal{R}^n$ such that $A\mathbf{x} = \mathbf{0}$. Similarly, A is called *left singular* if there exists a nonzero element $\mathbf{y} \in \mathcal{R}^m$ such that $\mathbf{y}^t A = \mathbf{0}^t$.

It is proved in [19, Theorem 9.1] that the notions of right and left singularity coincide for square matrices over commutative rings.

Moreover, a square matrix A is singular if and only if $\det A$ is a zero divisor in $\mathcal{R}$, where $\det A$ is the usual determinant of a square matrix over a commutative ring.

The notion of singularity for matrices over semirings can be defined in a similar way.

Definition 4.4. A matrix $A \in \mathcal{M}_{m,n}(\mathcal{S})$ is $\mathcal{S}$-*right singular* if $A\mathbf{x} = \mathbf{0}$ for some nonzero $\mathbf{x} \in \mathcal{S}^n$, and called $\mathcal{S}$-*left singular* if $\mathbf{y}^t A = \mathbf{0}^t$ for some nonzero $\mathbf{y} \in \mathcal{S}^m$.

We say A is $\mathcal{S}$-*singular* if A is either $\mathcal{S}$-left singular or $\mathcal{S}$-right singular.

The following example shows that there are $\mathcal{S}$-left singular matrices which are not $\mathcal{S}$-right singular, and vice versa, even over antinegative commutative semirings without zero divisors.

Example 4.5 ([13]). Let

$$A = \begin{bmatrix} 0 & 0 \\ 1 & 1 \end{bmatrix} \quad , \quad B = \begin{bmatrix} 1 & 0 \\ 1 & 0 \end{bmatrix} \in \mathcal{M}_2(\mathbb{Z}^+).$$

We have that $Ax = \mathbf{0}$ forces $\mathbf{x} = \mathbf{0}$ since $\mathbb{Z}^+$ is antinegative, while $[1,0]A = [0,0]$. Similarly $\mathbf{y}^t B = \mathbf{0}^t$ forces $\mathbf{y} = \mathbf{0}$ while $B \begin{bmatrix} 0, 1 \end{bmatrix}^t = \begin{bmatrix} 0, 0 \end{bmatrix}^t$.

Definition 4.6. A matrix $A \in \mathcal{M}_{m,n}(\mathcal{S})$ is $\mathcal{S}$-*nonsingular* if A is neither $\mathcal{S}$-left singular nor $\mathcal{S}$-right singular.

Lemma 4.7 ([13, Lemma 3.8]). *Let $\mathcal{S}$ be an antinegative semiring without zero divisors. Then the following conditions are equivalent for any matrix $A \in \mathcal{M}_{m,n}(\mathcal{S})$:*

  (i) *A is $\mathcal{S}$-singular,*

  (ii) *A has a zero row or a zero column.*

Note that if A is an $\mathcal{S}$-singular square matrix over the commutative semiring $\mathcal{S}$ then $(\|A\|^+, \|A\|^-) = (0,0)$. However, the following example shows that there are $\mathcal{S}$-nonsingular matrices with bideterminant equal to $(0,0)$.

Example 4.8. Over any commutative antinegative semiring,

$$\begin{Vmatrix} 0 & 0 & 1 \\ 1 & 1 & 0 \\ 0 & 0 & 1 \end{Vmatrix}^+ = 0 = \begin{Vmatrix} 0 & 0 & 1 \\ 1 & 1 & 0 \\ 0 & 0 & 1 \end{Vmatrix}^-.$$

If a semiring $\mathcal{S}$ is a subsemiring of an associative ring $\mathcal{R}$ without zero divisors, we can introduce another version of singularity as follows.

Definition 4.9. We say that a matrix $A \in \mathcal{M}_{m,n}(\mathcal{S})$ is $\mathcal{R}$-*right singular* if $Ax = \mathbf{0}$ for some nonzero $\mathbf{x} \in \mathcal{R}^n$, and $\mathcal{R}$-*left singular* if $\mathbf{y}^t A = \mathbf{0}^t$ for some nonzero $\mathbf{y} \in \mathcal{R}^m$.

Clearly, $\mathcal{R}$-right (respectively, left) singularity follows from $\mathcal{S}$-right (respectively, left) singularity. However, the following example shows that there are $\mathcal{S}$-nonsingular matrices which are $\mathcal{R}$-singular.

Example 4.10. For any n the matrix $J_n = \sum_{i,j=1}^{n} E_{i,j} \in \mathcal{M}_n(\mathbb{Z}_+)$ is $\mathbb{Z}$-left and $\mathbb{Z}$-right singular, but $\mathbb{Z}_+$-nonsingular.

Definition 4.11. We say that a matrix $A \in \mathcal{M}_{m,n}(\mathcal{S})$ is $\mathcal{R}$-*singular* if it is either $\mathcal{R}$-left singular or $\mathcal{R}$-right singular.

A matrix $A \in \mathcal{M}_{m,n}(\mathcal{S})$ is $\mathcal{R}$-*nonsingular* if it is not $\mathcal{R}$-singular.

Note that all non-square matrices are $\mathcal{R}$-singular.

If, in addition, we assume that $\mathcal{R}$ is a commutative ring, then the determinant

$$\det(A) = \sum_{\sigma \in S_n} (-1)^{\sigma} a_{1,\sigma(1)} \cdots a_{n,\sigma(n)} = \|A\|^+ - \|A\|^- \quad (A \in \mathcal{M}_n(\mathcal{S}))$$

is well-defined in $\mathcal{R}$, and the following analogue of [19, Theorem 9.1] can be obtained.

Lemma 4.12 ([13, Lemma 5.3]). *If $\mathcal{R}$ is an associative commutative ring without zero divisors then the following conditions are equivalent for a matrix $A \in \mathcal{M}_n(\mathcal{S})$:*

   (i) *A is $\mathcal{R}$-right singular;*
   (ii) *A is $\mathcal{R}$-left singular;*
   (iii) $\det A = 0$.

To conclude this section we mention some results on invertibility of matrices over semirings.

Definition 4.13. Let $\mathcal{G}$ be a multiplicative algebraic system with identity element $1_{\mathcal{G}}$. An element $g \in \mathcal{G}$ is *left invertible* if there is an element $f \in \mathcal{G}$ such that $fg = 1_{\mathcal{G}}$, *right invertible* if there is an element $h \in \mathcal{G}$ such that $gh = 1_{\mathcal{G}}$, and *invertible* if it is both left and right invertible.

Definition 4.14. A matrix $A \in \mathcal{M}_n(\mathcal{S})$ is called *monomial* if it has exactly one nonzero element in each row and column.

It follows from [46] that a matrix over an antinegative semiring is invertible (respectively, left invertible, right invertible) if and only if it is a monomial matrix such that all its nonzero elements are invertible (respectively, left invertible, right invertible).

5 Linear transformations that preserve matrix invariants

During the past century much effort has been devoted to the problem of classifying those linear transformations that preserve matrix invariants (linear preservers) over various algebraic structures. The first result of this type was obtained by G. Frobenius [34] at the end of the 19th century. He characterized all bijective determinant-preserving linear transformations of the algebra of complex matrices $\mathcal{M}_n(\mathbb{C})$.

Theorem 5.1 ([34]). *Let* $T : \mathcal{M}_n(\mathbb{C}) \to \mathcal{M}_n(\mathbb{C})$ *be a bijective linear transformation such that* $\det(T(X)) = \det(X)$ *for all* $X \in \mathcal{M}_n(\mathbb{C})$. *Then there exist invertible matrices* $P, Q \in \mathcal{M}_n(\mathbb{C})$ *with* $\det(PQ) = 1$, *such that* T *has the form*

$$T(X) = PXQ \text{ for all } X \in \mathcal{M}_n(\mathbb{C})$$

or

$$T(X) = PX^tQ \text{ for all } X \in \mathcal{M}_n(\mathbb{C}).$$

The matrices P *and* Q *are unique up to a scalar factor.*

This investigation was continued by J. Dieudonné [32], who characterized all bijective linear transformations that map the set of singular matrices into itself. Starting from these investigations, many researchers have studied the problem of characterizing the linear transformations on the $n \times n$ matrix algebra $\mathcal{M}_n(\mathcal{F})$ over a field $\mathcal{F}$ that preserve certain matrix relations, subsets, or properties. A detailed and complete exposition on this subject can be found in the surveys [50, 51, 55].

The complexity of a given matrix invariant depends on the structure and especially on the quantity of its linear preservers, which determine the number of arithmetical operations necessary to compute the invariant. Indeed, the majority of methods for computing the determinant, the rank and other matrix invariants reduce a matrix to a certain manageable form by means of transformations that do not change the invariant under consideration: that is, these methods are based on the linear preservers of the matrix invariant.

For example, it is known that a square matrix with coefficients in a field can be reduced to a diagonal form with only 1 and 0 appearing on the main diagonal, without changing the rank. This fact provides a simple algorithm for computing the rank of a square matrix of order n, which requires $O(n^3)$ operations. The same is true for the determinant function. However, the simplest known method of computing the permanent of a square $n \times n$-matrix (Ryser's formula) requires $(n - 1)(2^n - 1)$ multiplicative operations. Such a difference in computational complexity reflects the fact that there are only a few linear preservers of permanent, see the paper [52] by M. Marcus and F. May. These preservers are the transposition and the pre- and post-multiplications with certain invertible monomial matrices P and Q, while in the case of linear preservers of determinant and rank, P and Q are almost arbitrary nonsingular matrices.

During the last few decades much work has been devoted to linear preservers over semirings, in particular, over antinegative semirings, see for example [8, 13, 12, 11, 14, 18, 43, 59, 55, 61]. A number of results concerning linear preservers over fields have parallel versions for matrices over semirings. We provide a brief overview of these results for some important matrix invariants such as ranks, bideterminant, singularity and invertibility.

In the sequel we will use the following terminology.

Definition 5.2. Let S be a semiring, not necessarily commutative. An operator $T : \mathcal{M}_{m,n}(S) \to \mathcal{M}_{m,n}(S)$ is called *linear* if it is additive and satisfies

$$T(\alpha X) = \alpha T(X), \; T(X\alpha) = T(X)\alpha \quad \text{for all } X \in \mathcal{M}_{m,n}(S), \, \alpha \in S.$$

Definition 5.3. We say that an operator T *preserves* a set $\mathcal{P}$ if $X \in \mathcal{P}$ implies $T(X) \in \mathcal{P}$. It *strongly preserves* $\mathcal{P}$ if $X \in \mathcal{P} \iff T(X) \in \mathcal{P}$.

The preservers of matrix relations, functions, and invariants are defined in a similar way.

It often turns out that linear operators on $\mathcal{M}_{m,n}(S)$ that preserve a certain matrix invariant have one of the following forms.

Definition 5.4. An operator $T : \mathcal{M}_{m,n}(S) \to \mathcal{M}_{m,n}(S)$ is called a (W, V)-*operator* if there exist matrices W and V of appropriate orders such that $T(X) = WXV$ for all $X \in \mathcal{M}_{m,n}(S)$, or, in the case $m = n$, $T(X) = WX^t V$ for all $X \in \mathcal{M}_{m,n}(S)$.

Definition 5.5. An element of a semiring S is called *central* if it commutes with all elements of S. The *centre* of S is the set of all central elements in S.

Definition 5.6. The matrix $X \circ Y$ denotes the *Hadamard* or *Schur product*, i.e. the (i, j) entry of $X \circ Y$ is $x_{i,j} y_{i,j}$.

Definition 5.7. An operator T is called a (P, Q, B)-*operator* if there exist permutation matrices $P \in \mathcal{M}_m(S)$ and $Q \in \mathcal{M}_n(S)$ and a matrix $B = [b_{i,j}] \in \mathcal{M}_{m,n}(S)$ where all $b_{i,j}$ are nonzero central elements, such that $T(X) = P(X \circ B)Q$ for all $X \in \mathcal{M}_{m,n}(S)$, or, in the case $m = n$, $T(X) = P(X \circ B)^t Q$ for all $X \in \mathcal{M}_{m,n}(S)$.

Definition 5.8. An operator T is called *standard* if there exist permutation matrices $P \in \mathcal{M}_m(\mathcal{S})$ and $Q \in \mathcal{M}_n(\mathcal{S})$, and diagonal matrices $D \in \mathcal{M}_m(\mathcal{S})$ and $E \in \mathcal{M}_n(\mathcal{S})$ with invertible central elements on the main diagonal, such that $T(X) = PDXEQ$ for all $X \in \mathcal{M}_{m,n}(\mathcal{S})$, or, in the case $m = n$, $T(X) = PDX^t EQ$ for all $X \in \mathcal{M}_{m,n}(\mathcal{S})$.

Rank preserving linear transformations

Definition 5.9. Let $\mathcal{R}$ be a commutative ring. An element $r \in \mathcal{R}$ is called *irreducible* if r is not invertible in $\mathcal{R}$ and for any factorization $r = r_1 r_2$, $r_1, r_2 \in \mathcal{R}$, either r_1 or r_2 is invertible in $\mathcal{R}$.

Definition 5.10. A commutative ring $\mathcal{R}$ is called a *unique factorization domain* if the following conditions are satisfied.

(i) $\mathcal{R}$ has no zero divisors.

(ii) For any noninvertible $r \in \mathcal{R}$ there exist irreducible elements $r_1, \ldots, r_k \in \mathcal{R}$ such that $r = r_1 \cdots r_k$.

(iii) For any other factorization $r = q_1 \cdots q_l$, where $q_1, \ldots, q_l$ are irreducible in $\mathcal{R}$, it holds that $l = k$, and for any i, $1 \le i \le k$ there exist j, $1 \le j \le k$ such that $q_i = u_i r_j$ for a certain invertible element $u_i \in \mathcal{R}$.

Theorem 5.11 ([7, 16]). *Let $\mathcal{S}$ consist of nonnegative elements of a unique factorization domain in $\mathbb{R}$. Let $T : \mathcal{M}_{m,n}(\mathcal{S}) \to \mathcal{M}_{m,n}(\mathcal{S})$ be a linear operator, $\min\{m, n\} \ge 2$. Then the following statements are equivalent:*

(i) *T preserves the sets of matrices of factor rank 1 and 2;*

(ii) *T preserves the sets of matrices of rank 1 and 2, where rank is the usual rank of a real matrix;*

(iii) *T is an injective (W, V)-operator.*

Note that W and V need not be invertible over $\mathcal{S}$.

Theorem 5.12 ([16]). *Let $\mathcal{S}$ be a chain semiring, $T : \mathcal{M}_{m,n}(\mathcal{S}) \to \mathcal{M}_{m,n}(\mathcal{S})$ be a linear operator, $\min\{m, n\} \ge 2$. Then the following statements are equivalent:*

(i) *for every k, $1 \le k \le \min\{m, n\}$, T preserves the set of matrices of factor rank k;*

(ii) *T preserves the sets of matrices of factor rank 1 and 2;*

(iii) *T is bijective and preserves the set of matrices of factor rank 1;*

(iv) *T is a (W, V)-operator, where the matrices $W \in \mathcal{M}_m(\mathcal{S})$ and $V \in \mathcal{M}_n(\mathcal{S})$ are invertible.*

Theorem 5.13 ([16]). *Let $\mathcal{S}$ be an arbitrary semiring, $T : \mathcal{M}_{m,n}(\mathcal{S}) \to \mathcal{M}_{m,n}(\mathcal{S})$ be a linear operator. Then the following statements are equivalent:*

(i) *for every k, $1 \le k \le \min\{m, n\}$, T preserves the set of matrices of term rank k;*

(ii) *T preserves the sets of matrices of term rank 1 and 2;*

(iii) *T strongly preserves the set of matrices of term rank 1;*

(iv) *T is a (P, Q, B)-operator.*

In the case when $\mathcal{S}$ is a field the above conditions are also equivalent to the condition that

• *T is nonsingular and preserves the set of matrices of term rank 1.*

Similar results are obtained for linear preservers and strong linear preservers of column ranks and zero-term ranks, see [16] and references therein.

Semiring versions of Frobenius' theorem

The following are semiring analogues of Theorem 5.1.

Theorem 5.14 ([13]). *Let $\mathcal{S}$ be a commutative antinegative semiring without zero divisors and $T : \mathcal{M}_n(\mathcal{S}) \to \mathcal{M}_n(\mathcal{S})$ be a surjective linear transformation. Then $(\|T(X)\|^+, \|T(X)\|^-) = (\|X\|^+, \|X\|^-)$ for all $X \in \mathcal{M}_n(\mathcal{S})$ if and only if T is standard in the sense of Definition 5.8 with $(\|PQ\|^+, \|PQ\|^-) = (\|DE\|^+, \|DE\|^-) = (1, 0)$. Here the matrices P, Q are defined uniquely and the matrices D, E are defined uniquely up to an invertible scalar factor.*

If $\mathcal{S}$ is a subsemiring of a certain commutative ring $\mathcal{R}$, for example, $\mathcal{S} = \mathbb{R}^+$, $\mathbb{Q}^+$, or $\mathbb{Z}^+$, then the usual determinant function $\det : \mathcal{M}_n(\mathcal{S}) \to \mathcal{R}$ is well-defined. In this case the following theorem is true.

Theorem 5.15 ([13, 43]). *Let $\mathcal{S}$ be an arbitrary antinegative subsemiring of a commutative ring $\mathcal{R}$ without zero divisors, $T : \mathcal{M}_n(\mathcal{S}) \to \mathcal{M}_n(\mathcal{S})$ be a surjective linear transformation. Then $\det T(X) = \det X$ for all $X \in \mathcal{M}_n(\mathcal{S})$ if and only if T is standard in the sense of Definition 5.8, where $\det(PQ) = \det(DE) = 1$. Here P, Q are defined uniquely and D, E are defined uniquely up to an invertible scalar factor.*

Semiring versions of Dieudonné's theorem

The following analogues of Dieudonné's theorem (see [32]) can be obtained for matrices over semirings.

Theorem 5.16. *Let S be an antinegative semiring without zero divisors and $T : \mathcal{M}_{m,n}(S) \to \mathcal{M}_{m,n}(S)$ be a surjective linear operator. Then the following statements are equivalent:*

(i) *T preserves the set of S-singular matrices;*
(ii) *T preserves the set of S-nonsingular matrices;*
(iii) *T is a (P, Q, B)-operator where the entries of B are invertible elements.*

Proof. Similar to the proof of [13, Theorem 3.11]. □

Theorem 5.17. *Let S be an antinegative semiring without zero divisors and $T : \mathcal{M}_n(S) \to \mathcal{M}_n(S)$ be a surjective linear transformation. Then the following statements are equivalent:*

(i) *T preserves the set of $\mathcal{R}$-nonsingular matrices;*
(ii) *T preserves the set of $\mathcal{R}$-singular matrices;*
(iii) *T is standard in the sense of Definition 5.8.*

Here P, Q are defined uniquely and D, E are defined uniquely up to an invertible scalar factor.

Proof. Similar to the proof of [13, Theorem 5.7]. □

Although the fact that the set of invertible matrices over an antinegative semiring is in a sense very small, it is still possible to characterize the corresponding linear preservers.

Theorem 5.18. *Let S be an antinegative semiring without zero divisors, $T : \mathcal{M}_n(S) \to \mathcal{M}_n(S)$ be a bijective linear transformation. The following conditions are equivalent:*

(i) *T preserves the set of invertible matrices;*
(ii) *T preserves the set of left invertible matrices;*
(iii) *T preserves the set of right invertible matrices;*
(iv) *T is a (P, Q, B)-operator where the entries of B are invertible elements.*

Our final example shows that there are non-bijective invertibility preservers which are not (P, Q, B)-operators.

Example 5.19. We consider the transformation $T : \mathcal{M}_n(\mathcal{S}) \to \mathcal{M}_n(\mathcal{S})$ defined by

$$T(A) = \text{diag} \left(\sum_{i=1}^{n} a_{i,1}, \ldots, \sum_{i=1}^{n} a_{i,n} \right)$$

for all $A = [a_{i,j}] \in \mathcal{M}_n(\mathcal{S})$. Then T evidently preserves invertibility in $\mathcal{M}_n(\mathcal{S})$, but T is neither surjective nor a (P, Q, B)-operator.

Acknowledgements

This work was partially supported by a grant of the London Mathematical Society. The author is grateful to Marianne Akian, LeRoy Beasley, Peter Butkovič, Stephane Gaubert and Mikhail Kochetov for interesting discussions during the preparation of this paper. He also wishes to thank Antonio Belotti, Anna-Maria and Polina Castro, Jeremy Filding, Ilia Nuretdinov and Nicholas Young for careful reading of the manuscript and numerous suggestions on the presentation of the results.

Bibliography

[1] M. Akian, R. Bapat, and S. Gaubert. Max-plus algebras. In L. Hogben, editor, *Handbook of Linear Algebra*, Discrete Mathematics and its Applications. Chapman & Hall/CRC, Boca Raton, FL, 2006.

[2] M. Akian and S. Gaubert. Spectral theorems for convex monotone homogeneous maps, and ergodic control. *Nonlinear Anal.*, 52(2):637–679, 2003.

[3] N. Bacaer. Modéls mathématiques pour l'optimisation des rotations. *Comptes Rendus de l'Académie d'Agriculture de France*, 89(3):52, 2003.

[4] F. L. Baccelli, G. Cohen, G. J. Olsder, and J.-P. Quadrat. *Synchronization and linearity. An algebra for discrete event systems*. John Wiley & Sons Ltd., Chichester, 1992.

[5] A. Barvinok, D. S. Johnson, G. J. Woeginger, and R. Woodroofe. The maximum traveling salesman problem under polyhedral norms. In *Integer programming and combinatorial optimization (Houston, TX, 1998)*, volume 1412 of *Lecture Notes in Comput. Sci.*, pages 195–201. Springer, Berlin, 1998.

[6] A. I. Barvinok. Combinatorial optimization and computations in the ring of polynomials. DIMACS Technical Report 93-13, 1993.

[7] L. B. Beasley, D. A. Gregory, and N. J. Pullman. Nonnegative rank preserving operators. *Linear Algebra Appl.*, 65:207–223, 1985.

[8] L. B. Beasley and A. E. Guterman. LP-problems for rank inequalities over semirings: factorization rank. *Sovrem. Mat. Prilozh. Algebra*, 13:53–70, 2004. In Russian.

[9] L. B. Beasley and A. E. Guterman. Rank inequalities over semirings. *Journal of Korean Math. Soc.*, 42(2):223–241, 2005.

[10] L. B. Beasley, A. E. Guterman, Y.-B. Jun, and S.-Z. Song. Linear preservers of extremes of rank inequalities over semirings: row and column ranks. *Linear Algebra Appl.*, 413(2-3):495–509, 2006.

[11] L. B. Beasley, A. E. Guterman, S.-G. Lee, and S.-Z. Song. Linear transformations preserving the Grassmannian over $\mathcal{M}_n(\mathbb{Z}_+)$. *Linear Algebra Appl.*, 393:39–46, 2004.

[12] L. B. Beasley, A. E. Guterman, S.-G. Lee, and S.-Z. Song. Linear preservers of zeros of matrix polynomials. *Linear Algebra Appl.*, 401:325–340, 2005.

[13] L. B. Beasley, A. E. Guterman, S.-G. Lee, and S.-Z. Song. Frobenius and Dieudonné theorems over semirings. *Linear and Multilinear Algebra*, 55(1):19–34, 2007.

[14] L. B. Beasley, A. E. Guterman, and S.-C. Yi. Linear preservers of extremes of rank inequalities over semirings: term rank and zero-term rank. *Fundam. Prikl. Mat.*, 10(2):3–21, 2004. In Russian.

[15] L. B. Beasley, S. J. Kirkland, and B. L. Shader. Rank comparisons. *Linear Algebra Appl.*, 221:171–188, 1995.

[16] L. B. Beasley, C.-K. Li, and S. Pierce. Miscellaneous preserver problems. *Linear and Multilinear Algebra*, 33:109–119, 1992.

[17] L. B. Beasley and N. J. Pullman. Semiring rank versus column rank. *Linear Algebra Appl.*, 101:33–48, 1988.

[18] A. Berman, D. Hershkowitz, and C. R. Jonson. Linear transformations that preserve certain positivity classes of matrices. *Linear Algebra Appl.*, 68:79–106, 1986.

[19] W. C. Brown. *Matrices over commutative rings*, volume 169 of *Monographs and Textbooks in Pure and Applied Mathematics*. Marcel Dekker Inc., New York, 1993.

[20] R. A. Brualdi and H. J. Ryser. *Combinatorial matrix theory*, volume 39 of *Encyclopedia of Mathematics and its Applications*. Cambridge University Press, Cambridge, 1991.

[21] R. E. Burkard and P. Butkovič. Max algebra and the linear assignment problem. *Math. Program. Ser. B*, 98:415–429, 2003.

[22] P. Butkovič. Max-algebra: the linear algebra of combinatorics? *Linear Algebra Appl.*, 367:315–335, 2003.

[23] G. Cohen, D. Dubois, J.-P. Quadrat, and M. Viot. A linear-system-theoretic view of discrete-event processes and its use for performance evaluation in manufacturing. *IEEE Trans. Automat. Control*, 30(3):210–220, 1985.

[24] J. E. Cohen and U. G. Rothblum. Nonnegative ranks, decompositions, and factorizations of nonnegative matrices. *Linear Algebra Appl.*, 190:149–168, 1993.

[25] J. H. Conway. *Regular Algebra and Finite Machines*. Chapman and Hall, London, 1971.

[26] R. A. Cuninghame-Green. *Minimax algebra*, volume 166 of *Lecture Notes in Economics and Mathematical Systems*. Springer-Verlag, Berlin, 1979.

[27] R. A. Cuninghame-Green and P. Butkovič. Bases in max-algebra. *Linear Algebra Appl.*, 389:107–120, 2004.

[28] D. de Caen, D. A. Gregory, and N. J. Pullman. The Boolean rank of zero-one matrices. In *Proceedings of the Third Caribbean Conference on Combinatorics and Computing (Bridgetown, 1981)*, pages 169–173, Cave Hill Campus, Barbados, 1981. Univ. West Indies.

[29] R. Dedekind. Über die Theorie der ganzen algebraischen Zahlen. Supplement XI to P. G. Lejeune Dirichlet: *Vorlesung über Zahlentheorie 4 Aufl.*, Druck und Verlag, Braunschweig, 1894.

[30] M. Develin, F. Santos, and B. Sturmfels. On the rank of a tropical matrix. In *Combinatorial and computational geometry*, volume 52 of *MSRI Publications*, pages 213–242. Cambridge Univ. Press, Cambridge, 2005.

[31] M. Develin and B. Sturmfels. Tropical convexity. *Documenta Math.*, 9:1–27, 2004.

[32] J. Dieudonné. Sur une généralisation du groupe orthogonal à quatre variables. *Arch. Math.*, 1:282–287, 1949.

[33] S. Eilenberg. *Automata, languages, and machines. Vol. A.* Academic Press, New York, 1974. Pure and Applied Mathematics, Vol. 58.

[34] G. Frobenius. Über die Darstellung der endlichen Gruppen durch lineare Substitutionen. Sitzungsber., 994–1015. Preuss. Akad. Wiss (Berlin), Berlin, 1897.

[35] S. Gaubert and P. Butkovič. Sign-nonsingular matrices and matrices with unbalanced determinant in symmetrised semirings. *Linear Algebra Appl.*, 301:195–201, 1999.

[36] S. Ghosh. Matrices over semirings. *Inform. Sci.*, 90:221–230, 1996.

[37] K. Glazek. *A Guide to the Literature on Semirings and their Applications in Mathematics and Information Sciences.* Kluwer Academic Publishers, Dordrecht, 2002.

[38] J. S. Golan. *Semirings and their applications.* Kluwer Academic Publishers, Dordrecht, 1999. Updated and expanded version of *The theory of semirings, with applications to mathematics and theoretical computer science*, Longman Sci. Tech., Harlow, 1992.

[39] J. S. Golan. *Semirings and affine equations over them: theory and applications*, volume 556 of *Mathematics and its Applications*. Kluwer Academic Publishers Group, Dordrecht, 2003.

[40] M. Gondran and M. Minoux. *Graphs and Algorithms.* Wiley-Interscience, New York, 1984. Translated from the French by Steven Vajda.

[41] M. Gondran and M. Minoux. Linear algebra in dioids: a survey of recent results. *Ann. Discrete Math.*, 19:147–164, 1984.

[42] D. A. Gregory and N. J. Pullman. Semiring rank: Boolean rank and nonnegative rank factorization. *J. Combin. Inform. System Sci.*, 8:223–233, 1983.

[43] A. E. Guterman. Transformations of non-negative integer matrices preserving the determinant. *Uspekhi Mat. Nauk*, 58(6):147–148, 2003. Translation in *Russian Math. Surveys*, 58(6):1200–1201, 2003.

[44] U. Hebisch and H. J. Weinert. *Semirings: algebraic theory and applications in computer science*, volume 5 of *Series in Algebra*. World Scientific Publishing Co. Inc., River Edge, NJ, 1998. Translated from the 1993 German original.

[45] L. Henry. Nuptiality. *Theoret. Population Biol*, 3:135–152, 1972.

[46] S. N. Il'in. Invertible matrices over non-associative rings. In *Universal Algebra and its Applications*, pages 81–89. Peremena, Volgograd, 2000. In Russian.

[47] K. H. Kim. *Boolean Matrix Theory and Applications*, volume 70 of *Pure and Applied Mathematics*. Marcel Dekker, New York, 1982.

[48] V. N. Kolokol'tsov and V. Maslov. *Idempotent Analysis and Applications.* Kluwer, Dordrecht, 1997.

[49] J. Kuntzmann. *Théorie des réseaux (graphes)*. Dunod, Paris, 1972.

[50] C.-K. Li and S. Pierce. Linear preserver problems. *Amer. Math. Monthly*, 108(7):591–605, 2001.

[51] C.-K. Li and N. K. Tsing. Linear preserver problems: a brief introduction and some special techniques. *Linear Algebra Appl.*, 162–164:217–235, 1992.

[52] M. Marcus and F. May. On a theorem of I. Schur concerning matrix transformations. *Arch. Math.*, 11:401–404, 1960.

[53] G. Mikhalkin. Real algebraic curves, the moment map and amoebas. *Ann. of Math. (2)*, 151(1):309–326, 2000.

[54] H. Minc. *Permanents*. Addison-Wesley, Reading, Mass., 1978.

[55] S. Pierce *et al.* A survey of linear preserver problems. *Linear and Multilinear Algebra*, 33:1–119, 1992.

[56] P. L. Poplin and E. E. Hartwig. Determinantal identities over commutative semirings. *Linear Algebra Appl.*, 387:99–132, 2004.

[57] C. Reutenauer and H. Straubing. Inversion of matrices over a commutative semiring. *J. Algebra*, 88:350–360, 1984.

[58] V. N. Sachkov and V. E. Tarakanov. *Combinatorics of nonnegative matrices*, volume 213 of *Translations of Mathematical Monographs*. American Mathematical Society, Providence, RI, 2002. Translated from the 2000 Russian original by Valentin F. Kolchin.

[59] S.-Z. Song and S.-G. Hwang. Spanning column ranks and their preservers of nonnegative matrices. *Linear Algebra Appl.*, 254:485–495, 1997.

[60] B. Sturmfels. *Solving systems of polynomial equations*, volume 97 of *CBMS Regional Conference Series in Mathematics*. Amer. Math. Soc., Providence R.I., 2002.

[61] B. S. Tam. On the semiring of cone preserving maps. *Linear Algebra Appl.*, 35:79–108, 1981.

[62] H. Vandiver. Note on a simple type of algebra in which the cancellation law of addition does not hold. *Bull. Amer. Math. Soc.*, 40:914–920, 1934.

[63] O. Viro. Dequantization of real algebraic geometry on logarithmic paper. In *European Congress of Mathematics, Vol. I (Barcelona, 2000)*, volume 201 of *Progr. Math.*, pages 135–146. Birkhäuser, Basel, 2001.

[64] V. L. Watts. Boolean rank of Kronecker products. *Linear Algebra Appl.*, 336:261–264, 2001.

[65] M. Yannakakis. Expressing combinatorial optimization problems by linear programs. In *Proceedings of the 20th Annual ACM Symposium on Theory of Computing*, pages 223–228, 1998.

Faculty of Algebra
Department of Mathematics and Mechanics
Moscow State University,
Moscow 119992, GSP-2, Russia
guterman@list.ru

Algebraic geometry over Lie algebras

Ilya V. Kazachkov

Introduction

What is algebraic geometry over algebraic systems? Many important relations between elements of a given algebraic system $\mathcal{A}$ can be expressed by systems of equations over $\mathcal{A}$. The solution sets of such systems are called *algebraic sets* over $\mathcal{A}$. Algebraic sets over $\mathcal{A}$ form a category, if we take for morphisms polynomial functions in the sense of Definition 6.1 below. As a discipline, algebraic geometry over $\mathcal{A}$ studies structural properties of this category. The principal example is, of course, algebraic geometry over fields. The foundations of algebraic geometry over groups were laid by Baumslag, Myasnikov and Remeslennikov [3, 27]. The present paper transfers their ideas to algebraic geometry over Lie algebras.

Let A be a fixed Lie algebra over a field k. We introduce the category of A-Lie algebras in Sections 1 and 2. Sections 3–7 are built around the notion of a free A-Lie algebra $A[X]$, which can be viewed as an analogue of a polynomial algebra over a unitary commutative ring. We introduce a Lie-algebraic version of the concept of an algebraic set and study connections between algebraic sets, radical ideals of $A[X]$ and coordinate algebras (the latter can be viewed as analogues of factor algebras of a polynomial algebra over a commutative ring by a radical ideal). These concepts allow us to describe the properties of algebraic sets in two different languages:

- the language of radical ideals, and
- the language of coordinate algebras.

One of the most important results here is Corollary 7.3, which shows that the categories of coordinate algebras and algebraic sets are equivalent.

In Sections 8–12, we apply some ideas of universal algebra and model theory and introduce the notions of A-prevariety, A-variety, A-quasi-variety and A-universal closure. We transfer some methods of Myasnikov and Remeslennikov [27] to Lie algebras and solve Plotkin's problem on geometric equivalence of Lie algebras. Our exposition is based on a preprint by Daniyarova [8].

In the two final sections of this survey we describe applications of the general theorems from Section 1–12 to concrete classes of Lie algebras. In Section 13 we survey the papers [10, 11]:

- We study the universal closure of a free metabelian Lie algebra of finite rank $r \geq 2$ over a finite field k and find two convenient sets of axioms, Φ_r and Φ'_r for its description; the former is written in the first order language of Lie algebras L, the latter in the language $L_{\mathfrak{F}_r}$ enriched by constants from $\mathfrak{F}_r$.
- We describe the structure of finitely generated algebras from the universal closures $\mathfrak{F}_r$-$\mathrm{ucl}(\mathfrak{F}_r)$ and $\mathrm{ucl}(\mathfrak{F}_r)$ in languages $L_{\mathfrak{F}_r}$ and L.
- We prove that in both languages L and $L_{\mathfrak{F}}$ the universal theory of a free metabelian Lie algebra over a finite field is decidable.

Then we apply these results to algebraic geometry over the free metabelian Lie algebra $\mathfrak{F}_r$, $r \geq 2$, over a finite field k, as follows:

- we give a structural description of coordinate algebras of irreducible algebraic sets over $\mathfrak{F}_r$;
- we describe the structure of irreducible algebraic sets;
- we construct a theory of dimension in the category of algebraic sets over $\mathfrak{F}_r$;

Section 14 summarizes results by Daniyarova and Remeslennikov [12] on diophantine geometry over a free Lie algebra F. The objective of algebraic geometry is to classify irreducible algebraic sets and their coordinate algebras. We believe that the general classification problem for algebraic sets and coordinate algebras over a free Lie algebra is very complicated, and will therefore only treat the following two cases:

- algebraic sets defined by systems of equations in one variable;
- bounded algebraic sets (that is, algebraic sets contained in a finite dimensional affine subspace of F, see Definition 14.8).

In both cases we reduce the problem of classification of algebraic sets and coordinate algebras to problems in diophantine geometry over the ground field k.

We refer to [2, 6, 13, 26] for background facts on Lie algebras, model theory, category theory and universal algebra.

1 The category of A-Lie algebras

We work with a fixed algebra A of coefficients and introduce the notion of an A-Lie algebra, which is a Lie-algebraic analogue of an (associative) algebra over an associative ring.

Definition 1.1. Let A be a fixed Lie algebra over a field k. A Lie algebra B over k is called an *A-Lie algebra* if it contains a designated copy of A, which we shall usually identify with A. More precisely, an A-Lie algebra B is a Lie algebra together with an embedding $\alpha : A \to B$. A *morphism* or *A-homomorphism* φ from an A-Lie algebra B_1 to an A-Lie algebra B_2 is a homomorphism of Lie algebras which is the identity on A (or, in more formal language, $\alpha_1 \varphi = \alpha_2$ where α_1 and α_2 are the corresponding embeddings of the Lie algebra A into the A-Lie algebras B_1 and B_2).

Obviously, A-Lie algebras and A-homomorphisms form a category. In the special case $A = \{0\}$, the category of A-Lie algebras is the category of Lie algebras over k. Note that if A is a nonzero Lie algebra then the category of A-Lie algebras does not possess a zero object.

Note that A is itself an A-Lie algebra.

We denote by $\mathrm{Hom}_A(B_1, B_2)$ the set of all A-homomorphisms from B_1 to B_2; $\cong_A$ denotes isomorphism in the category of A-Lie algebras (A-*isomorphism*). The usual notions of free, finitely generated and finitely presented algebras carry over to the category of A-Lie algebras.

We say that the set X *generates* an A-Lie algebra B in the category of A-Lie algebras if the algebra B is generated by the set $A \cup X$ as a Lie algebra, i.e. $B = \langle A, X \rangle$. We use the notation $B = \langle X \rangle_A$.

Note that an A-Lie algebra B can be finitely generated in the category of A-Lie algebras without being finitely generated as a Lie algebra.

Definition 1.2. Let $X = \{x_1, \ldots, x_n\}$ be a finite set. An A-Lie algebra $A[X] = \langle x_1, \ldots, x_n \rangle_A$ is said to be *free* in the category of A-Lie algebras if, for any A-Lie algebra $B = \langle b_1, \ldots, b_n \rangle_A$ and any map ψ from $A[X]$ to B which is the identity on A and satisfies $\psi(x_i) = b_i$, $i = 1, \ldots, n$, the map ψ extends to an A-epimorphism $A[X] \longrightarrow B$. We sometimes say that X is a (*free*) base of $A[X]$.

A standard argument from universal algebra yields an equivalent form of this definition.

Definition 1.3. A *free A-Lie algebra with the free base* X is the free Lie product of the free (in the category of Lie k-algebras) Lie algebra $F(X)$ and algebra A, i.e. $A[X] = A * F(X)$.

2 The first order language

In this section we show that the category of A-Lie algebras is axiomatizable.

The standard first order language L of the theory of Lie algebras over a fixed field k contains a symbol for multiplication '$\circ$', a symbol for addition '$+$', a symbol for subtraction '$-$', a set of symbols $\{k_\alpha \mid \alpha \in k\}$ for multiplication by coefficients from the field k and a constant symbol '0' for zero. The category of A-Lie algebras requires a bigger language L_A; it consists of L together with the set of constant symbols for elements in A

$$L_A = L \cup \{c_a \mid a \in A\}.$$

It is clear that an A-Lie algebra B can be treated as a model of the language L_A if the new constant symbols are interpreted in the algebra B as $c_a = \alpha(a)$. For brevity, we sometimes omit the multiplication symbol '$\circ$'.

Therefore the class of all A-Lie algebras over a field k in the language L_A is given by the following two groups of axioms.

(i) The standard series of axioms that define the class of all Lie algebras over the field k.

(ii) Additional axioms **A**–**D** which describe the behaviour of constant symbols:

A $0 = c_0$;

B $c_{\alpha_1 a_1 + \alpha_2 a_2} = k_{\alpha_1}(c_{a_1}) + k_{\alpha_2}(c_{a_2})$
(for all $a_1, a_2 \in A$, $\alpha_1, \alpha_2 \in k$);

C $c_{a_1 a_2} = c_{a_1} \circ c_{a_2}$ (for all $a_1, a_2 \in A$);

D $c_a \neq 0$ (for all nonzero $a \in A$).

Axioms **A**, **B**, **C** and **D** imply that in an A-Lie algebra B constant symbols c_a are interpreted as distinct elements and form a subalgebra of B isomorphic to A.

3 Elements of algebraic geometry

Our next objective is to introduce Lie-algebraic counterparts to the classical concepts of algebraic geometry.

Let A be a fixed Lie algebra over a field k. Let $X = \{x_1, \ldots, x_n\}$ be a finite set of variables and $A[X]$ be the free A-Lie algebra with the base X. We view $A[X]$ as an analogue of a polynomial algebra in finitely many variables over a unitary commutative ring, and think of elements of $A[X]$ as polynomials with coefficients in A. We use functional notation

$$f = f(x_1, \ldots, x_n) = f(x_1, \ldots, x_n, a_1, \ldots, a_r)$$

thereby expressing the fact that the Lie polynomial f in $A[X]$ involves *variables* $x_1, \ldots, x_n$ and, if needed, *constants* $a_1, \ldots, a_r \in A$.

A formal equality $f = 0$ can be treated, in an obvious way, as an equation over A. Therefore every subset $S \subset A[X]$ can be treated as a *system of equations* with coefficients in A. In parallel with the commutative case, the set of solutions of S depends on the algebra used to solve the system. We are specially interested in *diophantine problems*, that is, solving systems in A, but for the time being we work in a more general situation and solve S in an arbitrary A-Lie algebra B.

Let B be an A-Lie algebra and S a subset of $A[X]$. Then the set

$$B^n = \{(b_1, \ldots, b_n) | \; b_i \in B\}$$

is called the *affine n-dimensional space over the algebra B*. A point $p = (b_1, \ldots, b_n) \in B^n$ is called a *root of a polynomial* $f \in A[X]$ if

$$f(p) = f(b_1, \ldots, b_n, a_1, \ldots, a_r) = 0.$$

We also say that the polynomial f *vanishes* at the point p. A point $p \in B^n$ is called a *root* or a *solution of the system* $S \subseteq A[X]$ if every polynomial from S vanishes at p.

Definition 3.1. Let B be an A-Lie algebra and let S be a subset of $A[X]$. Then the set

$$V_B(S) = \{p \in B^n | \; f(p) = 0 \; \forall f \in S\}$$

is called the (*affine*) *algebraic set over B defined by S*.

Definition 3.2. Let B be an A-Lie algebra, S_1 and S_2 subsets of $A[X]$. Then the systems S_1 and S_2 are called *equivalent over B* if $V_B(S_1) = V_B(S_2)$. A system S is called *inconsistent over B* if $V_B(S) = \emptyset$ and *consistent* otherwise.

Example 3.3 (Typical examples of algebraic sets).

(i) Every element $a \in A$ forms an algebraic set, $\{a\}$. Indeed if one takes $S = \{x - a\}$ then $V_B(S) = \{a\}$. In this example $n = 1$ and $X = \{x\}$.

(ii) Every element $\{(a_1, \ldots, a_n)\} \in A^n$ is an algebraic set: if $S = \{x_1 - a_1, \ldots, x_n - a_n\}$ then $V_B(S) = \{(a_1, \ldots, a_n)\}$.

(iii) The centralizer $C_B(M)$ of an arbitrary set of elements M from A is an algebraic set defined by the system $S = \{x \circ m \mid m \in M\}$.

(iv) The whole affine space B^n is the algebraic set defined by the system $S = \{0\}$.

(v) Let $A = \{0\}$. Then the empty set $\emptyset$ is not algebraic, since every algebraic set in B^n contains the point $(0, \ldots, 0)$.

We need more definitions.

- A polynomial $f \in A[X]$ is a *consequence of the system* $S \subseteq A[X]$ if $V(f) \supseteq V(S)$.
- Let Y be an arbitrary (not necessarily algebraic) subset of B^n. The set

$$\mathrm{Rad}_B(Y) = \{f \in A[X] \mid f(p) = 0 \ \forall p \in Y\}$$

is called the *radical* of the set Y. If $Y = \emptyset$ then, by the definition, its radical is the algebra $A[X]$.

If Y is an algebraic set $(Y = V_B(S))$ then we also refer to its radical as the radical of the system of equations S: $\mathrm{Rad}_B(S) = \mathrm{Rad}_B(V_B(S))$, i.e.

- a polynomial $f \in A[X]$ is a consequence of a system S if and only if $f \in \mathrm{Rad}_B(S)$;
- a polynomial f is a consequence of a system S if and only if the system $S' = S \cup \{f\}$ is equivalent to S.

Therefore, $\mathrm{Rad}_B(S)$ is the maximal (by inclusion) system of equations equivalent to S.

Proposition 3.4. *The radical of a set is an ideal of the algebra $A[X]$.*

Proof. Let $f, g \in \mathrm{Rad}_B(Y)$ and $h \in A[X]$ and let $y \in Y$. By definition,

$$(\alpha f + \beta g)(y) = \alpha f(y) + \beta g(y) = 0$$

and $(hf)(y) = h(y) \cdot f(y) = h(y) \cdot 0 = 0$, where $\alpha, \beta \in k$. $\square$

Lemma 3.5.

(i) *The radical of a system $S \subseteq A[X]$ contains the ideal $\mathrm{id}\langle S \rangle$ generated by the set S, $\mathrm{Rad}_B(S) \supseteq \mathrm{id}\langle S \rangle$.*

(ii) *Let Y_1 and Y_2 be subsets of B^n and S_1, S_2 subsets of $A[X]$.*

If $Y_1 \subseteq Y_2$ then $\mathrm{Rad}_B(Y_1) \supseteq \mathrm{Rad}_B(Y_2)$.
If $S_1 \subseteq S_2$ then $\mathrm{Rad}_B(S_1) \supseteq \mathrm{Rad}_B(S_2)$.

(iii) *For any family of sets $\{ Y_i \mid i \in I \}$, $Y_i \subseteq B^n$ we have*

$$\mathrm{Rad}_B \left(\bigcup_{i \in I} Y_i \right) = \bigcap_{i \in I} \mathrm{Rad}_B(Y_i).$$

(iv) *An ideal I of the algebra $A[X]$ is the radical of an algebraic set over B if and only if*

$$\mathrm{Rad}_B(V_B(I)) = I.$$

(v) *A set $Y \subseteq B^n$ is algebraic over B if and only if*

$$V_B(\mathrm{Rad}_B(Y)) = Y.$$

(vi) *Let $Y_1, Y_2 \subseteq B^n$ be two algebraic sets, then*

$$Y_1 = Y_2 \ \text{ if and only if } \ \mathrm{Rad}_B(Y_1) = \mathrm{Rad}_B(Y_2).$$

Thus the radical of an algebraic set describes it uniquely.

Proof. The proofs follow immediately from the definitions. As an example, we prove the fourth statement. If $\mathrm{Rad}_B(V_B(I)) = I$ then I is obviously a radical. Conversely, if I is the radical of an algebraic set then there exists a system S such that $I = \mathrm{Rad}_B(S)$. Then

$$V_B(I) = V_B(\mathrm{Rad}_B(S)) = V_B(S);$$

consequently, $\mathrm{Rad}_B(V_B(I)) = \mathrm{Rad}_B(V_B(S)) = \mathrm{Rad}_B(S) = I$. $\square$

Another crucial concept in algebraic geometry is that of *coordinate algebra*.

Definition 3.6. Let B be an A-Lie algebra, S a subset of $A[X]$ and $Y \subseteq B^n$ the algebraic set defined by the system S. Then the factor algebra

$$\Gamma_B(Y) = \Gamma_B(S) = A[X]\big/ \mathrm{Rad}_B(Y)$$

is called the *coordinate algebra* of the algebraic set Y (or of the system S).

Remark 3.7. Observe that if a system S is inconsistent over B then $\Gamma_B(S) = 0$. Note that the coordinate algebras of consistent systems of equations are A-Lie algebras and form a full subcategory of the category of all A-Lie algebras.

The coordinate algebra of an algebraic set can be viewed from a different perspective, as an algebra of polynomial functions. Indeed if $Y = V_B(S)$ is an algebraic set in B^n then the coordinate algebra $\Gamma_B(Y)$ can be identified with the A-Lie algebra of all polynomial functions on Y; the latter are the functions from Y into B of the form

$$\bar{f} : Y \to B, \quad p \to f(p) \quad (p \in Y),$$

where $f \in A[x_1, \ldots, x_n]$ is a polynomial.

It is clear that two polynomials $f, g \in A[X]$ define the same polynomial function if and only if $f - g \in \operatorname{Rad}_B(Y)$. The set of all polynomial functions $P_B(Y)$ from Y to B admits the natural structure of an A-Lie algebra. We formulate our observation as the following proposition.

Proposition 3.8. *Let B be an A-Lie algebra and Y a nonempty algebraic set. Then the coordinate algebra $\Gamma_B(Y)$ of Y is the A-Lie algebra of all polynomial functions on Y:*

$$\Gamma_B(Y) \cong_A P_B(Y).$$

Example 3.9. If $a \in A$ and $Y = \{a\}$ then $\Gamma_B(Y) \cong A$.

Similarly to the commutative case, points of an algebraic set can be viewed as certain Lie algebra homomorphisms.

Proposition 3.10. *Every algebraic set Y over B can be identified with the set $\operatorname{Hom}_A(\Gamma_B(Y), B)$ (see Section 1 for notation) by the rule*

$$\theta : \operatorname{Hom}_A(\Gamma_B(Y), B) \leftrightarrow Y.$$

Consequently, for any algebraic set Y there is a one-to-one correspondence between points of Y and A-homomorphisms from $\Gamma_B(Y)$ to B.

Proof. Indeed the coordinate algebra $\Gamma_B(Y)$ is the factor algebra

$$\Gamma_B(Y) = {}^{A[X]}\!\big/\!{}_{\operatorname{Rad}_B(Y)}$$

for $X = \{x_1, \ldots, x_n\}$. Defining an A-homomorphism from $\Gamma_B(Y)$ to B is equivalent to defining the images in B of the elements of X, that is, to fixing a point $(b_1, \ldots, b_n) \in B^n$. Therefore, if $\varphi \in \operatorname{Hom}_A(\Gamma_B(Y), B)$, we set $\theta(\varphi) = (b_1, \ldots, b_n)$, where $(b_1, \ldots, b_n)$ is the image of X under φ. The point $(b_1, \ldots, b_n)$ must satisfy the condition

$$f(b_1, \ldots, b_n) = 0 \text{ for all } f \in \mathtt{Rad}_B(Y).$$

Clearly, this condition holds only at the points of the algebraic set Y. Obviously, distinct homomorphisms correspond to distinct points and every point in Y has a pre-image in $\mathtt{Hom}_A(\Gamma_B(Y), B)$. □

Developing these ideas further, we come to the following method for computing the radical $\mathtt{Rad}_B(Y)$ of an algebraic set Y. Let $p \in B^n$ be an arbitrary point and denote by φ_p the A-homomorphism

$$\varphi_p : A[X] \to B, \quad f \in A[X], \quad \varphi_p(f) = f(p) \in B. \qquad (3.1)$$

Clearly, $\mathtt{Rad}_B(\{p\}) = \ker \varphi_p$. In view of Lemma 3.5 we have

$$\mathtt{Rad}_B(Y) = \bigcap_{p \in Y} \ker \varphi_p.$$

This equality clarifies the structure of the radical of an algebraic set and the structure of its coordinate algebra. Indeed, by Remak's Theorem we have an embedding

$$\Gamma_B(Y) \to \prod_{p \in Y} A[X] \big/ \ker \varphi_p.$$

The factor algebra $A[X] \big/ \ker \varphi_p$ is isomorphic to $\mathrm{im}\, \varphi_p$ and thus embeds into B. This implies that the coordinate algebra $\Gamma_B(Y)$ embeds into a cartesian power of the algebra B; we state this observation as the following proposition.

Proposition 3.11. *The coordinate algebra of an algebraic set over B embeds into an unrestricted cartesian power of B,*

$$\Gamma_B(Y) \hookrightarrow B^Y.$$

In particular, Proposition 3.11 implies that all the identities and quasi-identities which are true in B are also true in $\Gamma_B(S)$ (see Section 9 for definitions). In particular, if B is an abelian, metabelian or nilpotent Lie algebra, the coordinate algebra $\Gamma_B(Y)$ of an arbitrary algebraic set Y over B is abelian, metabelian or nilpotent, respectively; in the last case, the nilpotency class of $\Gamma_B(Y)$ does not exceed the nilpotency class of B.

4 The Zariski topology

To introduce the Zariski toppology on the n-dimensional affine space B^n, we first formulate the following auxiliary result.

Lemma 4.1. *If A is a nonzero Lie algebra then:*

(i) *the empty set $\emptyset$ is an algebraic set over any nonzero A-Lie algebra B (it suffices to take $S = \{a\}$ for some nonzero $a \in A$);*

(ii) *the whole affine space B^n is an algebraic set (see Example 3.3);*

(iii) *the intersection of a family of algebraic sets over B is also an algebraic set over B, viz.*

$$\bigcap_{i \in I} V_B(S_i) = V_B \left(\bigcup_{i \in I} S_i \right) \quad \text{for } S_i \in A\,[X].$$

In view of Lemma 4.1, we can define a topology in B^n by taking algebraic sets in B^n as a sub-basis for the collection of closed sets. We call this topology the *Zariski topology*. For later use we introduce the following notation for three families of subsets of B^n:

- Υ is the collection of all algebraic sets over B (the sub-basis of the topology);
- Υ_1 is the collection of all finite unions of sets from Υ (the basis of the topology);
- Υ_2 is the collection of all intersections of sets from Υ_1, i.e. Υ_2 is the set of all Zariski closed subsets of the space B^n.

Definition 4.2. An A-Lie algebra B is called *A-equationally Noetherian* if for any $n \in \mathbb{N}$ and any system $S \subseteq A\,[x_1, \ldots, x_n]$ there exists a finite subsystem $S_0 \subseteq S$ such that $V_B(S) = V_B(S_0)$.

Recall that a topological space (T, τ) is called *Noetherian* if and only if every strictly descending chain (with respect to inclusion) of closed subsets terminates. Provided that a sub-basis σ of τ is closed under intersections, one can give an equivalent formulation: a topological space (T, τ) is Noetherian if and only if every strictly descending chain of subsets from σ terminates [18].

Lemma 4.3. *The algebra B is A-equationally Noetherian if and only if, for every positive integer n, the affine space B^n is Noetherian.*

Proof. Let us assume for the time being that B is A-equationally Noetherian. Consider a decreasing chain $Y_1 \supset Y_2 \supset \ldots \supset Y_t \supset \ldots$ of algebraic sets. By Lemma 3.5, the radicals of these sets form an increasing chain $\mathrm{Rad}_B(Y_1) \subset \mathrm{Rad}_B(Y_2) \subset \dots$ Let

$$S = \bigcup_i \mathrm{Rad}_B(Y_i).$$

Since the system S is equivalent to its finite subsystem S_0, the chain of radicals contains only finite number of pairwise distinct ideals, therefore the both chains terminate after finitely many steps.

Now suppose that for every positive integer n the space B^n is topologically Noetherian. We wish to show that the A-Lie algebra B is A-equationally Noetherian. Let $S \subseteq A[x_1, \dots, x_n]$ be an arbitrary system of equations and s_1 a polynomial from S. If the system S is equivalent (over B) to its subsystem $S_0 = \{s_1\}$, then the statement is straightforward. Otherwise, there exists an element $s_2 \in S \smallsetminus \{s_1\}$ such that $V_B(\{s_1\}) \supset V_B(\{s_1, s_2\})$. Continuing the construction in an obvious way, we build a decreasing chain

$$V_B(\{s_1\}) \supset V_B(\{s_1, s_2\}) \supset \dots \supset V_B(\{s_1, \dots, s_r)\}) \supset \dots$$

Since the space B^n is topologically Noetherian the chain contains only a finite number of pairwise distinct sets. Therefore the system S is equivalent (over B) to a finite subsystem.   □

A closed set Y is called *irreducible* if $Y = Y_1 \cup Y_2$, where Y_1 and Y_2 are closed, implies that either $Y = Y_1$ or $Y = Y_2$.

Theorem 4.4. *Any closed subset Y of B^n over an A-equationally Noetherian A-Lie algebra B can be expressed as a finite union of irreducible algebraic sets:*

$$Y = Y_1 \cup \dots \cup Y_l.$$

This decomposition is unique if we assume, in addition, that $Y_i \nsubseteq Y_j$ for $i \neq j$; in that case, Y_i are referred to as the irreducible components of Y.

The proof is standard; see, for example, [17].

The dimension of an irreducible algebraic set Y is defined in the usual way. Let Y be an irreducible algebraic set. The supremum, if it exists, of all integers m such that there exists a chain of irreducible algebraic sets

$$Y = Y_0 \gneqq Y_1 \gneqq \dots \gneqq Y_m$$

is called the *dimension* of Y and is denoted by $\dim(Y)$. If the supremum does not exist then, by definition, we set $\dim(Y) = \infty$.

We can generalize this definition when Y is an arbitrary (not necessarily irreducible) algebraic set over an A-equationally Noetherian A-Lie algebra B: we define its dimension $\dim(Y)$ to be the supremum of the dimensions of its irreducible components.

5 A-Domains

For groups, the notion of a *domain* (a group that has no zero divisors) was introduced in [3]; it is useful for formulating irreducibility criteria for algebraic sets over groups. Here, we introduce a similar concept for Lie algebras.

Definition 5.1. Let B be an A-Lie algebra, $x \in B$.

- The principal ideal of the subalgebra $\langle A, x \rangle$ generated by x is called the A-*relative ideal generated by* x and denoted by $\langle x \rangle^A$.
- A nonzero element $x \in B$ is called an A-*zero divisor* if there exists $y \in B$, $y \neq 0$ such that $\langle x \rangle^A \circ \langle y \rangle^A = 0$.
- An A-Lie algebra B is an A-*domain* if it contains no nonzero A-zero divisors.

Example 5.2. Let A be a nilpotent Lie algebra. Then every element $x \in A$ is a zero divisor.

(Indeed, take a nonzero element y from the centre of A. Then $\mathrm{id}\,\langle y \rangle$ is a k-vector space with the basis $\{y\}$. The pair (x, y) is a pair of zero divisors.)

Recall that the *Fitting radical* of a Lie algebra A is the ideal generated by the set of all elements from the nilpotent ideals of A. Obviously, if A is an arbitrary Lie algebra over k, then every nonzero element x of the Fitting radical $\mathtt{Fit}(A)$ of the algebra A is a zero divisor since the ideal $\mathrm{id}\,\langle x \rangle$ is nilpotent (see [10]).

The next lemma shows that in an A-domain B every closed subset of the space B^n is algebraic.

Lemma 5.3. *Let B be an A-domain. Then $\Upsilon = \Upsilon_1 = \Upsilon_2$.*

Proof. By Lemma 4.1 it will suffice to show that if B is an A-domain then any finite union of algebraic sets is also algebraic.

Let $S_1, S_2 \subset A\,[X]$ be two consistent systems of equations and assume that $V_B(S_1)$ and $V_B(S_2)$ are the algebraic sets defined by the systems

S_1 and S_2, respectively. Then $V_B(S_1) \cup V_B(S_2) = V_B(S)$, where

$$S = \left\{ f_1 \circ f_2 \mid f_i \in \langle s_i \rangle^A, \ s_i \in S_i, \ i = 1, 2 \right\}.$$

$\square$

Lemma 5.4. *Let B be an A-domain and Y be an arbitrary subset of B^n. Then the closure of Y in the Zariski topology coincides with $V_B(\mathrm{Rad}_B(Y))$.*

Proof. Clearly the set $V_B(\mathrm{Rad}_B(Y))$ is closed and contains Y. We show that $V_B(\mathrm{Rad}_B(Y))$ is contained in every closed set Z such that $Y \subseteq Z$. According to Lemma 3.5, $\mathrm{Rad}_B(Y) \supseteq \mathrm{Rad}_B(Z)$ and thus $V_B(\mathrm{Rad}_B(Y)) \subseteq V_B(\mathrm{Rad}_B(Z))$. By Lemma 5.3, every closed set in B^n is algebraic over B, hence $V_B(\mathrm{Rad}_B(Z)) = Z$ and the statement follows. $\square$

6 The category of algebraic sets

In this section we introduce the category $\mathbf{AS}_{A,B}$ of algebraic sets over an A-Lie algebra B. Throughout this section we assume that B is an A-Lie algebra, $A[X] = A[x_1, \ldots, x_n]$ and that B^n is the affine n-space over B.

Definition 6.1. The objects of $\mathbf{AS}_{A,B}$ are algebraic sets in all affine n-spaces B^n, $n \in \mathbb{N}$. If $Y \subseteq B^n$ and $Z \subseteq B^m$ are algebraic sets, then a map $\psi : Y \to Z$ is a *morphism* if there exist $f_1, \ldots, f_m \in A[x_1, \ldots, x_n]$ such that, for any $(b_1, \ldots, b_n) \in Y$,

$$\psi(b_1, \ldots, b_n) = (f_1(b_1, \ldots, b_n), \ldots, f_m(b_1, \ldots, b_n)) \in Z.$$

Occasionally we refer to morphisms as *polynomial maps*.

We denote by $\mathrm{Hom}(Y, Z)$ the set of all morphisms from Y to Z.

Following usual conventions of category theory, algebraic sets Y and Z are called *isomorphic* if there exist morphisms $\psi : Y \to Z$ and $\theta : Z \to Y$ such that $\theta\psi = \mathrm{id}_Y$ and $\psi\theta = \mathrm{id}_Z$.

Lemma 6.2. *Let $Y \subseteq B^n$ and $Z \subseteq B^m$ be algebraic sets over B and let ψ be a morphism from Y to Z. Then*

(i) *ψ is a continuous map in the Zariski topology;*

(ii) *if Y is an irreducible algebraic set and ψ is an epimorphism then Z is also irreducible. In particular, both reducibility and irreducibility of algebraic sets are preserved by isomorphisms.*

Proof.

(i) By definition, ψ is continuous if and only if the pre-image $Y_1 \subseteq B^n$ of any algebraic set $Z_1 \subseteq B^m$ is also algebraic. Let $Z_1 = V_B(S)$; then

$$\psi^{-1}(Z_1) = \{p \in B^n \mid S(f_1(p), \ldots, f_m(p)) = 0\}.$$

Clearly this subset is algebraic over B.

(ii) Suppose that Z is reducible and $Z = Z_1 \cup Z_2$. Then

$$Y = \psi^{-1}(Z_1) \cup \psi^{-1}(Z_2)$$

and $\psi^{-1}(Z_1)$, $\psi^{-1}(Z_2)$ are proper closed subsets of Y.

$\square$

Lemma 6.3. *Let $Z_1 \subseteq B^n$ and $Z_2 \subseteq B^m$ be algebraic sets over B. Then the set $Z_1 \times Z_2 \subseteq B^{n+m}$ is also algebraic over B.*

Proof. Let $Z_1 = V_B(S_1)$, $S_1 \in A[X]$, $X = \{x_1, \ldots, x_n\}$ and let $Z_2 = V_B(S_2)$, $S_2 \in A[Y]$, $Y = \{y_1, \ldots, y_m\}$, where X and Y are disjoint. Observe that the set $Z_1 \times Z_2$ is defined by the system of equations $S_1 \cup S_2 \subseteq A[X \cup Y]$. $\square$

Example 6.4. Let A be a Lie k-algebra with trivial multiplication and assume that $B = A$. An elementary reinterpretation of some basic results of linear algebra yields the following results.

(i) Every consistent system of equations over A is equivalent to a triangular system of equations.

(ii) The morphisms in the category $\mathbf{AS}_{A,A}$ are affine transformations.

(iii) Every algebraic set $Y \subseteq A^n$ is isomorphic to an algebraic set of the form

$$(\underbrace{A, A, \ldots, A}_{s}, 0, \ldots, 0), \quad 0 \leq s \leq n.$$

(iv) Every coordinate algebra $\Gamma(Y)$ is A-isomorphic to

$$A \oplus \mathtt{lin}_k \{x_1, \ldots, x_s\},$$

where $0 \leq s \leq n$, and $\mathtt{lin}_k \{x_1, \ldots, x_s\}$ is the linear span of the elements $\{x_1, \ldots, x_s\}$ over k.

7 The Equivalence Theorem

One of the main problems in algebraic geometry is the problem of classifying algebraic sets up to isomorphism. In this section we prove that this problem is equivalent to the classification of coordinate algebras.

Denote by $\mathsf{CA}_{A,B}$ the category of all coordinate algebras of algebraic sets from $\mathsf{AS}_{A,B}$ (morphisms in $\mathsf{CA}_{A,B}$ are A-homomorphisms). As we observed in Section 3, coordinate algebras of nonempty algebraic sets over B form a full subcategory of the category of A-Lie algebras (see Remark 3.7).

Note that if the empty set is an algebraic set then the zero algebra is a coordinate algebra, and $\mathrm{Hom}_A(0, \Gamma_B(S)) = \mathrm{Hom}_A(\Gamma_B(S), 0) = \emptyset$.

For the time being we declare (and later show) that the categories $\mathsf{AS}_{A,B}$ and $\mathsf{CA}_{A,B}$ are equivalent but not isomorphic.

A pair (X, S), where here $X = \{x_1, \ldots, x_n\}$ and $S \subseteq A[X]$, where S is a radical ideal, is called a *co-presentation* of the coordinate algebra

$$\Gamma_B(S) = {}^{A[X]}\!\big/_{\mathrm{Rad}_B(S)}.$$

Let (X, S_1), (Y, S_2) be co-presentations, where $X = \{x_1, \ldots, x_n\}$, $S_1 \subseteq A[X]$ and $Y = \{y_1, \ldots, y_m\}$, $S_2 \in A[Y]$. An A-homomorphism $\varphi : A[X] \to A[Y]$ such that $\varphi(\mathrm{Rad}_B(S_1)) \subseteq \mathrm{Rad}_B(S_2)$ is called a morphism from the co-presentation (X, S_1) to the co-presentation (Y, S_2).

Naturally, co-presentations (X, S_1) and (Y, S_2) are called *isomorphic* if the respective coordinate algebras $\Gamma_B(S_1)$ and $\Gamma_B(S_2)$ are A-isomorphic.

The collection of all co-presentations together with the morphisms defined above form a category, which we call the *category of co-presentations of coordinate algebras* and denote by $\mathsf{CPA}_{A,B}$.

Theorem 7.1. *The categories* $\mathsf{AS}_{A,B}$ *and* $\mathsf{CPA}_{A,B}$ *are isomorphic.*

Proof. We construct two contravariant functors

$$\mathsf{F} : \mathsf{AS}_{A,B} \to \mathsf{CPA}_{A,B} \text{ and } \mathsf{G} : \mathsf{CPA}_{A,B} \to \mathsf{AS}_{A,B}$$

such that $\mathsf{FG} = \mathrm{id}_{\mathsf{CPA}_{A,B}}$ and $\mathsf{GF} = \mathrm{id}_{\mathsf{AS}_{A,B}}$. We shall define the functors F and G first on the objects and then on the morphisms of respective categories.

Every algebraic set Y, as well as every co-presentation, is defined by the cardinality n of the set of variables $X = \{x_1, \ldots, x_n\}$ and by the radical $S = \mathrm{Rad}(Y) \subset A[X]$ treated as a system of equations. We therefore set

$$\mathsf{F}(V_B(S)) = (X, S), \quad \mathsf{G}((X, S)) = V_B(S).$$

Next we define the functors on the morphisms. To this end, let $Z_1 = V_B(S_1)$ and $Z_2 = V_B(S_2)$ be algebraic sets over B, where $S_1 \subseteq A[X]$, $X = \{x_1, \ldots, x_n\}$ and $S_2 \in A[Y]$, $Y = \{y_1, \ldots, y_m\}$ are radicals. Suppose that (X, S_1), (Y, S_2) are the respective co-presentations. By definition, a contravariant functor is a morphism-reversing functor, i.e.

$$\psi \in \text{Hom}(Z_1, Z_2) \to \mathsf{F}(\psi) \in \text{Hom}((Y, S_2), (X, S_1))$$

$$\varphi \in \text{Hom}((Y, S_2), (X, S_1)) \to \mathsf{G}(\varphi) \in \text{Hom}(Z_1, Z_2).$$

Choose $\psi \in \text{Hom}(Z_1, Z_2)$ and polynomials $f_1, \ldots, f_m \in A[x_1, \ldots, x_n]$ so that

$$\psi(b_1, \ldots, b_n) = (f_1(b_1, \ldots, b_n), \ldots, f_m(b_1, \ldots, b_n)), \quad (b_1, \ldots, b_n) \in Z_1.$$

We define $\mathsf{F}(\psi) : A[Y] \to A[X]$ by $\mathsf{F}(\psi)(y_i) = f_i(x_1, \ldots, x_n)$, $i = 1, \ldots, m$. By definition, $\mathsf{F}(\psi)(S_2) \subseteq S_1$ and therefore $\mathsf{F}(\psi)$ is a morphism in the category $\text{CPA}_{A,B}$.

Now suppose that $\varphi \in \text{Hom}((Y, S_2), (X, S_1))$, i.e. $\varphi : A[Y] \to A[X]$ is an A-homomorphism satisfying $\varphi(S_2) \subseteq S_1$. We define the image $\mathsf{G}(\varphi) : Z_1 \to Z_2$ of φ under G by polynomial maps $f_1, \ldots, f_m \in A[x_1, \ldots, x_n]$, where $f_1, \ldots, f_m$ are the images of elements of Y under φ. It is easy to check the inclusion $\mathsf{G}(\varphi)(Z_1) \subseteq Z_2$; the equalities

$$\mathsf{FG} = \text{id}_{\text{CPA}_{A,B}} \text{ and } \mathsf{GF} = \text{id}_{\text{AS}_{A,B}}$$

follow from the definitions.

To complete the proof of the theorem we need to show that F and G are in fact functors, i.e. that

- for an arbitrary algebraic set Z, $\mathsf{F}(\text{id}_Z) = \text{id}_{\mathsf{F}(Z)}$;
- for any two morphisms ψ and θ in $\text{AS}_{A,B}$, $\mathsf{F}(\psi\theta) = \mathsf{F}(\theta)\mathsf{F}(\psi)$.

and similarly for G. The verification of these conditions is straightforward and left to the reader. $\square$

Corollary 7.2. *Two algebraic sets are isomorphic if and only if the respective co-presentations are. Two co-presentations are isomorphic if and only if the respective coordinate algebras are A-isomorphic.*

In informal terms, the categories $\text{CPA}_{A,B}$ and $\text{CA}_{A,B}$ look very much alike. The correspondence between the objects and the morphisms of the categories $\text{CPA}_{A,B}$ and $\text{CA}_{A,B}$ establishes a correspondence between the categories $\text{AS}_{A,B}$ and $\text{CA}_{A,B}$. However, this correspondence is not one-to-one: the same coordinate algebra corresponds to different algebraic sets because it has different co-presentations.

Corollary 7.3. *The category* $\mathrm{AS}_{A,B}$ *of algebraic sets over an A-Lie algebra B is equivalent to the category* $\mathrm{CA}_{A,B}$ *of coordinate algebras over B.*

We note also a corollary from the proof of Theorem 7.1:

Corollary 7.4. *Let Y and Z be algebraic sets over an algebra A and* $\Gamma(Y)$ *and* $\Gamma(Z)$ *the respective coordinate algebras. Then we have a one-to-one correspondence between* $\mathrm{Hom}(Y, Z)$ *and* $\mathrm{Hom}(\Gamma(Y), \Gamma(Z))$.

Moreover, every embedding of algebraic sets $Y \subseteq Z$ *corresponds to an A-epimorphism* $\varphi : \Gamma(Y) \to \Gamma(Z)$ *of the respective coordinate algebras. If, in addition,* $Y \subsetneq Z$ *then* $\ker \varphi \neq 1$.

8 Prevarieties

We have already mentioned in Section 3 that coordinate algebras of algebraic sets map into a cartesian power of B. This observation suggests that this object deserves a detailed study and, indeed, yields results about coordinate algebras. In this section we develop this approach. For that purpose, we introduce the concept of a prevariety and study connections between prevarieties and certain classes of coordinate algebras.

Given a class $\mathcal{K}$ of A-Lie algebras over a field k, we denote by $S_A(\mathcal{K})$ and $P_A(\mathcal{K})$, correspondingly, the classes of all A-Lie subalgebras and all unrestricted A-cartesian products of algebras from $\mathcal{K}$.

Definition 8.1. Let $\mathcal{K}$ be a class of A-Lie algebras over a field k. The class $\mathcal{K}$ is called an *A-prevariety* (or, for brevity, just a *prevariety*) if $\mathcal{K} = S_A P_A(\mathcal{K})$.

Assume that $\mathcal{K}$ is a class of A-Lie algebras over a field k. Then the least A-prevariety containing $\mathcal{K}$ is called the *A-prevariety generated by the class* $\mathcal{K}$ and is denoted by $A-\mathrm{pvar}(\mathcal{K})$. It is easy to see that $A-\mathrm{pvar}(\mathcal{K}) = S_A P_A(\mathcal{K})$.

Observe that the definition of the radical of an algebraic set can be given in terms of intersections of kernels of a certain collection of homomorphisms; restricting this collection to homomorphisms with images in a particular class of algebras, we arrive at the concept of a *radical with respect to a class*.

Definition 8.2. Let $\mathcal{K}$ be a class of A-Lie algebras over k, C an A-Lie algebra and S a subset of C. Consider the family of all A-homomorphisms

$\varphi_i : C \to D$ with $D \in \mathcal{K}$ and such that $\ker(\varphi_i) \supseteq S$. The intersection of their kernels is called the *radical of the set S with respect to the class $\mathcal{K}$*,

$$\mathrm{Rad}_{\mathcal{K}}(S) = \mathrm{Rad}_{\mathcal{K}}(S, C) = \bigcap_{S \subseteq \ker(\varphi_i)} \ker(\varphi_i).$$

Note that if $C = A[X]$ and $\mathcal{K} = \{B\}$ then $\mathrm{Rad}_B(S) = \mathrm{Rad}_{\mathcal{K}}(S, C)$.

Lemma 8.3. *Let $\mathcal{K}$ be a class of A-Lie algebras, B an arbitrary A-Lie algebra and $S \subseteq B$. Then*

(i) $\mathrm{Rad}_{\mathcal{K}}(S) \supseteq \mathrm{id}\langle S \rangle$.

(ii) $B/_{\mathrm{Rad}_{\mathcal{K}}(S)} \in A\text{-pvar}(\mathcal{K})$.

(iii) $\mathrm{Rad}_{\mathcal{K}}(S, B)$ *is the smallest ideal I of the algebra B containing S and such that $B/_I \in A\text{-pvar}(\mathcal{K})$.*

(iv) $\mathrm{Rad}_{\mathcal{K}}(S) = \mathrm{Rad}_{A\text{-pvar}(\mathcal{K})}(S)$.

Proof.

(i) The first statement is straightforward.

(ii) By Remak's theorem $B/_{\mathrm{Rad}_{\mathcal{K}}(S)}$ A-embeds into the cartesian power $\prod_{i \in I} B/_{\ker \varphi_i}$. For each $i \in I$ the algebra $B/_{\ker \varphi_i}$ A-embeds into the algebra $D_{\varphi_i} \in \mathcal{K}$. Therefore the algebra $B/_{\mathrm{Rad}_{\mathcal{K}}(S)}$ is an A-subalgebra of $\prod_{i \in I} D_{\varphi_i}$.

(iii) Let J be an ideal of B, $J \supseteq S$ and $\in A\text{-pvar}(\mathcal{K}) B/_J$. Then $B/_J$ is an A-subalgebra of $\prod_{i \in I} D_i$, where $D_i \in \mathcal{K}$. Consequently, for every $i \in I$ there exists an A-homomorphism $\varphi_i : B \to D_i$ such that $J \subseteq \ker \varphi_i$. Furthermore, $J = \bigcap_{i \in I} \ker \varphi_i$ and thus $J \supseteq \mathrm{Rad}_{\mathcal{K}}(S)$.

(iv) The inclusion $\mathrm{Rad}_{\mathcal{K}}(S) \supseteq \mathrm{Rad}_{A\text{-pvar}(\mathcal{K})}(S)$ is obvious. Suppose that $\mathrm{Rad}_{\mathcal{K}}(S) \supsetneq \mathrm{Rad}_{A\text{-pvar}(\mathcal{K})}(S)$; then

$$B/_{\mathrm{Rad}_{A\text{-pvar}(\mathcal{K})}(S)} \in A\text{-pvar}(\mathcal{K}),$$

a contradiction with statement (iii) of the lemma.

$\square$

Since the concept of a prevariety is extremely important, it will be useful to have an alternative definition.

An A-Lie algebra B is called *A-approximated* by a class $\mathcal{K}$, if for any $b \in B$, $b \neq 0$ there exists an A-homomorphism $\varphi_b : B \to C$ for some $C \in \mathcal{K}$ and such that $\varphi_b(b) \neq 0$. The set of all A-Lie algebras that are A-approximated by the class $\mathcal{K}$ is denoted by $\mathrm{Res}_A(\mathcal{K})$.

Lemma 8.4. *For any class of A-Lie algebras* $\mathcal{K}$

$$A-\mathrm{pvar}(\mathcal{K}) = \mathrm{Res}_A(\mathcal{K}).$$

Proof. Clearly $P_A(\mathcal{K}) \subseteq \mathrm{Res}_A(\mathcal{K})$, therefore $A-\mathrm{pvar}(\mathcal{K}) \subseteq \mathrm{Res}_A(\mathcal{K})$. To prove the converse, take an arbitrary A-Lie algebra $B \in \mathrm{Res}_A(\mathcal{K})$. According to Definition 8.2, $\mathrm{Rad}_{\mathcal{K}}(\{0\}, B) = 0$; therefore $B \in A-\mathrm{pvar}(\mathcal{K})$. $\square$

Another very important feature of prevarieties is that every prevariety has a theory of generators and relations.

Lemma 8.5. *Let* $\mathcal{K}$ *be an A-prevariety of A-Lie algebras over a field* k. *Let* B *be an A-Lie algebra and* $B = \langle X \mid R \rangle$ *its presentation in the category of all A-Lie algebras,* $R \subseteq A[X]$. *Then the algebra* C *lies in* $\mathcal{K}$ *if and only if* $\mathrm{id}\langle R \rangle = \mathrm{Rad}_{\mathcal{K}}(R, A[X])$.

Proof. Let $\mathrm{id}\langle R \rangle = \mathrm{Rad}_{\mathcal{K}}(R, A[X])$. Since

$$B \cong_A A[X]\big/\mathrm{id}\langle R \rangle\,,$$

then by Lemma 8.3 we have $B \in \mathcal{K}$.

On the other hand $\mathrm{id}\langle R \rangle \subseteq \mathrm{Rad}_{\mathcal{K}}(R, A[X])$. Therefore, if $B \in \mathcal{K}$, then by Lemma 8.3 $\mathrm{id}\langle R \rangle = \mathrm{Rad}_{\mathcal{K}}(R, A[X])$. $\square$

By Lemma 8.5, for every algebra $B \in \mathcal{K}$ there exists a presentation $B = \langle X \mid R \rangle_{\mathcal{K}}$, $R \subseteq A[X]$ of the form

$$B \cong_A A[X]\big/\mathrm{Rad}_{\mathcal{K}}(R, A[X])\cdot$$

Indeed, if we treat B as an A-Lie algebra then $B \cong_A A[X]\big/\mathrm{id}\langle R \rangle$. Next, since $B \in \mathcal{K}$ we see that $\mathrm{id}\langle R \rangle = \mathrm{Rad}_{\mathcal{K}}(R, A[X])$.

The algebra

$$A_{\mathcal{K}}[X] = A[X]\big/\mathrm{Rad}_{\mathcal{K}}(0, A[X])$$

is called the *free object of the prevariety* $\mathcal{K}$.

It might occur that $\mathrm{id}\langle R \rangle$ is not finitely generated, and yet

$$\mathrm{Rad}_{\mathcal{K}}(R, A[X]) = \mathrm{Rad}_{\mathcal{K}}(R_0, A[X])$$

where R_0 is a finite set. In this case the corresponding A-Lie algebra B is not finitely presented in the category of all A-Lie algebras but is finitely presented in $\mathcal{K}$. Moreover, two prevarieties $\mathcal{K}_1$ and $\mathcal{K}_2$ coincide if and only if we have, for any set $R \subseteq A[X]$ and any set X,

$$\mathrm{Rad}_{\mathcal{K}_1}(R, A[X]) = \mathrm{Rad}_{\mathcal{K}_2}(R, A[X]).$$

We denote by $\mathcal{K}_\omega$ the class of all finitely generated A-Lie subalgebras of $\mathcal{K}$.

Lemma 8.6. *Let $S_A(\mathcal{K}) = \mathcal{K}$ and B a finitely generated A-Lie algebra. Then $\mathrm{Rad}_\mathcal{K}(R, B) = \mathrm{Rad}_{\mathcal{K}_\omega}(R, B)$ for any $R \subseteq B$.*

Proof. Any A-homomorphism $\varphi : B \to D$, $D \in \mathcal{K}$ induces an A-homomorphism $\varphi_0 : B \to D_0$ such that $D_0 = \mathrm{im}\,\varphi$ and $\ker \varphi = \ker \varphi_0$. Since D_0 is an A-subalgebra of D, we have $K_0 \in \mathcal{K}$. Finally, since $D_0 = \mathrm{im}\,\varphi$, the algebra K_0 is a finitely generated A-algebra. $\qquad\square$

The following lemma is a direct corollary of Lemma 8.5.

Lemma 8.7. *Let $\mathcal{K}_1$ and $\mathcal{K}_2$ be prevarieties of A-Lie algebras. Then $(\mathcal{K}_1)_\omega = (\mathcal{K}_2)_\omega$ if and only if for any finite set X and any finite subset $R \subseteq A[X]$ the radicals $\mathrm{Rad}_{\mathcal{K}_1}(R, A[X])$ and $\mathrm{Rad}_{\mathcal{K}_2}(R, A[X])$ coincide.*

The next theorem connects the theory of prevarieties with algebraic geometry.

Theorem 8.8. *Let B be an A-Lie algebra over a field k. Then all coordinate algebras over B lie in the prevariety $A-\mathrm{pvar}(B)$ and, conversely, every finitely generated A-Lie algebra from the prevariety $A-\mathrm{pvar}(B)$ is a coordinate algebra of an algebraic set over B.*

Proof. Let $C = \langle X \mid S \rangle_A$ be a finitely generated A-Lie algebra. By Lemma 8.5, $C \in A-\mathrm{pvar}(B)$ if and only if $\mathrm{id}\,\langle S \rangle = \mathrm{Rad}_{A-\mathrm{pvar}(B)}(S, A[X])$. According to Lemma 8.3, $\mathrm{Rad}_{A-\mathrm{pvar}(B)}(S) = \mathrm{Rad}_{A-\mathrm{pvar}(B)}(S, A[X])$. Moreover, $\mathrm{Rad}_B(S) = \mathrm{Rad}_{A-\mathrm{pvar}(\mathcal{K})}(S, A[X])$.

Consequently, $C \in A-\mathrm{pvar}(B)$ if and only if $\mathrm{id}\,\langle S \rangle = \mathrm{Rad}_B(S)$. Therefore C is a coordinate algebra of an algebraic set of $V_B(S)$. $\qquad\square$

Corollary 8.9. *A finitely generated A-Lie algebra C is a coordinate algebra over an A-Lie algebra B if and only if C is A-approximated by B.*

Remark 8.10. In Section 7 we introduced the notion of a co-presentation. We can now reformulate this definition as follows: a co-presentation (X, S) of a coordinate algebra $\Gamma_B(S)$ is its presentation in $A-\mathrm{pvar}(B)$.

If C is a coordinate algebra of an irreducible algebraic set over B then it possesses yet another, stronger property, which is very important in the study of universal closures (see Sections 9 and 11).

Definition 8.11. An A-Lie algebra C is said to be A-*discriminated* by an A-Lie algebra B if for every finite subset $\{c_1, \ldots, c_m\}$ of nonzero elements from the algebra C there exists an A-homomorphism $\varphi : C \to B$ such that $\varphi(c_i) \neq 0$, where $i = 1, \ldots, m$. The set of all A-Lie algebras discriminated by the algebra B is denoted by $\mathrm{Dis}_A(B)$.

Lemma 8.12. *Let B be an A-Lie algebra and C the coordinate algebra of an irreducible algebraic set over B. Then $C \in \mathrm{Dis}_A(B)$.*

Proof. Let $C = \Gamma_B(Y) = \Gamma_B(S)$, where Y is an irreducible algebraic set over B. Assume that the statement of the lemma is not true. Then there exists a tuple

$$f_1 + \mathrm{Rad}_B(S), \ldots, f_m + \mathrm{Rad}_B(S) \in \Gamma_B(S), \quad f_i \notin \mathrm{Rad}_B(S), \ i = 1, \ldots, m,$$

such that the image of at least one of these elements under every A-homomorphism $\varphi : C \to B$ is zero. Let Y_i be the algebraic set over B defined by the system of equations $S \cup \{f_i\}$, $Y_i = V_B(S \cup \{f_i\})$. In this notation

$$Y = Y_1 \cup \cdots \cup Y_m, \quad Y \neq Y_i, \quad i = 1, \ldots, m.$$

This leads to a contradiction, since Y is an irreducible algebraic set. $\square$

Let $\bar{B} = \prod_{i \in I} B^{(i)}$ be the unrestricted cartesian power of the algebra B. We turn the algebra $\bar{B}$ into an A-Lie algebra by using the diagonal embedding of A into B. Suppose that the cardinality of the set I is the maximum of the cardinalities of B and $\aleph_0$. Then Theorem 8.8, Proposition 3.11 and Definition 8.1 yield the following result.

Theorem 8.13. *Let $Y \subseteq B^n$ be an algebraic set over B. Then the coordinate algebra $\Gamma(Y)$ A-embeds into the algebra $\bar{B}$. Conversely, every finitely generated A-subalgebra of the algebra $\bar{B}$ is the coordinate algebra of an algebraic set over B.*

Definition 8.14. An n-generated A-subalgebra $C = \langle c_1, \ldots, c_n \rangle_A$ of $\bar{B}$ is called the *realization of $\Gamma_B(Y)$ in $\bar{B}$* if the complete set of relations $R \subset A[X]$ (for the generators $c_1, \ldots, c_n \in \bar{B}$) coincides with $\mathrm{Rad}(Y)$.

Note that in the above definition, the number n of generators coincides with the dimension of the affine space $B^n \supseteq Y$, and the generators are chosen in such a way that $R = \mathrm{Rad}(Y)$. However, the realization of $\Gamma_B(Y)$ in $\bar{B}$ is not unique.

Let $C = \langle c_1, \ldots, c_n \rangle_A$ and $\tilde{C} = \langle \tilde{c}_1, \ldots, \tilde{c}_n \rangle_A$ be two n-generated A-subalgebras of $\bar{B}$ and $R, \tilde{R} \subset A[X]$ be the respective complete sets of relations. The algebras C and $\tilde{C}$ are realizations of the coordinate algebra of the same algebraic set if and only if $R = \tilde{R}$.

Therefore Theorem 8.13 can be refined as follows.

Theorem 8.15. *Let* $Y \subseteq B^n$ *be an algebraic set. Its coordinate algebra* $\Gamma_B(Y)$ *has a realization in* $\bar{B}$, $\Gamma_B(Y) = \langle c_1, \ldots, c_n \rangle_A$, $c_i \in \bar{B}$. *Furthermore, generators* $c_1, \ldots, c_n$ *can be chosen in such a way that*

$$Y = \left\{ (c_1^{(i)}, \ldots, c_n^{(i)}) \mid i \in I \right\}.$$

Conversely, if C *is a finitely generated A-subalgebra of* $\bar{B}$, *with generators* $c_1, \ldots, c_n$ *say, then there exists a unique algebraic set* Y *such that* C *is a realization of* $\Gamma_B(Y)$ *in* $\bar{B}$. *In this case*

$$Y \supseteq \left\{ (c_1^{(i)}, \ldots, c_n^{(i)}) \mid i \in I \right\};$$

moreover, the closure

$$\overline{\left\{ (c_1^{(i)}, \ldots, c_n^{(i)}) \mid i \in I \right\}}$$

in the Zariski topology coincides with Y.

Theorem 8.8 demonstrates the importance of the prevariety $A-\mathtt{pvar}(B)$ in studying algebraic geometry over an A-Lie algebra. Unfortunately, the language of prevarieties is not very convenient, since prevarieties are not necessarily axiomatizable classes. In the next section we introduce some additional logical classes which have the advantage of being axiomatizable, and discuss some of their connections with prevarieties.

9 Universal classes

Given a class $\mathcal{K}$ of A-Lie algebras over a field k, we construct several model-theoretical classes of A-Lie algebras. First we need a number of definitions.

- An *A-universal sentence* of the language L_A is a formula of the type

$$\forall x_1 \cdots \forall x_n \left(\bigvee_{j=1}^{s} \bigwedge_{i=1}^{t} (u_{ij}(\bar{x}, \bar{a}_{ij}) = 0 \ \wedge \ w_{ij}(\bar{x}, \bar{c}_{ij}) \neq 0) \right),$$

where $\bar{x} = (x_1, \ldots, x_n)$ is an n-tuple of variables, $\bar{a}_{ij}$ and $\bar{c}_{ij}$ are sets of constants from the algebra A and u_{ij}, w_{ij} are terms in the language L_A in variables $x_1, \ldots, x_n$.

If an A-universal sentence involves no constants from the algebra A this notion specializes into the standard notion of a universal sentence in the language L.

- An *A-identity* of the language L_A is the formula of the form

$$\forall x_1 \cdots \forall x_n \left(\bigwedge_{i=1}^{m} r_i(\bar{x}, \bar{a}_{ij}) = 0 \right),$$

where $r_i(\bar{x})$ are terms in the language L_A in variables $x_1, \ldots, x_n$. If an A-identity involves no constants from A we come to the standard notion of an identity of the language L.

- An *A-quasi-identity* of the language L_A is a formula of the form

$$\forall x_1 \cdots \forall x_n \left(\bigwedge_{i=1}^{m} r_i(\bar{x}, \bar{a}_{ij}) = 0 \rightarrow s(\bar{x}, \bar{b}) = 0 \right),$$

where $r_i(\bar{x})$ and $s(\bar{x})$ are terms. A coefficient-free analogue is the notion of a quasi-identity.

Now we are ready to introduce the main definitions of this section: they are standard concepts of universal algebra and model theory.

- A class of Lie algebras $\mathcal{K}$ is called a *variety* if it can be axiomatized by a set of identities.
- A class of Lie algebras $\mathcal{K}$ is called a *quasivariety* if it can be axiomatized by a set of quasi-identities.
- A class of Lie algebras $\mathcal{K}$ is called a *universal class* if it can be axiomatized by a set of universal sentences.

Replacing identity by A-identity, etc., we come to 'A-versions' of these definitions: A-variety, A-quasivariety and A-universal class.

- The A-variety $A-\text{var}(\mathcal{K})$ generated by the class $\mathcal{K}$ is the class of all A-Lie algebras that satisfy all the identities of the language L_A satisfied by all algebras from $\mathcal{K}$.
- The A-quasivariety $A-\text{qvar}(\mathcal{K})$ generated by the class $\mathcal{K}$ is the class of all A-Lie algebras that satisfy all the quasi-identities of the language L_A satisfied by all algebras from $\mathcal{K}$.
- The A-universal closure $A-\text{ucl}(\mathcal{K})$ generated by the class $\mathcal{K}$ is the class of all A-Lie algebras that satisfy all the universal sentences of the language L_A satisfied by all algebras from $\mathcal{K}$.

For later use (see Section 13) we also need to consider the classes $\mathtt{var}(\mathcal{K})$, $\mathtt{qvar}(\mathcal{K})$ and $\mathtt{ucl}(\mathcal{K})$, which by definition are the variety, quasi-variety and universal closure generated by the class $\mathcal{K}$ in the first order language L. The classes $\mathtt{var}(\mathcal{K})$, $\mathtt{qvar}(\mathcal{K})$ and $\mathtt{ucl}(\mathcal{K})$ are special cases of the classes $A\!-\!\mathtt{var}(\mathcal{K})$, $A\!-\!\mathtt{qvar}(\mathcal{K})$ and $A\!-\!\mathtt{ucl}(\mathcal{K})$.

Note the following inclusions:

$$A\!-\!\mathtt{ucl}(\mathcal{K}) \subseteq A\!-\!\mathtt{qvar}(\mathcal{K}) \subseteq A\!-\!\mathtt{var}(\mathcal{K}).$$

The first inclusion is obvious, while the second one is implied by the fact that every identity is equivalent to a conjunction of a finite number of quasi-identities. For instance, the identity

$$\forall x_1 \cdots \forall x_n \left(\bigwedge_{i=1}^{m} r_i(\bar{x}) = 0 \right)$$

is equivalent to the set of m quasi-identities

$$\forall x_1 \cdots \forall x_n \forall y (y = y \to r_i(\bar{x}) = 0).$$

Obviously the classes $A\!-\!\mathtt{var}(\mathcal{K})$ and $A\!-\!\mathtt{qvar}(\mathcal{K})$ are prevarieties. Moreover,

$$A\!-\!\mathtt{pvar}(\mathcal{K}) \subseteq A\!-\!\mathtt{qvar}(\mathcal{K}) \subseteq A\!-\!\mathtt{var}(\mathcal{K}).$$

10 Quasivarieties

Every quasivariety is a prevariety and therefore quasivarieties allow for a theory of generators and relations. In this section we give a method of computation of radicals with respect to a quasivariety.

Let $\mathcal{K}$ be a quasivariety of A-Lie algebras over a field k, B an A-Lie algebra over k and S a subset of B.

Set R_0 to be the ideal of B generated by the set S and suppose that R_i is already defined. Denote by T_i the set of all elements in B of the form $s(b_1, \ldots, b_n)$, where $b_1, \ldots, b_n \in B$ and the quasi-identity

$$\forall x_1 \cdots \forall x_n \left(\bigwedge_{j=1}^{m} r_j(\bar{x}) = 0 \to s(\bar{x}) = 0 \right)$$

is true in all algebras from $\mathcal{K}$, $r_j(b_1, \ldots, b_n) \in R_i$, $j = 1, \ldots, m$. In this notation we define the ideal $R_{i+1} = \mathtt{id}\langle R_i \cup T_i \rangle$. As a result, we have an ascending chain of ideals in B:

$$R_0 \leq \cdots \leq R_i \leq R_{i+1} \leq \cdots$$

Lemma 10.1. *In this notation,* $\mathrm{Rad}_\mathcal{K}(S, B) = \bigcup_{i=0}^{\infty} R_i.$

Proof. Let $R = \bigcup_{i=0}^{\infty} R_i$. An easy induction on i shows that

$$\mathrm{Rad}_\mathcal{K}(S, B) \supseteq R.$$

To prove the reverse inclusion, it suffices to show $B/R \in \mathcal{K}$, that is, to verify that all quasi-identities which are true in every algebra from the class $\mathcal{K}$ are true in the algebra B/R. The latter easily follows from the definition of the set R. □

Lemma 10.2. *Let* $f \in \mathrm{Rad}_\mathcal{K}(S, B)$. *Then there exists a finite subset* $S_{0,f} \subseteq S$ *such that* $f \in \mathrm{Rad}_\mathcal{K}(S_{0,f}, B)$.

Proof. We use induction on i. If $f \in R_0$ then the statement is obvious. Assume that it holds for all the elements from R_i and consider an element $f \in R_{i+1}$. By definition of R_{i+1}, f has the form $s(b_1, \ldots, b_n)$. Using an argument similar to that of Lemma 10.1, we have

$$s(b_1, \ldots, b_n) \in \mathrm{Rad}_\mathcal{K}(\{r_j(b_1, \ldots, b_n), \ j = 1, \ldots m\}).$$

Since, by the inductive assumption, there exist finite sets $S_j \subseteq S$ such that $r_j(b_1, \ldots, b_n) \in \mathrm{Rad}_\mathcal{K}(S_j)$, we have $S_{0,f} = S_1 \cup \cdots \cup S_m$. □

As we mentioned earlier the prevariety $A-\mathrm{pvar}(B)$ is of exceptional importance for algebraic geometry over the algebra B. Note however that a prevariety is not in general an axiomatizable class. On the other hand, every quasivariety admits an axiomatic description and $A-\mathrm{pvar}(B) \subseteq A-\mathrm{qvar}(B)$. The situation is clarified by the following result of Malcev [25].

Proposition 10.3. *A prevariety is axiomatizable if and only if it is a quasivariety.*

This fact leads to the following question:

For which $\mathcal{K}$ is the prevariety $\mathrm{pvar}(\mathcal{K})$ *a quasivariety?*

This question is known in the theory of quasivarieties as *Malcev's Problem* [26]. Malcev himself gave the following sufficient condition for the prevariety $\mathrm{pvar}(\mathcal{K})$ to be a quasivariety [25].

Proposition 10.4. *Let* $\mathcal{K}$ *be an axiomatizable class of Lie algebras. Then* $\mathrm{pvar}(\mathcal{K})$ *is a quasivariety.*

A complete solution of Malcev's Problem has recently been found by Gorbunov [15].

Fortunately, if a system of equations $S(X) = 0$ has only finitely many variables (as is always the case in the algebraic geometry over B) the radical $\mathrm{Rad}_B(S)$ in $A[X]$ depends only on finitely generated A-Lie algebras from $A-\mathrm{pvar}(B)$. So we can refine the question.

The restricted Malcev problem. *For which $\mathcal{K}$ do the subclasses of finitely generated A-Lie algebras in $A-\mathrm{pvar}(\mathcal{K})$ and $A-\mathrm{qvar}(\mathcal{K})$ coincide?* We are particularly interested in $A-\mathrm{pvar}(B)$ and $A-\mathrm{qvar}(B)$.

A criterion in Lemma 10.8 gives an answer to this question. For its formulation we need to define A-q_ω-compactness.

Definition 10.5. An A-Lie algebra B is called A-q_ω-*compact* if for any $n \in \mathbb{N}$, any system of equations $S \subseteq A[x_1, \ldots, x_n]$ and every its consequence $f \in \mathrm{Rad}_B(S)$ there exists a finite subsystem $S_{0,f} \subseteq S$, $S_{0,f} = \{f_1, \ldots, f_m\}$ such that the following A-quasi-identity is true in the algebra B:

$$\forall x_1 \cdots \forall x_n \left(\bigwedge_{i=1}^{m} f_i(\bar{x}) = 0 \to f(\bar{x}) = 0 \right),$$

or, in other words, $f \in \mathrm{Rad}_B(S_{0,f})$.

The notion of a q_ω-compact algebra is a natural generalization of that of a Noetherian algebra. To say that an algebra is Noetherian is equivalent to requiring that the finite subsystem S_0 from Definition 10.5 is universal for all the consequences of S. Also, q_ω-compactness of an algebra implies that this finite subsystem S_0 depends on the choice of a consequence of S.

Remark 10.6. If an A-Lie algebra B is A-equationally Noetherian then B is A-q_ω-compact.

Recall that a set of formulas T is called *compact* (for the algebra B) if B satisfies all formulas from T whenever every finite submodel satisfies all formulas from T. The expression 'q_ω-compactness' comes from the following observation: an algebra B is A-q_ω-compact whenever B is compact with respect to the sentences of the language L_A of the form

$$T = \{s = 0 \mid s \in S\} \cup \{f \neq 0\}.$$

B. Plotkin and others use different terminology for saying that an algebra is q_ω-compact or equationally Noetherian: the corresponding expressions are *logically Noetherian* for q_ω-compact algebraic structures and *algebraically Noetherian* for equationally Noetherian structures.

Example 10.7. Let B be a nilpotent Lie algebra of class two given by the following presentation in the variety of class ≤ 2 nilpotent Lie algebras:

$$B = \langle a_i, b_i \; (i \in \mathbb{N}) \mid a_i \circ a_j = 0, b_i \circ b_j = 0, a_i \circ b_j = 0 \; (i \neq j) \rangle.$$

Then the infinite quasi-identity

$$\forall x \forall y \left(\bigwedge_{i \in \mathbb{N}} (x \circ a_i = 0 \bigwedge_{j \in \mathbb{N}} x \circ b_j = 0) \to x \circ y = 0 \right)$$

holds in B, but for any finite subsets I, J of $\mathbb{N}$ the following quasi-identity does not hold in B:

$$\forall x \forall y \left(\bigwedge_{i \in I} (x \circ a_i = 0 \bigwedge_{j \in J} x \circ b_j = 0) \to x \circ y = 0 \right).$$

Therefore the algebra B is not q_ω-compact.

Lemma 10.8. *An A-Lie algebra B is A-q_ω-compact if and only if*

$$(A\mathrm{-pvar}(B))_\omega = (A\mathrm{-qvar}(B))_\omega.$$

Proof. We observe first that

$$(A\mathrm{-pvar}(B))_\omega = (A\mathrm{-qvar}(B))_\omega$$

if and only if

$$\mathrm{Rad}_{A\mathrm{-pvar}(B)}(S) = \mathrm{Rad}_{A\mathrm{-qvar}(B)}(S)$$

for any system of equations $S \subseteq A[X]$ with a finite number of variables. Recall that, in view of Lemma 8.3, $\mathrm{Rad}_{A\mathrm{-pvar}(B)}(S) = \mathrm{Rad}_B(S)$. Hence it suffices to show that the algebra B is q_ω-compact if and only if $\mathrm{Rad}_{A\mathrm{-qvar}(B)}(S) = \mathrm{Rad}_B(S)$ for any system $S \subseteq A[X]$ with a finite number of variables. Using Lemma 10.1 and definition of the sets R_i, $i \in \mathbb{N}$ we conclude that the A-q_ω-compactness of B follows from the condition $\mathrm{Rad}_B(S) = \mathrm{Rad}_{A\mathrm{-qvar}(B)}(S)$ and hence from the equality $\mathrm{Rad}_{A\mathrm{-pvar}(B)}(S) = \mathrm{Rad}_{A\mathrm{-qvar}(B)}(S)$.

Now we have to prove the reverse implication. Let B be an A-q_ω-compact A-Lie algebra. We shall show that $\mathrm{Rad}_{A\mathrm{-qvar}(B)}(S) = \mathrm{Rad}_B(S)$ for an arbitrary system S of equations.

It follows from the definition of the radical with respect to a quasivariety that $\text{Rad}_B(S) \supseteq \text{Rad}_{A-\text{qvar}(B)}(S)$. Therefore it will suffice to prove that $f \in \text{Rad}_{A-\text{qvar}(B)}(S)$ for any $f \in \text{Rad}_B(S)$. Since B is q_ω-compact then there exists a finite subsystem $S_{0,f} \subseteq S$, $S_{0,f} = \{f_1, \ldots, f_m\}$ such that the quasi-identity

$$\forall x_1 \cdots \forall x_n \left(\bigwedge_{j=1}^{m} f_j(\bar{x}) = 0 \rightarrow f(\bar{x}) = 0 \right)$$

holds in B. Therefore this quasi-identity is satisfied by every algebra $C_i \in A-\text{qvar}(B)$ and thus $f \in \text{Rad}_{A-\text{qvar}(B)}(S)$. $\square$

Corollary 10.9. *Let B be an A-equationally Noetherian A-Lie algebra. Then the class of all coordinate algebras of algebraic sets over B coincides with the class $(A-\text{qvar}(B))_\omega$.*

Myasnikov and Remeslennikov [27, §9, Problem 2] asked if there exists a q_ω-compact group which is not equationally Noetherian; a similar question can be formulated for an arbitrary algebraic structure. The question was answered by Goebel and Shelah [14] who constructed such a group; the same construction works for Lie algebras. However, the author's conversations with B. Plotkin, A. Myasnikov and V. Remeslennikov led to an observation that all counterexamples known so far are not finitely generated.

Problem. *Are there finitely generated q_ω-compact A-Lie algebras which are not A-equationally Noetherian?*

11 Universal closure

If an A-Lie algebra B is A-equationally Noetherian then every algebraic set over B is a finite union of irreducible algebraic sets and, moreover, this presentation is unique (Theorem 4.4). This shifts the focus of the study of algebraic sets onto their irreducible components. It turns out that irreducible algebraic sets and the corresponding coordinate algebras are the algebraic counterparts to the universal closure of the algebra B (see Theorem 11.5 below).

Lemma 11.1. *Let B and C be A-Lie algebras such that $C \in \text{Dis}_A(B)$ (see Definition 8.11). Then $C \in A-\text{ucl}(B)$.*

Proof. Recall that the condition $C \in A-\text{ucl}(B)$ means that every finite submodel of the algebra C A-embeds into the algebra B. This is obvious, since C is A-discriminated by the algebra B. $\qquad\square$

Corollary 11.2. *Let B be an A-Lie algebra, and suppose that a finitely generated A-Lie algebra C is the coordinate algebra of an irreducible algebraic set over B. Then $C \in A-\mathtt{ucl}(B)$.*

Lemma 11.3. *Let B be an A-equationally Noetherian A-Lie algebra and C a finitely generated A-Lie algebra. If $C \in A-\mathtt{ucl}(B)$ then $C \in \mathtt{Dis}_A(B)$.*

Proof. Since B is A-equationally Noetherian, we have, in view of Remark 10.6 and Lemma 10.8, that

$$(A-\mathtt{pvar}(B))_\omega = (A-\mathtt{qvar}(B))_\omega.$$

And since $A-\mathtt{ucl}(B) \subseteq A-\mathtt{qvar}(B)$ the algebra C lies in the class $(A-\mathtt{pvar}(B))_\omega$, i.e. C is a coordinate algebra of an algebraic set over B and has the form $C = \Gamma_B(S)$ for $S \subseteq A[X]$, $X = \{x_1, \ldots, x_n\}$. Since B is A-equationally Noetherian we can assume without loss of generality that $S = \{s_1, \ldots, s_p\}$ is a finite system of equations.

Assume now that the algebra C is not discriminated by B. Then there exists an m-tuple

$$f_1 + \mathtt{Rad}_B(S), \ldots, f_m + \mathtt{Rad}_B(S) \in \Gamma_B(S), \quad f_i \notin \mathtt{Rad}_B(S), \ i = 1, \ldots, m$$

such that the image of at least one of these elements under every A-homomorphism $\varphi : C \to B$ is zero. In other words, this implies that the algebra B satisfies the universal formula

$$\Phi = \forall x_1 \cdots \forall x_n \left(\bigwedge_{j=1}^{p} s_j(\bar{x}) = 0 \to \bigvee_{i=1}^{m} f_i(\bar{x}) = 0 \right).$$

This yields a contradiction, since $C \in A-\mathtt{ucl}(B)$ and the formula Φ is not true in the algebra C. Indeed, after substituting the elements

$$x_1 + \mathtt{Rad}_B(S), \ldots, x_n + \mathtt{Rad}_B(S) \in \Gamma_B(S)$$

into Φ we see that the consequence is a false statement while the condition is true. $\qquad\square$

We denote by $\mathtt{LDis}_A(B)$ the set of all A-Lie algebras such that every finitely generated A-Lie subalgebra of an algebra from $\mathtt{LDis}_A(B)$ is A-discriminated by B. The following result follows directly from the previous discussion.

Theorem 11.4. *If B is an A-equationally Noetherian A-Lie algebra then*

$$A-\mathrm{ucl}(B) = \mathrm{LDis}_A(B).$$

Theorem 11.5. *Let B be an A-equationally Noetherian A-Lie algebra and C a finitely generated A-Lie algebra. Then C is the coordinate algebra of an irreducible algebraic set over B if and only if $C \in A-\mathrm{ucl}(B)$.*

Proof. If C is a coordinate algebra of an irreducible algebraic set over B then, in view of Corollary 11.2, $C \in A-\mathrm{ucl}(B)$.

Conversely, let C be a finitely generated A-Lie algebra from the class $A-\mathrm{ucl}(B)$. By Remark 10.6 and Lemma 10.8,

$$(A-\mathrm{pvar}(B))_\omega = (A-\mathrm{qvar}(B))_\omega.$$

Since $A-\mathrm{ucl}(B) \subseteq A-\mathrm{qvar}(B)$, C lies in the class $(A-\mathrm{pvar}(B))_\omega$: that is, C is the coordinate algebra of an algebraic set over B, Y say. By virtue of Theorem 11.4, C is A-discriminated by the algebra B. We are left to show that the algebraic set Y is irreducible.

We argue towards a contradiction and assume that the algebraic set Y is reducible: $Y = Y_1 \cup \cdots \cup Y_m$, $Y \neq Y_i$, $i = 1, \ldots, m$. Then

$$\mathrm{Rad}_B(Y) = \mathrm{Rad}_B(Y_1) \cap \cdots \cap \mathrm{Rad}_B(Y_m),$$

$\mathrm{Rad}_B(Y) < \mathrm{Rad}_B(Y_i)$, $i = 1, \ldots, m$.

Let $f_i \in \mathrm{Rad}_B(Y_i) \smallsetminus \mathrm{Rad}_B(Y)$, $i = 1, \ldots, m$ and let $p \in Y$. For the time being we treat the elements of C as polynomial functions (see Proposition 3.8). Since every A-homomorphism $\varphi : C \to B$ can be regarded as a substitution of a point $p \in Y$ into the elements of C (see Equation (3.1)), the polynomial f_i vanishes at $p \in Y_i$. Since every point of Y is contained in at least one of its irreducible components, we come to a contradiction with the fact that $C = \Gamma_B(Y)$ is A-discriminated by the algebra B. $\qquad\square$

Theorem 8.13 shows that the description of coordinate algebras is equivalent to the description of finitely generated subalgebras of $\bar{B}$. It turns out that any coordinate algebra of an irreducible algebraic set embeds into the ultrapower $\prod_{i \in I} B^{(i)}/\mathcal{D}$ with respect to an ultrafilter $\mathcal{D}$ over the set I (for definitions see [6, 7]). We turn the algebra $\prod_{i \in I} B^{(i)}/\mathcal{D}$ into an A-Lie algebra by using the diagonal embedding of A.

The following theorem follows from a classical theorem of Malcev [26].

Theorem 11.6.

(i) *Let Y be an irreducible algebraic set over B. Then the coordinate algebra $\Gamma_B(Y)$ A-embeds into an ultrapower $\prod_{i \in I} B^{(i)}/\mathcal{D}$, where $|I| = |B|$.*

(ii) *If B is A-equationally Noetherian, then any finitely generated A-subalgebra of any ultrapower $B^J/\mathcal{D}$ is a coordinate algebra of an irreducible algebraic set over B.*

12 Geometric equivalence

B. Plotkin [28] introduced an important notion of geometrical equivalence for algebraic structures. Myasnikov and Remeslennikov [27] discuss this notion in the case of groups and observe that all their results can be transferred to an arbitrary algebraic structure. In this section, we transfer their results to Lie algebras.

At the intuitive level of understanding, two Lie algebras are geometrically equivalent if they produce identical algebraic geometries. More precisely, A-Lie algebras B and C are called *geometrically equivalent* if for every positive integer n and every system $S \subseteq A[X]$, $X = \{x_1, \ldots, x_n\}$ the radicals $\mathrm{Rad}_B(S)$ and $\mathrm{Rad}_C(S)$ coincide.

Since radicals completely determine algebraic sets and their coordinate algebras, the study of algebraic geometry over B is equivalent to the study of algebraic geometry over C, provided that B and C are geometrically equivalent. In the latter case, the respective categories of algebraic sets are isomorphic.

Lemma 12.1. *A-Lie algebras B and C are geometrically equivalent if and only if*

$$(A-\mathrm{pvar}(B))_\omega = (A-\mathrm{pvar}(C))_\omega.$$

Proof. Assume that B and C are geometrically equivalent. Then the families of coordinate algebras over B and C coincide and, consequently, the classes $(A-\mathrm{pvar}(B))_\omega$ and $(A-\mathrm{pvar}(C))_\omega$ also coincide by virtue of Theorem 8.8.

Now suppose that $(A-\mathrm{pvar}(B))_\omega = (A-\mathrm{pvar}(C))_\omega$. Let $S \subseteq A[X]$ be an arbitrary system and consider its radical $\mathrm{Rad}_B(S)$. Applying Lemma 8.3, we have

$$\mathrm{Rad}_B(S) = \mathrm{Rad}_{A-\mathrm{pvar}(B)}(S) = \mathrm{Rad}_{A-\mathrm{pvar}(C)}(S) = \mathrm{Rad}_C(S).$$

By definition, B and C are geometrically equivalent. $\qquad\square$

Corollary 12.2. *If A-Lie algebras B and C are geometrically equivalent then*

$$A-\mathrm{qvar}(B) = A-\mathrm{qvar}(C).$$

Proof. By definition,

$$A-\mathrm{qvar}(B) = A-\mathrm{qvar}(A-\mathrm{pvar}(B)) = A-\mathrm{qvar}((A-\mathrm{pvar}(B))_\omega).$$

Now, in view of Lemma 12.1, $A-\mathrm{qvar}(B) = A-\mathrm{qvar}(C)$. □

Observe that geometric equivalence of A-Lie algebras B and C does not follow from equality of the quasivarieties generated by B and C. As Lemma 12.3 demonstrates, a counterexample can be found only for non q_ω-compact A-Lie algebras. Such counterexamples indeed exist, see Example 10.7. We refer to [27] for more detail.

Nevertheless, for q_ω-compact Lie algebras, equality of quasivarieties is equivalent to geometric equivalence.

Lemma 12.3. *Let B and C be q_ω-compact A-Lie algebras. Then B and C are geometrically equivalent if and only if $A-\mathrm{qvar}(B) = A-\mathrm{qvar}(C)$.*

Proof. By Corollary 12.2, the geometric equivalence of the algebras B and C yields $A-\mathrm{qvar}(B) = A-\mathrm{qvar}(C)$.

Since B and C are q_ω-compact, then

$$(A-\mathrm{pvar}(B))_\omega = (A-\mathrm{qvar}(B))_\omega \text{ and } (A-\mathrm{pvar}(C))_\omega = (A-\mathrm{qvar}(C))_\omega$$

by Lemma 10.8. Therefore the families of coordinate algebras over B and C coincide and the algebras B and C are geometrically equivalent. □

13 Algebraic geometry over free metabelian Lie algebras

The objective of this section is to give a brief account of recent results in algebraic geometry over free metabelian Lie algebras. We shall follow [9, 10, 11].

Throughout this section $\mathfrak{F}$ will denote a free metabelian Lie algebra over a field k. We also use the notation $\mathfrak{F}_r = \langle a_1, \ldots, a_r \rangle$ for the free metabelian Lie algebra of rank r with the set of free generators (the free base) $a_1, \ldots, a_r$.

Recall that a metabelian Lie algebra is a Lie algebra A which satisfies the identity:

$$(a \circ b) \circ (c \circ d) = 0.$$

We denote by $\texttt{Fit}(A)$ the Fitting radical of the algebra A. It is well-known that $\texttt{Fit}(A)$ has a natural structure of a module over the ring of polynomials, see [10].

13.1 The Δ-localization and the direct module extension of the Fitting radical

In this subsection we introduce U-algebras; for any U-algebra A over k we describe two extensions of the Fitting radical of A (introduced in [10]). These constructions play an important role in the study of the universal closure of the algebra A (see Sections 9 and 11).

Following [10], we call a metabelian Lie algebra A over a field k a *U-algebra* if

- $\texttt{Fit}(A)$ is abelian;
- $\texttt{Fit}(A)$ is a torsion-free module over the ring of polynomials.

Let A be a U-algebra over a field k and $\{z_\alpha,\ \alpha \in \Lambda\}$ a maximal set of elements from A which are linearly independent modulo $\texttt{Fit}(A)$. We denote by V the linear span of this set over k.

Let $\Delta = \langle x_\alpha, \alpha \in \Lambda \rangle$ be a maximal ideal of the ring $R = k\,[x_\alpha, \alpha \in \Lambda]$. Denote by R_Δ the localization of the ring R with respect to Δ and by $\texttt{Fit}_\Delta(A)$ the localization of the module $\texttt{Fit}(A)$ with respect to Δ, that is, the closure of $\texttt{Fit}(A)$ under the action of the elements of R_Δ (for definitions see [5] and [21]). Consider the direct sum $V \oplus \texttt{Fit}_\Delta(A)$ of vector spaces over k. By definition, the multiplication on V is inherited from A and the multiplication on $\texttt{Fit}_\Delta(A)$ is trivial. Set

$$u \circ z_\alpha = u \cdot x_\alpha, \quad u \in \texttt{Fit}_\Delta(A),\ z_\alpha \in V, \quad u \cdot x_\alpha \in \texttt{Fit}_\Delta(A).$$

and extend multiplication on $\texttt{Fit}_\Delta(A)$ to V by linearity.

In this notation, the algebra A_Δ is called the Δ-*localization* of the algebra A.

Proposition 13.1. *Let A be a U-algebra and A_Δ its Δ-localization. Then* $\texttt{ucl}(A) = \texttt{ucl}(A_\Delta)$, *i.e. the algebras A and A_Δ are universally equivalent.*

Let M be a torsion-free module over the ring of polynomials R. By definition, the algebra $A \oplus M$ is the direct sum of k-vector spaces $V \oplus \texttt{Fit}(A) \oplus M$. To define the structure of an algebra on $V \oplus \texttt{Fit}(A) \oplus M$ we need to introduce multiplication. The multiplication on V is inherited from A. The multiplication on $\texttt{Fit}(A) \oplus M$ is trivial. If $b \in M$ and

$z_\alpha \in V$, we set $b \circ z_\alpha = b \cdot x_\alpha$ and extend this multiplication by b to elements from V using the distributivity law. This operation turns $A \oplus M$ into a metabelian Lie algebra over k.

Proposition 13.2. *Let A be a U-algebra and M a finitely generated module over R. Then*

- $\mathtt{ucl}(A) = \mathtt{ucl}(A \oplus M)$;
- $A-\mathtt{ucl}(A) = A-\mathtt{ucl}(A \oplus M)$;
- $\mathtt{Fit}(A \oplus M) = \mathtt{Fit}(A) \oplus M$.

The operations of Δ-localization and direct module extension commute. Following [11], we denote by $\mathfrak{F}_{r,s}$ the direct module extension of the Fitting radical of $\mathfrak{F}_r$ by the free R-module T_s of the rank s, $\mathfrak{F}_{r,s} = \mathfrak{F}_r \oplus T_s$.

13.2 The case of a finite field

Our aim is to construct diophantine algebraic geometry over the free metabelian Lie algebra $\mathfrak{F}_r$. When $r = 1$, algebraic geometry over $\mathfrak{F}_r$ degenerates to Example 6.4. Therefore we restrict ourselves to the non-degenerate case $r \geq 2$.

The universal axioms for the Φ_r-algebras

In this section we formulate two collections of universal axioms Φ_r and Φ'_r in the languages L and $L_{\mathfrak{F}_r}$, which axiomatize the universal classes $\mathtt{ucl}(\mathfrak{F}_r)$ and $\mathfrak{F}_r-\mathtt{ucl}(\mathfrak{F}_r)$. Most of these formulas are in the first order language L and consequently belong to both Φ_r and Φ'_r. We shall write the two series of formulas simultaneously, at every step indicating the differences between Φ_r and Φ'_r.

Axiom $\Phi 1$ below is the metabelian identity, axiom $\Phi 2$ postulates that there are no non-abelian nilpotent subalgebras and axiom $\Phi 3$ is the CT-axiom:

$$\forall x_1, x_2, x_3, x_4 \quad (x_1 x_2)(x_3 x_4) = 0. \tag{$\Phi 1$}$$

$$\forall x \forall y \quad xyx = 0 \wedge xyy = 0 \rightarrow xy = 0. \tag{$\Phi 2$}$$

$$\forall x \forall y \forall z \quad x \neq 0 \wedge xy = 0 \wedge xz = 0 \rightarrow yz = 0. \tag{$\Phi 3$}$$

We next introduce an universal formula $\mathtt{Fit}(x)$ of the language L and in one variable which defines the Fitting radical

$$\mathtt{Fit}(x) \equiv (\forall y \; xyx = 0). \tag{13.1}$$

An analogue of formula (13.1) in the language $L_{\mathfrak{F}_r}$ is

$$\mathsf{Fit}'(x) \equiv \bigwedge_i (xa_ix = 0). \tag{13.2}$$

From now on we restrict ourselves to the case of a finite field k. In particular, the vector space $\mathfrak{F}_r/_{\mathtt{Fit}(\mathfrak{F}_r)}$ is finite, and its dimension over k is r.

Lemma 13.3. *Let k be a finite field and $n \in \mathbb{N}$, $n \le r$. Then the formula*

$$\varphi(x_1,\ldots,x_n) \equiv \bigwedge_{(\alpha_1,\ldots,\alpha_n)\neq\bar{0}} \neg\mathsf{Fit}(\alpha_1x_1 + \cdots + \alpha_nx_n)$$

of the language L is true on the elements $\{b_1,\ldots,b_n\}$ of $\mathfrak{F}_r$ if and only if $b_1,\ldots,b_n$ are linearly independent modulo $\mathtt{Fit}(\mathfrak{F}_r)$.

We can express the dimension axiom using formula φ:

$$\forall x_1 \cdots \forall x_{r+1} \, \neg\varphi(x_1,\ldots,x_{r+1}). \tag{$\Phi 4$}$$

Since φ is an existential formula, the formula $\neg\varphi$ and hence axiom $\Phi 4$ are universal. This axiom postulates that the dimension of the factor space $B/_{\mathtt{Fit}(B)}$ is at most r, provided that B satisfies $\Phi 1 - \Phi 3$.

Recall that $\mathtt{Fit}(\mathfrak{F}_r)$ has a structure of a module over the ring of polynomials $R = k[x_1,\ldots,x_r]$. The series of axioms $\Phi 5$, $\Phi'5$, $\Phi 6$, $\Phi 7$ and $\Phi'7$ express module properties of $\mathtt{Fit}(\mathfrak{F}_r)$. In these axioms we use module (over R) notation. In particular, this involves rewriting the polynomial $f(x_1,\ldots,x_n)$, $n \le r$ in the signature of metabelian Lie algebra; see [10] for a more detailed description of transcription from the module signature to the Lie algebra signature.

The Fitting radical of the free metabelian Lie algebra is a torsion-free module over the ring R. We use this observation to write the following infinite series of axioms. For every nonzero polynomial $f \in k[x_1,\ldots,x_n]$, $n \le r$, write

$$\Phi 5 : \forall z_1 \forall z_2 \forall x_1 \cdots \forall x_n (z_1z_2 \cdot f(x_1,\ldots,x_n) = 0 \wedge z_1z_2 \neq 0)$$
$$\to (\neg\,\varphi(x_1,\ldots,x_n)). \tag{$\Phi 5$}$$

Since $\varphi(x_1,\ldots,x_n)$ is a $\exists$-formula the formula $\Phi 5$ is a $\forall$-formula.

This property can be expressed in the language $L_{\mathfrak{F}_r}$ as follows. For any nonzero polynomial $f \in k[x_1,\ldots,x_r]$ write:

$$\forall z_1 \forall z_2 \quad (z_1z_2 \cdot f(a_1,\ldots,a_r) = 0 \to z_1z_2 = 0). \tag{$\Phi'5$}$$

The main advantage of this formula is that it does not involve the formula φ, thus the restriction on the cardinality of the field k is not significant.

For every nonzero Lie polynomial $l(a_1, \ldots, a_n)$, $n \le r$ in variables $a_1, \ldots, a_r$ from the free base of $\mathfrak{F}_r$, we write

$$\forall x_1 \cdots \forall x_n \quad \varphi(x_1, \ldots, x_n) \to (l(x_1, \ldots, x_n) \neq 0). \qquad (\Phi 6)$$

Since $\varphi(x_1, \ldots, x_n)$ is a $\exists$-formula the formula $\Phi 6$ is a $\forall$-formula.

The series of axioms $\Phi 7$ and $\Phi' 7$ are quite sophisticated. We first introduce higher-dimensional analogues of Formulas (13.1) and (13.2):

$$\mathsf{Fit}(y_1, \ldots, y_l; x_1, \ldots, x_n)$$

$$\equiv \left(\bigwedge_{i=1}^n (y_1 x_i y_1 = 0) \right) \wedge \cdots \wedge \left(\bigwedge_{i=1}^n (y_l x_i y_l = 0) \right),$$

$$\mathsf{Fit}'(y_1, \ldots, y_l) \equiv \mathsf{Fit}'(y_1) \wedge \cdots \wedge \mathsf{Fit}'(y_l).$$

We begin with the series of axioms $\Phi' 7$ in the language $L_{\mathfrak{F}_r}$. Let S be a fixed finite system of module equations with variables $y_1, \ldots, y_l$ over the module $\mathsf{Fit}(\mathfrak{F}_r)$. Every equation from S has the form

$$h = y_1 f_1(\bar{x}) + \cdots + y_l f_l(\bar{x}) - c = 0, \quad c = c(a_1, \ldots, a_r) \in \mathsf{Fit}(\mathfrak{F}_r),$$

where $\bar{x} = \{x_1, \ldots, x_r\}$ is a vector of variables and $f_1, \ldots, f_l \in R = k[\bar{x}]$. Suppose that S is inconsistent over $\mathsf{Fit}(\mathfrak{F}_r)$. This fact can be written, in an obvious way, as a logical formula in the signature of a module. The system S gives rise to a system of equations S_1 over $\mathfrak{F}_r$. Replace every module equation $h_i = 0$ from S by the equation $h_i' = 0$, $i = 1, \ldots, m$ in the signature of $L_{\mathfrak{F}_r}$ (see [10]). This procedure results in a system of equations S_1 over $\mathfrak{F}_r$. For every inconsistent system of module equations S, write

$$\psi_S' \equiv \forall y_1 \cdots \forall y_l \quad \mathsf{Fit}'(y_1, \ldots, y_l) \to \bigvee_{i=1}^m h_i'(y_1, \ldots, y_l) \neq 0. \qquad (\Phi' 7)$$

Note that we have not used the restriction on the cardinality of the field k.

Now we turn our attention to the collection Φ_r in the language L. Let S be a system of m module equations inconsistent over the Fitting radical $\mathsf{Fit}_\Delta(\mathfrak{F}_n)$ of Δ-local Lie algebra $(\mathfrak{F}_n)_\Delta$. We write, for every

$n \in \mathbb{N}$, $n \leq r$ and every such system S:

$$\psi_{n,S} \equiv \forall x_1 \cdots \forall x_n \, \forall y_1 \cdots \forall y_l \quad \varphi(x_1, \ldots, x_n) \wedge \mathsf{Fit}(y_1, \ldots, y_l) \rightarrow$$

$$\rightarrow \bigvee_{i=1}^{m} h_i(y_1, \ldots, y_l; x_1, \ldots, x_n) \neq 0.$$

$$(\Phi 7)$$

The Lie polynomials h_i, $i = 1, \ldots, m$ are constructed from the system S according to the following procedure. Consider the i-th equation of S. It has the form

$$h_i' = y_1 f_1(x_1, \ldots, x_n) + \cdots + y_l f_l(x_1, \ldots, x_n) - c = 0,$$

$$f_i \in R, \; c = c(a_1, \ldots, a_n) \in \mathsf{Fit}(\mathfrak{F}_n).$$

After replacing every occurrence of a_j in $c(a_1, \ldots, a_n)$ by x_j for $j = 1, \ldots, n$, and rewriting the polynomial h_i' in the signature of Lie algebras (see [10]), the resulting Lie polynomial is h_i.

Denote by Φ_r and by Φ_r' the universal classes axiomatized by $\Phi 1$-$\Phi 7$ and by $\Phi 1$-$\Phi 4$, $\Phi' 5$, $\Phi 6$, $\Phi' 7$. respectively. The algebras from these classes are called, correspondingly, Φ_r-*algebras* and Φ_r'-*algebras*.

13.3 Main results

Assume that the ground field k is finite.

Theorem 13.4. *Let A be an arbitrary finitely generated metabelian Lie algebra over a finite field k. Then the following conditions are equivalent:*

- $A \in \mathsf{ucl}(\mathfrak{F}_r)$;
- *there exists $s \in \mathbb{N}$ so that A is a subalgebra of $\mathfrak{F}_{r,s}$;*
- *A is a Φ_r-algebra.*

Corollary 13.5. *The universal closure $\mathsf{ucl}(\mathfrak{F}_r)$ of the free metabelian Lie algebra $\mathfrak{F}_r$ is axiomatized by Φ_r.*

Theorem 13.6. *Let A be an arbitrary finitely generated metabelian $\mathfrak{F}_r$-Lie algebra over a finite field k. Then the following conditions are equivalent:*

- *$A \in \mathfrak{F}_r-\mathsf{ucl}(\mathfrak{F}_r)$;*
- *A is a Φ_r'-algebra;*
- *A is $\mathfrak{F}_r$-isomorphic to the algebra $\mathfrak{F}_r \oplus M$, where M is a torsion-free module over $R = k[x_1, \ldots, x_n]$.*

It is shown in [10] that for every torsion-free R-module M the algebra $\mathfrak{F}_r \oplus M$ $\mathfrak{F}_r$-embeds into $\mathfrak{F}_{r,s}$ for some $s \in \mathbb{N}$. It follows that all Φ'_r-algebras can be treated as $\mathfrak{F}_r$-subalgebras of $\mathfrak{F}_{r,s}$.

Corollary 13.7. *The universal closure* $\mathrm{ucl}(\mathfrak{F}_r)$ *of the free metabelian Lie algebra* $\mathfrak{F}_r$ *is axiomatized by* Φ'_r.

Theorem 13.8. *Axioms* Φ_r *(and* Φ'_r*) form a recursive set. The universal theory in the language* L *(in the language* $L_{\mathfrak{F}_r}$*, respectively) of the algebra* $\mathfrak{F}_r$ *(treated as an* $\mathfrak{F}_r$*-Lie algebra, respectively) over a finite field* k *is decidable.*

Theorem 13.9. *The compatibility problem for a system of equations over the free metabelian Lie algebra* $\mathfrak{F}_r$ *is decidable.*

This result contrasts with a result of Roman'kov [29] on the compatibility problem over metabelian groups: he proved that in the case of free metabelian groups of a sufficiently large rank the problem is undecidable. His argument holds for free metabelian Lie rings and free metabelian Lie algebras, provided that the compatibility problem for the ground field is undecidable (and thus the ground field is infinite).

We next classify all irreducible algebraic sets over $\mathfrak{F}_r$. Combining Theorems 11.5 and 13.6 we state

Proposition 13.10. *Let* Γ *be an* $\mathfrak{F}_r$*-Lie algebra. Then* Γ *is a coordinate algebra of an irreducible algebraic set over* $\mathfrak{F}_r$ *if and only if* Γ *is* $\mathfrak{F}_r$*-isomorphic to* $\mathfrak{F}_r \oplus M$*, where* M *is a torsion-free module over* $k[x_1, \ldots, x_r]$.

Let $R = k[x_1, \ldots, x_r]$ and let M be a finitely generated torsion-free module over R. Let $\mathrm{Hom}_R(M, \mathrm{Fit}(\mathfrak{F}_r))$ be the set of all R-homomorphisms from M to $\mathrm{Fit}(\mathfrak{F}_r)$ treated as a module over R. The lemma below describes the canonical implementation of an algebraic set.

Lemma 13.11. *In this notation, we have a one-to-one correspondences*

$$\mathrm{Hom}_R(M, \mathrm{Fit}(\mathfrak{F}_r)) \leftrightarrow \mathrm{Hom}_{\mathfrak{F}_r}(\mathfrak{F}_r \oplus M, \mathfrak{F}_r) \leftrightarrow Y$$

where Y *is an irreducible algebraic set over* $\mathfrak{F}_r$ *such that* $\Gamma(Y) = \mathfrak{F}_r \oplus M$.

Theorem 13.12. *Up to isomorphism, every irreducible algebraic set over* $\mathfrak{F}_r$ *is either* $\mathrm{Hom}_R(M, \mathrm{Fit}(\mathfrak{F}_r))$ *for some finitely generated torsion-free module* M *over the ring* R*, or a point.*

Corollary 13.13. *Every irreducible algebraic set in the affine space* $\mathfrak{F}_r^n$*, $n = 1$, is, up to isomorphism, either a point or* $\mathrm{Fit}(\mathfrak{F}_r)$.

Recall that the rank $r(M)$ of the module M over a ring R is the supremum of cardinalities of linearly independent over R sets of elements from M. We set $r(\Gamma(Y)) = r(M)$ if $\Gamma(Y) = \mathfrak{F}_r \oplus M$.

Theorem 13.14. *For an irreducible algebraic set Y over $\mathfrak{F}_r$*

$$\dim(Y) = r(\Gamma(Y)) = r(M).$$

Remark 13.15. When this paper had already been written E. Daniyarova published her PhD thesis, where in particular she obtained a description of the quasivarieties $\mathtt{qvar}(\mathfrak{F}_r)$ and $\mathfrak{F}_r{-}\mathtt{qvar}(\mathfrak{F}_r)$.

14 Algebraic geometry over a free Lie algebra

In this section we outline results by Daniyarova and Remeslennikov [12] on diophantine geometry over the free Lie algebra.

The aim of algebraic geometry is to classify irreducible algebraic sets and their coordinate algebras. We expect that the classification problem of algebraic sets and coordinate algebras in the case of the free Lie algebra is very difficult if taken in full generality. We treat this problem only in the two following cases:

- for algebraic sets defined by systems of equations with one variable;
- for bounded algebraic sets (see Definition 14.8).

In these cases, we reduce the problem to the corresponding problem in the diophantine geometry over the ground field k.

The classification of algebraic sets and coordinate algebras over free groups was considered in a series of papers [1, 3, 7, 16, 22, 23, 24, 27], the earliest of which [24] dates back to 1960. In that paper Lyndon studied one-variable systems of equations over free groups. Recently, a satisfactory classification of irreducible algebraic sets over a free group $\mathbb{F}$ and their coordinate groups was given in [19, 20]. We begin this section by comparing results for free Lie algebras [12] and for free groups [7].

Let $\mathbb{F}$ be a free group and S any system of equations in one variable over $\mathbb{F}$ (i.e. $S \subseteq \mathbb{F} * \langle x \rangle$) such that $V_{\mathbb{F}}(S) \neq \emptyset$. The full description of all algebraic sets in $\mathbb{F}^1 = \mathbb{F}$ is given by the following two theorems.

Theorem 14.1. *Any coordinate group $\Gamma_{\mathbb{F}}(Y)$ of an irreducible algebraic set $Y \subseteq \mathbb{F}^1$ satisfies one of the following three conditions:*

- $\Gamma_{\mathbb{F}}(Y) \cong \mathbb{F}$;
- $\Gamma_{\mathbb{F}}(Y) \cong \mathbb{F} * \langle x \rangle$, *where $*$ denotes the free product of groups;*

- $\Gamma_{\mathbb{F}}(Y) \cong \langle \mathbb{F}, t \mid [u, t] = 1 \rangle$, *where* u *generates a maximal cyclic subgroup of* $\mathbb{F}$.

Theorem 14.2. *If* $V \neq \mathbb{F}^1$ *is an irreducible algebraic set defined by a system of equations with one variable then:*

- *either* V *is a point, or*
- *there exist elements* f, g, $h \in \mathbb{F}$ *such that* $V = fC_{\mathbb{F}}(g)h$, *where* $C_{\mathbb{F}}(g)$ *stands for the centralizer of* g *in* $\mathbb{F}$.

On the other hand, the classification of algebraic sets is much more complicated in the case of one-variable equations over the free Lie algebra.

Theorem 14.3. *Up to isomorphism, an algebraic set defined by a consistent system of equations with one variable over* F *is either bounded (see Definition 14.8) or coincides with* F.

Roughly speaking, bounded algebraic sets are the ones contained in a finite dimensional affine subspace of F. To make this definition explicit, we introduce the notion of a parallelepipedon (Section 14.1) and then show that algebraic geometry in a parallelepipedon is equivalent to diophantine geometry over the ground field (Section 14.3).

14.1 Parallelepipedons

We shall now show that every finite dimensional affine subspace V of a finitely generated free Lie algebra F is an algebraic set.

Let V be a finite dimensional subspace of F, $V = \text{lin}_k(v_1, \ldots, v_m)$ where lin_k is the linear span over k. Set

$$s_1(x) = x \circ v_1; \quad s_2(x) = s_1(x, v_1) \circ s_1(v_2, v_1) = (x \circ v_1) \circ (v_2 \circ v_1).$$

Recursively, set

$$s_m(x) = s_{m-1}(x, v_1, \ldots, v_{m-1}) \circ s_{m-1}(v_m, v_1, \ldots, v_{m-1}). \tag{14.1}$$

Note that equations $s_l(x) = 0$ are linear over k and thus define vector subspaces of the algebra F.

Proposition 14.4. *Any finite dimensional linear subspace of the algebra* F *is an algebraic set over* F. *Furthermore, an* l-*dimensional subspace of* F *can be defined by an equation of the type* s_l.

Moreover, affine shifts of linear subspaces (affine subspaces) of F are also algebraic sets. Let $c \in F$ be an arbitrary element. Then the subspace $V + c$ (here V is a linear subspace of dimension l) is, obviously, also an algebraic set over the free Lie algebra F in the obvious way: it is defined by the equation $s_l(x - c) = 0$.

Corollary 14.5. *Every affine shift of an arbitrary finite dimensional linear subspace of the algebra F is an algebraic set over F.*

Using the same argument for systems of equations with n variables, we get the following result.

Corollary 14.6. *Let V_i, $i = 1, \ldots, n$, be finite dimensional linear (or affine) subspaces of the free Lie algebra F. The direct product $\mathbb{V} = V_1 \times \cdots \times V_n \subset F^n$ is an algebraic set over F.*

We denote by $\mathbb{V}$ the direct product of affine finite dimensional subspaces of F, i.e.

$$\mathbb{V} = (V_1 + c_1) \times \cdots \times (V_n + c_n) \subset F^n,$$

$V_i = \text{lin}_k \left\{ v_1^i, \ldots, v_m^i \right\}$. We call such spaces *parallelepipedons*.

Let $\bar{F} = \prod_{i \in I} F^{(i)}$ and $\bar{k} = \prod_{i \in I} k^{(i)}$ be the unrestricted cartesian powers of F and k, where the cardinality of the set I coincides with the cardinality of F. We turn the algebra $\bar{F}$ into an F-Lie algebra by identifying the diagonal copy of the algebra F with F. We next apply Theorem 8.15 and obtain

Proposition 14.7. *Let $\Gamma(\mathbb{V}) = \langle \xi_1, \ldots, \xi_n \rangle_F$ be a realization (see Definition 8.14) of $\Gamma(\mathbb{V})$ in $\bar{F}$, $\xi_1, \ldots, \xi_n \in \bar{F}$. Then the generators ξ_i have the form*

$$\xi_i = t_1^i v_1^i + \cdots + t_{m_i}^i v_{m_i}^i, \quad t_1^i, \ldots, t_{m_i}^i \in \bar{k}; \qquad (14.2)$$

and the coefficients t_j^i satisfy the conditions

- $\left\{ (t_j^{i\,(l)}), \ l \in I \right\} = k^M$ *if the field k is finite, or*
- *the elements t_i^j's are algebraically independent over k, otherwise.*

14.2 Bounded algebraic sets and coordinate algebras

Definition 14.8. An algebraic set Y over F is called *bounded* if it is contained in a parallelepipedon.

We list some properties of bounded algebraic sets.

- An arbitrary bounded algebraic set Y is contained in infinitely many parallelepipedons.
- Let $Y, Z \subset F^n$ be algebraic sets and let Y be bounded. Suppose next that there exists an epimorphism $\phi : Y \to Z$. Then Z is also bounded.

This section is mostly concerned with the parallel concept of bounded coordinate algebra. Set

$$B(\bar{F}) = \left\{ \xi \in \bar{F} \mid \text{the degrees of the coordinates of } \xi \text{ are bounded above} \right\}.$$

Proposition 14.9. *In this notation,* $B(\bar{F}) \cong F \otimes_k \bar{k}$.

A coordinate algebra is called *bounded* if it has a realization in $B(\bar{F})$ (see Definition 8.14).

Proposition 14.10.

   (i) *An algebraic set* $Y \subseteq F^n$ *over* F *is bounded if and only if its coordinate algebra* $\Gamma_F(Y)$ *is bounded.*

   (ii) *If* $Y \subseteq \mathbb{V}$ *then any realization of* $\Gamma_F(Y) = \langle \xi_1, \dots, \xi_n \rangle$ *has the form* (14.2).

   (iii) *If the generators* ξ_i *in a realization of* $\Gamma_F(Y)$ *have the form* (14.2) *then* $Y \subset \mathbb{V}$.

14.3 The correspondences between algebraic sets, radicals and coordinate algebras

The objective of this section is to establish a correspondence between bounded algebraic geometry over the free Lie algebra and algebraic geometry over the ground field k.

We begin with bounded algebraic geometry in dimension 1, that is, consider algebraic sets $Y \subset F$. Fix a parallelepipidon $\mathbb{V} = V + c \subset F$, where $V = \text{lin}_k \{ v_1, \dots, v_m \}$, $\dim V = m$, $c \in F$. We shall show that there exists a one-to-one correspondence between bounded algebraic sets over F from $\mathbb{V}$ and algebraic sets over k that lie in the affine space k^m:

$$Y_F \subseteq \mathbb{V} \leftrightarrow Y_k \subseteq k^m.$$

Our correspondence takes the following form. Let $Y_F \subseteq \mathbb{V}$ then

$$Y_k = \{ (\alpha_1, \dots, \alpha_m) \in k^m \mid \alpha_1 v_1 + \cdots + \alpha_m v_m + c \in Y_F \} \subseteq k^m. \quad (14.3)$$

Conversely, let Y_k be an algebraic set in k^m. Set

$$Y_F = \{ \alpha_1 v_1 + \cdots + \alpha_m v_m \mid (\alpha_1, \dots, \alpha_m) \in Y_k \} \subseteq \mathbb{V}. \quad (14.4)$$

We wish to show that the sets defined by (14.3) and (14.4) are algebraic. We do this by constructing the corresponding systems of equations, S_F and S_k, such that $Y_F = V(S_F)$ and $Y_k = V(S_k)$.

As shown in Lemma 3.5, an algebraic set is uniquely determined by its radical, which will be more convenient for us to work with. First we define the correspondence between polynomials in $F[x]$ and $k[y_1, \ldots, y_m]$.

Consider an arbitrary Lie polynomial $f(x) \in \mathtt{Rad}_F(Y)$ and a point

$$p = \alpha_1 v_1 + \cdots + \alpha_m v_m + c \in \mathbb{V}.$$

We treat the coefficients α_i as variables and write $f(p)$ in the form

$$f(p) = g_1(\alpha_1, \ldots, \alpha_m)u_1 + \cdots + g_s(\alpha_1, \ldots, \alpha_m)u_s,$$

where $g_1, \ldots, g_s \in k[y_1, \ldots, y_m]$ and $u_1, \ldots, u_s \in F$ are linearly independent and do not depend on the point p. Clearly

$$f(p) = 0 \text{ if and only if } g_1(\alpha_1, \ldots, \alpha_m) = \ldots = g_s(\alpha_1, \ldots, \alpha_m) = 0$$

We can therefore set $S_f = \{g_1, \ldots, g_s\} \subset k[y_1, \ldots, y_m]$.

We next construct the inverse correspondence. Consider a polynomial

$$g(y_1, \ldots, y_m) = \sum_{\bar{i}} \alpha_{\bar{i}} y_1^{i_1} \cdots y_m^{i_m} = 0, \quad \alpha_{\bar{i}} \in k, \quad \bar{i} = (i_1, \ldots, i_m) \in \mathbb{N}^m$$

and set $M_1 = \max\{i_1\}, \ldots, M_m = \max\{i_m\}$. We associate with the polynomial $g(\bar{y})$ a Lie polynomial $f_g(x)$ such that

$$g(\bar{y}) \in \mathtt{Rad}(Y_k) \leftrightarrow f_g(x) \in \mathtt{Rad}(Y_F).$$

Set

$$
\begin{array}{rcl}
f_m(x) & = & s_{m-1}(x - c, v_1, \ldots, v_{n-1}) \\
f_{m-1}(x) & = & s_{m-1}(x - c, v_1, \ldots, v_{n-2}, v_n) \\
& \vdots & \\
f_1(x) & = & s_{n-1}(x - c, v_2, \ldots, v_{n-1}, v_n),
\end{array}
$$

where s_{n-1} is the Lie polynomial defined by Equation (14.1). Substituting the point p into the f_i's we obtain

$$f_m(p) = \alpha_m b_m; \quad f_{m-1}(p) = \alpha_{m-1} b_{m-1}; \quad \ldots; \quad f_1(p) = \alpha_1 b_1,$$

where $b_1, \ldots, b_m \in F$ are not zeros. Choose a Lie polynomial $a \in F$ so that the degree (see [2] for the definition) of a is greater than the degrees of the polynomials b_i. In that case all the products of the form

$ab_1 \cdots b_1 b_2 \cdots b_2 \cdots b_n \cdots b_n$ are nonzero (see [30] or [4]). We define the Lie polynomial $f_g(x)$ as

$$f_g(x) = \sum_{\bar{i}} \alpha_{\bar{i}} a \circ \underbrace{f_1(x) \circ \cdots \circ f_1(x)}_{i_1} \circ \underbrace{b_1 \circ \cdots \circ b_1}_{M_1 - i_1} \circ \cdots$$

$$\cdots \circ \underbrace{f_m(x) \circ \cdots \circ f_m(x)}_{i_m} \circ \underbrace{b_n \circ \cdots \circ b_n}_{M_m - i_m}.$$

We get, after substituting the point $x = p$ into $f_g(x)$,

$$f_g(p) = g(\alpha_1, \ldots, \alpha_n) a \circ \underbrace{b_1 \circ \cdots \circ b_1}_{M_1} \circ \cdots \circ \underbrace{b_m \circ \cdots \circ b_m}_{M_m}.$$

Therefore, $f_g(x) = 0$ if and only if $g(\alpha_1, \ldots, \alpha_n) = 0$.

We now define the correspondences

$$\mathrm{Rad}(Y_F) \to \mathrm{Rad}(S_k)$$

$$\mathrm{Rad}(Y_k) \to \mathrm{Rad}(S_F)$$

by setting

$$\mathrm{Rad}(Y_F) \to S_k = \{ S_f \mid f \in \mathrm{Rad}(Y_F) \}$$

and

$$\mathrm{Rad}(Y_k) \to S_F = \{ f_g(x) \in F[x] \mid g \in \mathrm{Rad}(Y_k) \} \cup s_m(x - c, v_1, \ldots, v_m),$$

where $V_F(s_m(x - c, v_1, \ldots, v_m)) = \mathbb{V}$; see Equation (14.1).

The previous argument holds not only in the case when $Y_F \subset V + c$ but also in a more general context $Y_F \subseteq F^n$. Furthermore,

$$\mathbb{V} = V_1 + c_1 \times \cdots \times V_n + c_n \subset F^n,$$

with

$$V_i = \mathrm{link} \{ v_1^i, \ldots, v_m^i \}.$$

Indeed, similarly to the dimension 1 case, we can define the correspondence between algebraic sets from $\mathbb{V}$ and affine subsets from k^M, where $M = m_1 + \cdots + m_n$, by setting

$$Y_F = \{ (\alpha_1^1 v_1^1 + \cdots + \alpha_{m_1}^1 v_{m_1}^1, \ldots, \alpha_1^n v_1^n + \cdots + \alpha_{m_1}^n v_{m_1}^n + c_n) \} \subseteq \mathbb{V}$$

$$\leftrightarrow Y_k = \{ \underbrace{\alpha_1^1, \ldots, \alpha_{m_1}^1}_{m_1}, \ldots, \underbrace{\alpha_1^n, \ldots, \alpha_{m_n}^n}_{m_n} \} \subseteq k^M.$$

$$(14.5)$$

The following theorem demonstrates the relation between this construction and the correspondence (14.5).

Theorem 14.11.

- Let $Y_k \subseteq k^M$ be an algebraic set over k. Then the corresponding set $Y_F \subset \mathbb{V}$ is algebraic over F. Moreover, $Y_F = V(S_F)$.
- Let $Y_F \subseteq \mathbb{V}$ be an algebraic set over F. Then the corresponding set Y_k is algebraic over k. Moreover, $Y_k = V(S_k)$.
- The maps $Y_F \to Y_k$ and $Y_k \to Y_F$ define a one-to-one correspondence $Y_F \leftrightarrow Y_k$ between algebraic (over F) sets from $\mathbb{V} \subset F$ and algebraic (over k) sets from k^M. Consequently, $Y_F \to Y_k \to Y_F = \mathrm{id}_{\mathrm{AS}(F)}$ and $Y_k \to Y_F \to Y_k = \mathrm{id}_{\mathrm{AS}(k)}$.
- The correspondence $\mathrm{Rad}(Y_F) \leftrightarrow \mathrm{Rad}(Y_k)$ defined above is a one-to-one correspondence between radical ideals from

$$k\left[y_1^1, \ldots, y_{m_1}^1, \ldots, y_1^n, \ldots, y_{m_n}^n\right]$$

and those radical ideals from $\mathfrak{F}[x_1, \ldots, x_n]$ that contain

$$s_{m_1}(x_1 - c_1), \ldots, s_{m_n}(x_n - c_n).$$

Corollary 14.12.

- Let S be a system of equations over F. If S defines a bounded algebraic set, then S is equivalent to a finite system S_0.
- If the ground field k is finite, then any subset $M \subset \mathbb{V}$ is algebraic over F.

A realization of a coordinate ring over k can be defined by analogy with Definition 8.14. The correspondence (14.5) can be reformulated in terms of coordinate algebras as follows.

Theorem 14.13. Let $Y_F \subset \mathbb{V}$ be a bounded algebraic set over F and let $Y_k \subseteq k^M$ be the corresponding algebraic set over k. Consider the F-algebra

$$C_F = \langle \xi_1, \ldots, \xi_n \rangle_F,$$

where each ξ_i has the form (14.2). Then C_F is a realization of $\Gamma_F(Y_F)$ if and only if the k-subring

$$C_k = \langle k, t_1^1, \ldots, t_{m_1}^1, \ldots, t_1^n, \ldots, t_{m_n}^n \rangle$$

is a realization of the coordinate ring $\Gamma_k(Y_k)$.

We have to address some of the differences between the case of dimension 1 and the case of higher dimensions. If $V = V_1 \times \cdots \times V_m$ is a fixed variety in k^N, we interpret it in k^N by partitioning the N variables into m groups, $N = n_1 + \cdots + n_m$. Note that the affine space k^N, $N > 1$, represented as a sum of m sub-spaces is not equivalent to the whole k^N, since the notion of a 'decomposed affine space' cannot be expressed in terms of morphisms of algebraic sets.

Note that the interpretations of bounded algebraic geometry over the algebra F in dimension 1 and in higher dimensions (as diophantine geometry over k) are very similar.

Finally, we classify algebraic sets defined by systems of equations with one variable. By Proposition 14.10 the coordinate algebra of any bounded algebraic set has a realization in $B(\bar{F})$. Moreover, according to Theorem 8.15, $\Gamma(Y)$ is isomorphic to the subalgebra $\langle F, \xi \rangle$ of $B(\bar{F})$. Using the machinery of combinatorics in Lie algebras [4, 30], one can show that the only alternatives are:

- $\xi \in B(\bar{F})$, in which case Y is bounded; or
- $\xi \notin B(\bar{F})$, in which case $\Gamma(Y) \cong F * \langle x \rangle$.

Theorem 14.14. *Every algebraic set defined by a system of equations in one variable over the free Lie algebra F is (up to isomorphism) either bounded, empty, or coincides with F.*

Note that the Jacobi identity is not essential for the proofs of most of the results of Section 14 and they can be generalized to free anti-commutative algebras [12, Appendix].

Acknowledgements

This work was supported by a grant from the London Mathematical Society. I am grateful to R. Bryant, E. Daniyarova and R. Stör for useful comments and remarks, as well as to S. Rees, A. Duncan, G. Megyesi, K. Goda, M. Batty for their support and to H. Khudaverdyan for lending me his office. My special thanks go to the London Mathematical Society for the unique opportunity to write this survey. This text would have never appeared without V. N. Remeslennikov, to whom I express my sincere gratitude.

Bibliography

[1] K. I. Appel. One-variable equations in free groups. *Proc. Amer. Math. Soc.*, 19:912–918, 1968.

[2] Y. A. Bahturin. *Identities in Lie Algebras*. Moscow, Nauka, 1985. In Russian.

[3] G. Baumslag, A. G. Myasnikov, and V. N. Remeslennikov. Algebraic geometry over groups I. Algebraic sets and ideal theory. *J. Algebra*, 219:16–79, 1999.

[4] L. A. Bokut and G. P. Kukin. *Algorithmic and Combinatorial Algebra*, volume 255 of *Mathematics and its Applications*. Kluwer Academic Publishers Group, Dordrecht, 1994.

[5] N. Bourbaki. *Elements of mathematics. Commutative algebra*. Hermann, Paris, 1972. Translated from the French.

[6] C. C. Chang and H. J. Keisler. *Model Theory*, volume 73 of *Studies in Logic and the Foundations of Mathematics*. North-Holland Publishing Co., Amsterdam, 1973.

[7] I. M. Chiswell and V. N. Remeslennikov. Equations in free groups with one variable 1. *J. Group Theory*, 3:445–466, 2000.

[8] E. Yu. Daniyarova. Elements of algebraic geometry over Lie algebras. Preprint.

[9] E. Yu. Daniyarova, I. V. Kazatchkov, and V. N. Remeslennikov. Algebraic geometry over free metabelian Lie algebra I: U-algebras and A-modules. Preprint N34, Omsk, OmGAU, 2001.

[10] E. Yu. Daniyarova, I. V. Kazatchkov, and V. N. Remeslennikov. Algebraic geometry over metabelian Lie algebras I: U-algebras and universal classes. *Fundam. Prikl. Mat.*, 9(3):37–63, 2003. Translation in *J. Math. Sci. (New York)*, 135(5):3290–3310, 2006.

[11] E. Yu. Daniyarova, I. V. Kazatchkov, and V. N. Remeslennikov. Algebraic geometry over metabelian Lie algebras II: The finite field case. *Fundam. Prikl. Mat.*, 9(3):65–87, 2003. Translation in *J. Math. Sci. (New York)*, 135(5):3311–3326, 2006.

[12] E. Yu. Daniyarova and V. N. Remeslennikov. Bounded algebraic geometry over a free Lie algebra. *Algebra Logika*, 44(3):269–304, 2005. Translation in *Algebra Logic*, 44(3):148–167, 2005.

[13] C. Faith. *Algebra: rings, modules and categories. I*. Die Grundlehren der mathematischen Wissenschaften, Band 190. Springer-Verlag, New York, 1973.

[14] R. Goebel and S. Shelah. Radicals and Plotkin's problem concerning geometrically equivalent groups. *Proc. Amer. Math. Soc.*, 130:673–674, 2002.

[15] V. A. Gorbunov. *Algebraic theory of quasivarieties*, volume 5 of *Sibirskaya Shkola Algebry i Logiki*. Nauchnaya Kniga, Novosibirsk, 1998.

[16] V. Guirardel. Limit groups and groups acting freely on $\mathbb{R}^n$-trees. *Geom. Topol.*, 8:1427–1470 (electronic), 2004. Cf. arXiv math.GR/0306306.

[17] R. Hartshorne. *Algebraic Geometry*, volume 52 of *Graduate Texts in Mathematics*. Springer-Verlag, New York-Heidelberg-Berlin, 1977.

[18] J. L. Kelley. *General topology*. D. Van Nostrand Company, Inc., Toronto-New York-London, 1955.

[19] O. Kharlampovich and A. Myasnikov. Irreducible affine varieties over a free group. I: Irreducibility of quadratic equations and Nullstellensatz. *J. Algebra*, 200:472–516, 1998.

[20] O. Kharlampovich and A. Myasnikov. Irreducible affine varieties over a free group. II: Systems in row-echelon form and description of residually free groups. *J. Algebra*, 200:517–570, 1998.

[21] S. Lang. *Algebra. Revised third edition.*, volume 211 of *Graduate Texts in Mathematics*. Springer-Verlag, New York, 2002.

[22] A. A. Lorents. The solution of systems of equations in one unknown in free groups. *Dokl. Akad. Nauk.*, 148:1253–1256, 1963.

[23] A. A. Lorents. Equations without coefficients in free groups. *Dokl. Akad. Nauk.*, 160:538–540, 1965.

[24] R. C. Lyndon. Equations in free groups. *Trans. Amer. Math. Soc*, 96:445–457, 1960.

[25] A. I. Malcev. Some remarks on quasi-varieties of algebraic structures. *Algebra and Logic*, 5(3):3–9, 1966.

[26] A. I. Malcev. *Algebraic Structures*. Moscow, Nauka, 1987.

[27] A. G. Myasnikov and V. N. Remeslennikov. Algebraic geometry over groups II. Logical Foundations. *J. Algebra*, 234:225–276, 2000.

[28] B. Plotkin. Varieties of algebras and algebraic varieties. Categories of algebraic varieties. *Siberian Advances in Math.*, 7(2):64–97, 1997.

[29] V. A. Roman'kov. On equations in free metabelian groups. *Siberian Math. J.*, 20(3):671–673, 1979.

[30] A. I. Shirshov. *Selected papers. Rings and Algebras.* "Nauka", Moscow, 1984. In Russian.

I. V. Kazachkov

805 Sherbrooke St. West

Mathematics and Statistics Department

McGill University

Montreal, QC,

Canada H3A 2K6

ilya.kazachkov@gmail.com

Destabilization of closed braids

Andrey V. Malyutin

Acknowledgements

I would like to thank Professor Peter Walters and Warwick Mathematics Institute for their warm hospitality.

Introduction

This paper concerns braid theory and several related subjects: knot theory, surface mapping class groups, hyperbolic geometry of surfaces, etc.

The central topic of the paper is the Markov destabilization of braids. Here we shortly describe some features of the 'destabilization problem' in the braid group and explain how the other topics of this paper relate to this one. There are many works related to destabilization, and we refer to only a few of them. See for instance [7, 6, 10, 5, 8, 9], [14, 17] and [19, 20, 21]; older references include [2, 3, 22, 23, 24].

The *Artin braid group* B_n on n strands is determined by the presentation

$$B_n := \langle \sigma_1, \ldots, \sigma_{n-1} \mid \sigma_i \sigma_j = \sigma_j \sigma_i, \ |i - j| \geq 2; \ \sigma_i \sigma_{i+1} \sigma_i = \sigma_{i+1} \sigma_i \sigma_{i+1} \rangle.$$

The group B_1 is trivial, while $B_2 \cong \mathbb{Z}$. Elements of B_n are called *braids*. The generators $\sigma_1, \ldots, \sigma_{n-1}$ are *Artin's generators*.

Closed braids

Suppose $L \subset \mathbb{R}^3$ is an oriented link in the oriented 3-space $\mathbb{R}^3$. Suppose $A \subset \mathbb{R}^3 \smallsetminus L$ is an infinite 'unknotted' curve in $\mathbb{R}^3$ such that the space $\mathbb{R}^3 \smallsetminus A$ is homeomorphic to the open solid torus. Recall that an open

82

2-disc D in an open solid torus T is said to be *meridional* if the space $T \setminus D$ is homeomorphic to the open 3-ball. The link L is called a *closed braid with axis A* if there is a fibration of the open solid torus $\mathbb{R}^3 \setminus A$ into meridional discs such that L intersects each disc of this fibration transversally (and orientations of the intersections are coherent). The number of intersection points of L and a disc of the fibration is the *index* of the closed braid (L, A). The closed braids of index n are called *n-braids*. We use the term 'closed braid' to refer to an isotopy class of closed braids as well.

The standard 'closing procedure' (see, e.g. [2]) transforms a braid $\beta \in B_n$ into a closed n-braid. There is a well-known one-to-one correspondence between conjugacy classes in the braid groups $B_1, B_2, B_3, \ldots$ and isotopy classes of closed braids. For this reason, we denote by $\widehat{\beta}$ both the conjugacy class of a braid β and the corresponding closed braid, which we identify with the conjugacy class.

We denote by $\mathcal{L}(\widehat{\alpha})$ (or just by $\mathcal{L}(\alpha)$) the link type represented by a closed braid $\widehat{\alpha}$.

The destabilization problem

We say that a closed $(n-1)$-braid (= a conjugacy class) $\widehat{\alpha} \subset B_{n-1}$ is obtained from a closed n-braid $\widehat{\beta} \subset B_n$ by a *positive* (respectively, *negative*) *destabilization* if there is a braid $\alpha_1 \in \widehat{\alpha}$ such that $\alpha_1 \sigma_{n-1} \in \widehat{\beta}$ (respectively, $\alpha_1 \sigma_{n-1}^{-1} \in \widehat{\beta}$):

$$\widehat{\beta} \ni \alpha_1 \sigma_{n-1}^{\pm 1} \quad \longmapsto \quad \alpha_1 \in \widehat{\alpha}.$$

We also say that $\widehat{\beta}$ is obtained from $\widehat{\alpha}$ by a *stabilization*.

Thus, $\widehat{\beta}$ *admits a destabilization*, or is *destabilizable*, if $\widehat{\beta}$ contains a braid of the form $\sigma_{n-1}^{\pm 1} W$, where W is a word in the letters $\{\sigma_1^{\pm 1}, \ldots, \sigma_{n-2}^{\pm 1}\}$. We observe that $\widehat{\beta}$ admits a positive destabilization if and only if $\widehat{\beta}^{-1}$ admits a negative destabilization. Hence, $\widehat{\beta}$ is destabilizable if and only if $\widehat{\beta}^{-1}$ is destabilizable.

Note that we can use the generator σ_1 as the 'outermost generator' instead of σ_{n-1}: a closed braid $\widehat{\beta}$ admits a destabilization if and only if $\widehat{\beta}$ contains a braid of the form $\sigma_1^{\pm 1} W$, where W is a word in the letters $\{\sigma_2^{\pm 1}, \ldots, \sigma_{n-1}^{\pm 1}\}$.

Stabilization and destabilization of braids were introduced by Markov in 1935 (see [22]). The celebrated Markov theorem states that two closed braids represent one and the same link type if and only if they are related by a finite sequence of stabilizations and destabilizations.

While a stabilization can be applied to any closed braid (furthermore, it can 'usually' be applied in many different ways), not each closed braid is destabilizable. The '*destabilization problem*' is the problem of recognizing whether a given closed braid admits a destabilization. The formulation of this problem can be found, for example, in [17, 4].

In this paper we present, in particular, an algorithm which solves the 'destabilization problem', i.e. an algorithm which detects whether a given closed braid is destabilizable. By an 'algorithm' we mean a 'computational procedure' that takes a finite number of steps.

Approaches to the destabilization problem and related problems

We consider the following three approaches to the destabilization problem.

The 'knot approach'. Under this approach, we regard braids (or rather closed braids) as objects in 3-space (see the remarks above). Here the destabilization problem has a clear knot-geometric meaning. The problems most closely related to the destabilization problem are the problems of admissibility of other 'braid moves', e.g. the *exchange move* and the *generalized flype*. Braid moves that are different from the stabilization and destabilization were introduced by Birman and Menasco, see [5, 6, 7, 8, 10].

The 'group approach'. This approach is focused on the group structure. We actually used this approach above in the definition of the destabilization. Admissibility of other braid moves can also be formulated in these terms.

From the point of view of group theory, the 'braid moves problems' belong to a vast family of recognition problems for discrete groups. The 'simplest' questions in this family are the word and conjugacy problems, the problems of (stable) commutator length, etc. Another sample problem is that of recognizing whether a given element of a group G belongs to the product of a given collection of subsets (say, subgroups) of G. For mapping class groups, we have, e.g. the problem of detecting how many reducible elements of a group we must multiply to obtain a given element.

The 'surface approach'. This approach treats the braid group B_n as the mapping class group of the n-punctured disc D_n, so that we regard braids as classes of automorphisms of D_n. It turns out that the destabilization problem can be formulated in terms of the action of the braid group (= the mapping class group) on the space of isotopy classes of curves (see Subsection 4.1 below).

Admissibility problems for other braid moves can also be formulated in these terms. From this point of view, recognition problems for braid moves are just a few problems in a long list that includes, among others, the problem of detecting the Nielsen-Thurston type of a surface automorphism, the problem of detecting the strong irreducibility (this problem is not solved yet), etc.

To construct an algorithm solving the destabilization problem, we use the 'surface approach'. It involves results of hyperbolic geometry and the Nielsen-Thurston theory. The methods developed can be applied to recognize (in 'generic' cases) whether a given braid admits an exchange move or a flype. Also, these methods enable us to detect the strong irreducibility in mapping class groups of certain punctured surfaces and surfaces with nonempty boundary.

Structure of the paper

In Section 1, we present results concerning closed braids that can be destabilized in an infinite number of different ways.

In Section 2, we give auxiliary definitions and results related to the hyperbolic geometry on a punctured disc.

Section 3 concerns the 'surface approach' to the braid group.

In Section 4, we describe an algorithm that solves the destabilization problem.

1 'Infinitely destabilizable' braids

Notation. Let $\widehat{\beta}$ be a closed n-braid. We denote by $\mathfrak{D}(\widehat{\beta})$ the set of all closed $(n-1)$-braids that are obtained from $\widehat{\beta}$ by a destabilization.

By $\mathfrak{S}(\widehat{\beta})$ we denote the set of all closed $(n+1)$-braids that are obtained from $\widehat{\beta}$ by a stabilization. Thus, we have

$$\widehat{\alpha} \in \mathfrak{D}(\widehat{\beta}) \iff \widehat{\beta} \in \mathfrak{S}(\widehat{\alpha}).$$

Proposition 1.1. *Let $\beta \in B_3$ be a 3-braid representing a knot (i.e. a one-component link). Let $\beta_\star$ be the 5-braid represented by the word $\beta\sigma_2^{-2}\sigma_3\sigma_4\sigma_2\sigma_3$:*

$$\beta_\star := \beta\sigma_2^{-2}\sigma_3\sigma_4\sigma_2\sigma_3 \in B_5.$$

(Obviously, $\beta_\star$ represents the same knot as β.) Then the set $\mathfrak{D}(\widehat{\beta}_\star)$ is infinite.

Proof. We set

$$\beta_s := \sigma_4^{-s}\beta_\star\sigma_4^s.$$

(It is clear that $\beta_s \in \widehat{\beta}_\star$.) We obviously have $\sigma_i^s\sigma_{i+1}\sigma_i = \sigma_{i+1}\sigma_i\sigma_{i+1}^s$. Using this relation, we obtain

$$
\begin{aligned}
\beta_s &= \sigma_4^{-s}\beta\sigma_2^{-2}\sigma_3\sigma_4\sigma_2\sigma_3\sigma_4^s \\
&= \beta\sigma_2^{-2}\sigma_4^{-s}\sigma_3\sigma_4\sigma_2\sigma_3\sigma_4^s \\
&= \beta\sigma_2^{-2}\sigma_3\sigma_4\sigma_3^{-s}\sigma_2\sigma_3\sigma_4^s \\
&= \beta\sigma_2^{-2}\sigma_3\sigma_4\sigma_2\sigma_3\sigma_2^{-s}\sigma_4^s \qquad (1.1)\\
&= \beta\sigma_2^{-2}\sigma_3\sigma_2\sigma_4\sigma_3\sigma_4^s\sigma_2^{-s} \\
&= \beta\sigma_2^{-2}\sigma_3\sigma_2\sigma_3^s\sigma_4\sigma_3\sigma_2^{-s} \\
&= \beta\sigma_2^{s-2}\sigma_3\sigma_2\sigma_4\sigma_3\sigma_2^{-s}.
\end{aligned}
$$

We destabilize β_s:

$$\beta_s = \beta\sigma_2^{s-2}\sigma_3\sigma_2\sigma_4\sigma_3\sigma_2^{-s} \ (\in B_5)$$

$$\xrightarrow{\text{destabilization}} \beta\sigma_2^{s-2}\sigma_3\sigma_2\sigma_3\sigma_2^{-s} \ (\in B_4)$$

$$= \beta\sigma_2^{s-2}\sigma_2\sigma_3\sigma_2\sigma_2^{-s} = \beta\sigma_2^{s-1}\sigma_3\sigma_2^{1-s}$$

and define

$$\alpha_i := \beta\sigma_2^i\sigma_3\sigma_2^{-i} \in B_4.$$

Taking the result of the above destabilization and assuming that $i = s - 1$, we get the inclusion

$$\{\widehat{\alpha}_i\}_{\mathbb{Z}} \subset \mathfrak{D}(\widehat{\beta}_\star).$$

Let us show that the set $\{\widehat{\alpha}_i\}_{\mathbb{Z}}$ of closed braids (i.e. conjugacy classes) is infinite. Consider the homomorphism

$$\varrho : B_4 \to B_3 \qquad (\sigma_3 \mapsto \sigma_1, \sigma_1 \mapsto \sigma_1, \sigma_2 \mapsto \sigma_2).$$

Recall that any group homomorphism induces a map on the level of conjugacy classes. We see that the map of conjugacy classes induced by the homomorphism ϱ maps $\{\widehat{\alpha}_i\}_{\mathbb{Z}}$ to $\{\widehat{\alpha}'_i\}_{\mathbb{Z}}$, where $\alpha'_i = \beta\sigma_2^i\sigma_1\sigma_2^{-i}$. In order to prove that the set $\{\widehat{\alpha}_i\}_{\mathbb{Z}}$ is infinite, it is sufficient to show that the set $\{\widehat{\alpha}'_i\}_{\mathbb{Z}}$ is infinite, which in turn follows from the fact that the set of link types $\{\mathcal{L}(\widehat{\alpha}'_i)\}_{\mathbb{Z}}$ is infinite (recall that $\mathcal{L}(\widehat{\alpha})$ denotes the link type represented by $\widehat{\alpha}$).

Let

$$ S : B_n \to S_n, \qquad \sigma_i \mapsto (i, i+1) $$

be the standard homomorphism from B_n to the symmetric group. Since β represents a knot, it follows that $S(\beta) \in \{(1, 2, 3); (3, 2, 1)\}$.

We consider the case where $S(\beta) = (1, 2, 3)$, the proof in the second case being similar. Suppose i is even (say, $i = 2k$). We see that $S(\sigma_2^{2k}\sigma_1\sigma_2^{-2k}) = (1, 2)$. Then we have

$$ S(\alpha'_{2k}) = S(\beta\sigma_2^{2k}\sigma_1\sigma_2^{-2k}) = (1, 2, 3) \circ (1, 2) = (2, 3). $$

This means that $\widehat{\alpha}'_{2k}$ represents a two-component link.

We easily see that the linking number of the two components of the link $\mathcal{L}(\widehat{\alpha}'_{2k}) = \mathcal{L}(\beta\sigma_2^{2k}\sigma_1\sigma_2^{-2k})$ is equal to $b + k$, where b is a constant depending on β only. This means that if $k_1 \neq k_2$, then the oriented links $\mathcal{L}(\widehat{\alpha}'_{2k_1})$ and $\mathcal{L}(\widehat{\alpha}'_{2k_2})$ are different, whence it follows that the conjugacy classes $\widehat{\alpha}'_{2k_1}$ and $\widehat{\alpha}'_{2k_2}$ are different. Therefore, the sets $\{\widehat{\alpha}'_i\}_{\mathbb{Z}}$ and $\{\widehat{\alpha}_i\}_{\mathbb{Z}}$ are infinite. This completes the proof. $\qquad\square$

Remark 1.2. Using Proposition 1.1, it is not hard to prove that any knot (or, more generally, any link) is represented by a braid β such that the set $\mathfrak{D}(\widehat{\beta})$ is infinite.

Definition 1.3. We say that a closed braid $\widehat{\beta}$ is *doubly destabilizable* if $\widehat{\beta}$ contains a braid of the form $\sigma_2\sigma_1\sigma_3\sigma_2 W$ or of the form $\sigma_2^{-1}\sigma_1^{-1}\sigma_3^{-1}\sigma_2^{-1} W$, where W is a word in the letters $\{\sigma_3^{\pm 1}, \ldots, \sigma_{n-1}^{\pm 1}\}$. In particular, it follows that if $\widehat{\beta}$ is doubly destabilizable, then $\mathfrak{D}(\widehat{\beta})$ contains a destabilizable braid. Clearly, the braid $\widehat{\beta}_*$ in Proposition 1.1 is doubly destabilizable.

Proposition 1.4. *Suppose $\widehat{\beta}$ is a closed braid such that the set $\mathfrak{D}(\widehat{\beta})$ is infinite. Then $\widehat{\beta}$ is doubly destabilizable.*

We omit the proof.

2 Preliminaries: the free group, punctured disc, and hyperbolic geometry

This section contains notation, definitions, and several statements that easily follow from well-known results and simple 'geometric' considerations. Proofs are mostly omitted.

2.1 The free group and its boundary

Let F be the free group of rank $n \geq 2$ with generators $\{u_1, \ldots, u_n\}$:

$$F = \langle u_1, \ldots, u_n \mid \rangle.$$

We denote by U the set of the generators and their inverses:

$$U := \{u_1, \ldots, u_n; u_1^{-1}, \ldots, u_n^{-1}\}.$$

Recall that a word $w_1 w_2 \ldots w_k$ in letters of the alphabet U is said to be *reduced* if $w_i \neq w_{i+1}^{-1}$ for each $i \in \{1, \ldots, k-1\}$; recall that an element $v \in F$ is represented by a unique reduced word. The number of letters in the reduced word representing an element $v \in F$ is called the *length* of v and we denote it by $|v|$.

The boundary ∂F. We denote by ∂F the *boundary* of the free group F. Recall that there exists a one-to-one correspondence between the set of points of ∂F and the set of all infinite (to the right) reduced words in generators and their inverses. We use the following metric on the set $F \cup \partial F$: let $v \neq w \in F \cup \partial F$, and let $V = v_1 v_2 \ldots$ and $W = w_1 w_2 \ldots$ be the reduced words (either finite or infinite) in generators and their inverses that represent v and w, respectively. Then we set $d(v, w) := 1/r$, where r is the smallest positive integer such that $v_r \neq w_r$ (we mean that the inequality $v_r \neq w_r$ holds true, in particular, if we have either $|v| = r - 1$ or $|w| = r - 1$). Clearly, the metric space $(F \cup \partial F, d)$ is compact.

Action on the boundary. The free group acts naturally on its boundary. We give two (slightly) different descriptions of this action.

On the one hand, any element $v \in F$ determines the left shift $\phi_v : a \mapsto va$ of the group F. Being an isometry of F (in the standard *word metric* on F), ϕ_v has a unique continuous (with respect to the metric d) extension

$$\overline{\phi}_v : F \cup \partial F \to F \cup \partial F.$$

On the other hand, an element $v \in F$ determines the inner automorphism $\psi_v : a \mapsto vav^{-1}$ of the group F. Being a quasi-isometry of F with word metric, ψ_v has a unique continuous extension

$$\overline{\psi}_v : F \cup \partial F \to F \cup \partial F.$$

Clearly, for any $v \in F$ the mappings $\overline{\psi}_v$ and $\overline{\phi}_v$ coincide on ∂F. We denote the result of the action of an element $v \in F$ on a point $w \in \partial F$ by vw:

$$vw := \overline{\phi}_v(w) = \overline{\psi}_v(w).$$

Infinite powers. We see that for each nontrivial element $v \in F$ the sequence of powers $(v, v^2, v^3, \dots)$ converges to a point of the boundary ∂F. We denote this point by $v^{+\infty}$. We also use the notation $v^{-\infty} := (v^{-1})^{+\infty}$.

2.2 The fundamental group of the punctured disc

The punctured disc D_n. Let $\mathbb{C}$ be the complex plane. We choose n distinct points on the real interval $(-1, 1) \subset \mathbb{C}$ and denote these points by $z_1, \dots, z_n$ from the left to the right. For convenience, we also denote by z_0 and by z_{n+1} the points -1 and 1 of $\mathbb{C}$, respectively.

We denote by D^2 the closed disc of radius 1 with center 0 in the complex plane $\mathbb{C}$, and we denote by D_n the disc D^2 with the points $z_1, \dots, z_n$ removed. We also fix the point $x_0 := -i \in \partial D_n$ on the boundary of the punctured disc D_n as a base point.

The fundamental group π_1. The fundamental group $\pi_1(D_n, x_0)$ is a free group of rank n. In what follows, we denote the group $\pi_1(D_n, x_0)$ just by π_1. Recall that an element of $\pi_1 = \pi_1(D_n, x_0)$ is a homotopy class of oriented loops

$$\gamma : ([0, 1], \{0, 1\}) \to (D_n, x_0).$$

We denote by u_k $(1 \leq k \leq n)$ the element of π_1 represented by a loop that intersects the interval $(-1, 1) \subset \mathbb{C}$ exactly twice: the first time (recall that we consider oriented loops) between the points z_{k-1} and z_k and the second time between the points z_k and z_{k+1}.

Obviously, the set $\{u_1, \dots, u_n\}$ generates the free group π_1.

Definition 2.1 (The distinguished element of π_1). Let $\mathbf{u} \in \pi_1$ be the homotopy class of the loop having endpoints at x_0 and tracing the circle

∂D_n once clockwise. Obviously, $\mathbf{u} = u_1 u_2 \cdots u_n$. In what follows, $\mathbf{u}$ is called the *distinguished element*.

We denote by $\pi_1^{(0)}$ the set of powers of $\mathbf{u}$, i.e.

$$\pi_1^{(0)} := \{\ldots, \mathbf{u}^{-1}, \mathbf{u}^0 = e, \mathbf{u}^1, \mathbf{u}^2, \ldots\};$$

and we denote by $\pi_1^{(1)}$ the set of all other elements of π_1, i.e.

$$\pi_1^{(1)} := \pi_1 \smallsetminus \pi_1^{(0)}.$$

2.3 Universal covering space

We denote by $D_n^\sim$ the universal covering space of the punctured disc D_n. By definition $D_n^\sim$ is a simply connected space. Hence the interior of $D_n^\sim$ is homeomorphic to an open 2-disc. The boundary $\partial D_n^\sim$ (it is the inverse image of ∂D_n) is a countable collection of open intervals.

Let $\mathcal{X}$ be the inverse image of the base point $x_0 \in \partial D_n$ (see the notation of the previous section) in the universal cover. Note that $\mathcal{X} \subset \partial D_n^\sim$. We fix, once and for all, a base point $\mathbf{x} \in \mathcal{X}$.

We establish a one-to-one correspondence between $\mathcal{X}$ and π_1 by assigning to an element $v \in \pi_1 = \pi_1(D_n, x_0)$ the endpoint of a curve $\gamma : [0,1] \to D_n^\sim$, where γ is the lift of any loop representing v such that $\gamma(0) = \mathbf{x}$.

Remark 2.2 (An important convention). In what follows, we identify the elements of the group π_1 with the corresponding points of $\mathcal{X} \subset \partial D_n$. In particular, the base point $\mathbf{x}$ corresponds to the trivial element e of π_1.

Finally, recall that the fundamental group of a topological space is isomorphic to the group of *deck transformations* of the universal cover of this space. A canonical isomorphism between the group π_1 and the group of deck transformations of $D_n^\sim$ takes an element $v \in \pi_1$ to the deck transformation that maps the base point $\mathbf{x}$ to the point $v \in \mathcal{X} = \pi_1$.

2.4 Hyperbolic structure

It is well known that D_n admits a hyperbolic metric (provided $n \geq 2$) in which the circle ∂D_n is a geodesic and the punctures are cusps. Now we fix, once and for all, such a hyperbolic metric on D_n. (There exists a vast family of such metrics, but the choice is irrelevant for the constructions below.)

Hyperbolic structure on the universal cover. The hyperbolic metric on D_n lifts to a hyperbolic metric on the universal cover $D_n^\sim$. It is not hard to prove that $D_n^\sim$ is convex with respect to this metric. Hence, $D_n^\sim$ can be isometrically embedded in the hyperbolic plane $\mathbb{H}^2$, and we identify $D_n^\sim$ with a subset of $\mathbb{H}^2$. Under this inclusion, the boundary of $D_n^\sim \subset \mathbb{H}^2$ is a union of disjoint geodesics in $\mathbb{H}^2$, and $\mathbb{H}^2 \setminus D_n^\sim$ is a union of disjoint open half-planes.

Points at infinity. We denote by $\overline{\mathbb{H}}^2$ the standard compactification of the hyperbolic plane, i.e. the closed disc obtained from $\mathbb{H}^2$ by adding a circle at infinity (the *absolute*). We denote by $D_n^{\overline{\sim}}$ the closure of $D_n^\sim$ in $\overline{\mathbb{H}}^2$ ($D_n^{\overline{\sim}}$ is a compactification of the space $D_n^\sim$); thus $D_n^{\overline{\sim}} \cap \mathbb{H}^2 = D_n^\sim$.

Since $D_n^\sim$ is convex, $D_n^{\overline{\sim}}$ is homeomorphic to a closed disc. We denote by $\mathcal{K}$ the set of the limit points of $D_n^\sim$ on the absolute:

$$\mathcal{K} := D_n^{\overline{\sim}} \setminus D_n^\sim = D_n^{\overline{\sim}} \cap \partial \overline{\mathbb{H}}^2.$$

The set $\mathcal{K}$ is a Cantor subset of the absolute.

2.5 The circle Θ

Since the compactification $D_n^{\overline{\sim}} \subset \overline{\mathbb{H}}^2$ of the universal cover is homeomorphic to a closed disc, the boundary $\partial D_n^{\overline{\sim}}$ is homeomorphic to a circle. We denote this boundary by Θ. Note that

$$\Theta = \partial D_n^\sim \cup \mathcal{K}.$$

The natural inclusion $\pi_1 \subset \Theta$. Since $\partial D_n^\sim \subset \Theta$, Remark 2.2 yields a natural inclusion

$$\pi_1 = \mathcal{X} \subset \Theta.$$

In particular, the base point $\mathbf{x} = e$ is now a base point of Θ.

Proposition 2.3. *The inclusion $\pi_1 \to \Theta$ has a unique continuous extension*

$$\pi_1 \cup \partial \pi_1 \to \Theta.$$

The image of the boundary $\partial \pi_1$ under this mapping is $\mathcal{K}$.

The proof of this is standard.

By the convention adopted in Remark 2.2 we have $\pi_1 \subset \Theta$. We 'extend' this convention to the boundary $\partial \pi_1$. If $w \in \partial \pi_1$, we denote the corresponding point of $\mathcal{K} \subset \Theta$ by the same symbol w. Note that under this convention, in particular, we have $v^{+\infty} \in \Theta$ for any $v \in \pi_1 \setminus e$.

Remark 2.4. It can be shown that the inverse image in $\partial \pi_1$ of a point $y \in \mathcal{K}$ is either one point or a pair of points, furthermore, we can completely describe the set of pairs of points in $\partial \pi_1$ that are 'glued' in $\mathcal{K}$: it is the set

$$\{\{vu_i^{+\infty}, vu_i^{-\infty}\} \mid v \in \pi_1, u_i \in \{u_1, \ldots, u_n\}\}.$$

(See the notation in Section 2.1)

Definition 2.5 (Equivariance). Suppose a group G acts on a space S_1 (an *action* of G on a topological space X is a homomorphism from G to the group $\mathrm{Homeo}(X)$ of homeomorphisms of X), and suppose G also acts on another space, say S_2. Recall that a mapping $m : S_1 \to S_2$ is *equivariant* with respect to these actions if for each $g \in G$ and each $s \in S_1$ we have $m(g(s)) = g(m(s))$.

Definition 2.6 (Natural action of π_1 on Θ). It is clear that each deck transformation of $D_n^{\sim}$ is an isometry in the hyperbolic metric and is uniquely extended to a homeomorphism of the disc $D_n^{\overline{\sim}}$. Therefore, each deck transformation induces a homeomorphism of the circle $\Theta = \partial D_n^{\overline{\sim}}$. The isomorphism between the group π_1 and the group of deck transformations of $D_n^{\sim}$ (see Section 2.3) determines a homomorphism

$$\pi_1 \to \mathrm{Homeo}_+(\Theta).$$

In other words, we obtain an action of the group π_1 on the space Θ. We denote by $v(y)$ (or just by vy) the result of the action of an element $v \in \pi_1$ on a point $y \in \Theta$.

The definition of this action easily implies the following result.

Proposition 2.7. *The inclusion $\pi_1 \to \Theta$ is equivariant with respect to the action of π_1 on Θ (as described just above) and the action of π_1 on itself by left multiplication.*

Furthermore, the continuous extension $\pi_1 \cup \partial \pi_1 \to \Theta$ (see Proposition 2.3) is equivariant with respect to the aforementioned action of π_1 on Θ and the extended action of π_1 on $\pi_1 \cup \partial \pi_1$ by left multiplication (see Section 2.1).

Proposition 2.8. *Let v be a nontrivial element of the fundamental group π_1. Then the corresponding homeomorphism $v : \Theta \to \Theta$ has either exactly one or exactly two fixed points. More precisely:*

- *if v is conjugate to a power of a generator (say, $v = yu_i^k y^{-1}$, where $y \in \pi_1$ and k is a nonzero integer), then the homeomorphism $v : \Theta \to$*

Θ *has exactly one fixed point* $v^{+\infty} = yu_i^{+\infty} = yu_i^{-\infty} = v^{-\infty} \in \Theta$ (see Remark 2.4).

- *if v is not conjugate to a power of a generator, then the homeomorphism* $v : \Theta \to \Theta$ *has exactly two fixed points* $v^{-\infty}$ *and* $v^{+\infty}$; *the fixed point* $v^{-\infty}$ *is repelling, while the fixed point* $v^{+\infty}$ *is attracting.*

2.6 Orderings

2.6.1 Definitions

The linear ordering '$\prec$'. We consider the space $\Theta \smallsetminus \mathbf{x}$ (we also write '$\Theta \smallsetminus e$' to refer to this space, see Remark 2.2). Being homeomorphic to the real line, this space bears two natural mutually inverse total orderings. We denote by $\prec$ the ordering for which

$$u_1 \prec u_2.$$

(See Section 2.5: under the convention adopted in Remark 2.2, the generators u_1 and u_2 of π_1 are also points of Θ.) We also use the symbol $\preceq$:

$$(\theta_1 \preceq \theta_2) \iff \text{either } (\theta_1 \prec \theta_2) \text{ or } (\theta_1 = \theta_2).$$

In what follows, the statement 'the condition $\theta_1 \prec \theta_2$ holds for a pair of points $\theta_1, \theta_2 \in \Theta$' means that:

$$\mathbf{x} \neq \theta_1 \neq \theta_2 \neq \mathbf{x} \quad \text{and} \quad \theta_1 \prec \theta_2.$$

Definition 2.9 (Intervals). Suppose that θ_1 and θ_2 are points of $\Theta \smallsetminus e$. We introduce the following natural notation:

$$(\theta_1, \theta_2) := \{\theta \in \Theta \smallsetminus e \mid \theta_1 \prec \theta \prec \theta_2\};$$

$$[\theta_1, \theta_2) := \{\theta \in \Theta \smallsetminus e \mid \theta_1 \preceq \theta \prec \theta_2\};$$

$$(\theta_1, \theta_2] := \{\theta \in \Theta \smallsetminus e \mid \theta_1 \prec \theta \preceq \theta_2\};$$

$$[\theta_1, \theta_2] := \{\theta \in \Theta \smallsetminus e \mid \theta_1 \preceq \theta \preceq \theta_2\}.$$

These sets are called *intervals*.

Cyclic ordering. Since the space Θ is homeomorphic to a circle, there are two natural mutually inverse complete cyclic orderings on Θ. We use the ordering, which corresponds to the ordering $\prec$. We denote this cyclic ordering by the symbol $\curvearrowright$. Namely, we write

$$x_1 \curvearrowright x_2 \curvearrowright x_3 \quad \text{and} \quad e \curvearrowright x_1 \curvearrowright x_2$$

whenever $x_1 \prec x_2 \prec x_3$.

We also write

$$x_2 \curvearrowright x_3 \curvearrowright x_1$$

whenever

$$x_1 \curvearrowright x_2 \curvearrowright x_3.$$

Note that these rules determine our cyclic ordering completely.

Remark. By the above definitions, the cyclic ordering $\curvearrowright$ is invariant with respect to the action of π_1 on Θ, i.e. for any $v \in \pi_1$ we have $vx_1 \curvearrowright vx_2 \curvearrowright vx_3$ if $x_1 \curvearrowright x_2 \curvearrowright x_3$.

Notation. Suppose $x_1, x_2, \ldots, x_j$ are points in Θ. The notation

$$(*) \qquad\qquad x_1 \curvearrowright x_2 \curvearrowright \cdots \curvearrowright x_j$$

means that

(i) the points $x_1, x_2, \ldots, x_j$ are pairwise distinct, and
(ii) if we move along the circle Θ clockwise from the point x_1, then first we meet the point x_2, after that we meet x_3, ..., then x_j, and then x_1 again.

Observe that $(*)$ is equivalent to

$$x_2 \curvearrowright \cdots \curvearrowright x_j \curvearrowright x_1.$$

We note also that if $f : \Theta \to \Theta$ is an orientation-preserving homeomorphism of Θ, then $(*)$ is equivalent to

$$f(x_1) \curvearrowright f(x_2) \curvearrowright \cdots \curvearrowright f(x_j).$$

2.6.2 Comparing the elements of π_1 with respect to $\prec$

In this subsection, we explain how to compare the elements of the fundamental group $\pi_1 \subset \Theta$ with respect to the ordering $\prec$.

First, observe that we have

$$u_1 \prec u_1^{-1} \prec u_2 \prec u_2^{-1} \prec \cdots \prec u_n \prec u_n^{-1}$$

or, equivalently,

$$e \curvearrowright u_1 \curvearrowright u_1^{-1} \curvearrowright u_2 \curvearrowright u_2^{-1} \curvearrowright \cdots \curvearrowright u_n \curvearrowright u_n^{-1}.$$

Now, let b and c be distinct nontrivial elements of the free group π_1. Let $B = b_1 b_2 \ldots$ and $C = c_1 c_2 \ldots$ be the reduced words representing b and c, respectively.

If their first letters b_1 and c_1 are distinct, then

$$b \prec c \iff b_1 \prec c_1;$$

$$b \succ c \iff b_1 \succ c_1.$$

If $b_1 = c_1$, let us denote by V the maximal common initial subword of the words B and C. This means that

$$V = v_1 \ldots v_i := b_1 \ldots b_i = c_1 \ldots c_i,$$

where either $|b| = i < |c|$, or $|c| = i < |b|$, or $b_{i+1} \neq c_{i+1}$. Furthermore, if the length of B or C is i, we set $b_{i+1} := e$ or $c_{i+1} := e$ respectively.

Thus, the triple $(v_i^{-1} = b_i^{-1} = c_i^{-1}, b_{i+1}, c_{i+1})$ is a triple of pairwise distinct letters, and we have

$$b \prec c \iff v_i^{-1} \curvearrowright b_{i+1} \curvearrowright c_{i+1};$$

$$b \succ c \iff v_i^{-1} \curvearrowright c_{i+1} \curvearrowright b_{i+1}.$$

2.7 The set Γ of geodesics

Let I be the connected component of $\partial D_n^{\sim}$ containing the base point $\mathbf{x}$. Since we consider $D_n^{\sim}$ as a subset of $\mathbb{H}^2$ (see Section 2.4), I is thus a geodesic in $\mathbb{H}^2$. Also, I is an open arc of the circle Θ. We define

$$\Gamma := \Theta \smallsetminus \mathrm{clos}(I).$$

Thus, Γ is an open arc of Θ, and Γ is homeomorphic to the real line $\mathbb{R}$. Being a subinterval of $\Theta \smallsetminus \mathbf{x}$, the set Γ inherits the ordering $\prec$.

Proposition 2.10. *There exists a natural homeomorphism between the interval Γ and the space A of all geodesics in D_n that start at the base point x_0 and go into the interior of D_n. (Here, by a geodesic in D_n, we mean a geodesic with respect to the hyperbolic metric on D_n that was fixed at the start of Section 2.4.)*

Remark. In the proposition, we regard the set A as a topological space because A has a natural metric: the distance between two geodesics is equal to the absolute value of the angle between these geodesics at the base point x_0.

Proof. Let γ be a geodesic from A. Let $\widetilde{\gamma}$ be the lift of γ to the universal cover that starts at the base point $\mathbf{x}$. Then $\widetilde{\gamma}$ connects $\mathbf{x}$ with a certain point $f(\gamma) \in \boldsymbol{\Gamma}$. Clearly, the mapping $\gamma \mapsto f(\gamma)$ is a homeomorphism between A and $\boldsymbol{\Gamma}$. $\square$

In what follows, we identify the points of $\boldsymbol{\Gamma}$ with the corresponding geodesics in D_n and call points of $\boldsymbol{\Gamma}$ *geodesics*.

Definition 2.11 (Fundamental geodesics). We say that a point (= a geodesic) $\gamma \in \boldsymbol{\Gamma}$ is *fundamental* if γ represents an element of the fundamental group $\pi_1 = \pi_1(D_n, x_0)$, i.e. if $\gamma \in \mathcal{X}$ (see Section 2.3 for the definition of $\mathcal{X}$). In other words, a geodesic $\gamma \subset D_n$ is fundamental if it ends at the base point x_0.

We denote the set $\boldsymbol{\Gamma} \cap \mathcal{X}$ of all fundamental geodesics by $\boldsymbol{\Gamma}_\pi$. Note that under the natural inclusion $\pi_1 = \mathcal{X} \subset \boldsymbol{\Theta}$ we have

$$\pi_1^{(1)} \subset \boldsymbol{\Gamma} \quad , \quad \pi_1^{(0)} \subset I.$$

Thus, the natural inclusion $\pi_1 \to \boldsymbol{\Theta}$ yields a one-to-one correspondence

$$\boldsymbol{\Gamma}_\pi \cong \pi_1^{(1)}.$$

Remark. As mentioned above , we may identify a point $\gamma \in \boldsymbol{\Gamma}$ with the corresponding geodesic in D_n. By the convention adopted in Remark 2.2, we may also identify a point $\gamma \in \mathcal{X}$ with the corresponding element of the fundamental group. Thus, we regard a fundamental geodesic as a trinity

'point of the interval $\boldsymbol{\Gamma}$ — geodesic — element of the free group'.

The set $\mathcal{K}_*$. Recall that $\mathbf{u}$ is the distinguished element. We define

$$\mathcal{K}_* := \mathcal{K} \cap \boldsymbol{\Gamma} = \mathcal{K} \setminus \{\mathbf{u}^{-\infty}, \mathbf{u}^{+\infty}\}.$$

Following the convention described earlier (after Proposition 2.3), we may identify $\partial \pi_1$ with the Cantor set $\mathcal{K}$; the points $\mathbf{u}^{-\infty}$ and $\mathbf{u}^{+\infty}$ are the endpoints of the open arc $\boldsymbol{\Gamma} \subset \boldsymbol{\Theta}$, and are also the endpoints of the open arc $I \subset \boldsymbol{\Theta}$. It easily follows that for each $v \in \boldsymbol{\Gamma}_\pi$ we have $v^{+\infty} \in \mathcal{K}_*$.

Definition 2.12. A subset of a topological space is said to be *perfect* if it is closed and has no isolated points.

Let R be a topological space homeomorphic to the real line $\mathbb{R}$. Recall that a nonempty subset H of R is called a *Cantor set* if H is perfect, nowhere dense, and compact. We say that a nonempty subset $E \subset R$ is a ∗-*Cantor set* if H is perfect, nowhere dense, and bilaterally unbounded.

Thus, $\mathcal{K}_*$ is a ∗-Cantor subset of $\boldsymbol{\Gamma}$.

Remark. It is clear that the points of the set $\mathcal{K}_*$ are infinite geodesics in D_n, while the points of the set $\boldsymbol{\Gamma} \smallsetminus \mathcal{K}_* = \partial D_n^{\sim} \smallsetminus I$ are finite geodesics in D_n. In particular, each fundamental geodesic is finite.

Simple geodesics. Recall that a geodesic is said to be *simple* if it has no transverse self-intersections. Usually, this notion concerns either two-sided infinite or closed geodesics. For geodesics of $\boldsymbol{\Gamma}$ we need a more precise definition of 'simplicity'.

We say that a fundamental geodesic $\gamma \in \boldsymbol{\Gamma}_\pi$ is *simple* if its image in D_n is homeomorphic to a circle. In other words, we do not regard the coincidence of the endpoints of a fundamental geodesic as a transverse self-intersection.

If a geodesic $\gamma \in \boldsymbol{\Gamma}$ is not fundamental, then any self-intersection of γ is transverse. Thus, there is no ambiguity about the definition of simplicity in this case.

We use the following notation:

- $\boldsymbol{\Sigma}$ denotes the set of all simple geodesics in $\boldsymbol{\Gamma}$;
- $\boldsymbol{\Sigma}_\pi$ denotes the set of all simple fundamental geodesics in $\boldsymbol{\Gamma}$;
- $\boldsymbol{\Sigma}_{\mathrm{inf}}$ denotes the set of all simple infinite geodesics in $\boldsymbol{\Gamma}$.

Types of geodesics in $\boldsymbol{\Gamma}$. Here is a table with types of geodesics in $\boldsymbol{\Gamma}$ (by the symbol '■' we mark the types that are 'not very important' in our considerations):

Types of geodesics in $\boldsymbol{\Gamma}$

	Fundamental ($\boldsymbol{\Gamma}_\pi$)	Finite non-fundamental	Infinite ($\mathcal{K}_*$)
Simple ($\boldsymbol{\Sigma}$)	$\boldsymbol{\Sigma}_\pi$	■	$\boldsymbol{\Sigma}_{\mathrm{inf}}$
Non-simple	$\boldsymbol{\Gamma}_\pi \smallsetminus \boldsymbol{\Sigma}_\pi$	■	■

2.8 Laminations

We recall several notions and facts of lamination theory (see [12, 26, 15]).

A (geodesic) *lamination* $L \subset D_n$ is a nonempty closed set that is a union of disjoint simple complete geodesics. These geodesics are the *leaves* of L, while the connected components of the set $D_n \setminus L$ are the *regions* of L.

A lamination $L \subset D_n$ is *perfect* if it has no isolated leaves. A perfect lamination contains an infinite number of leaves.

A perfect lamination L is *prime* if each leaf of L is dense in L. Each perfect lamination is a union of a finite number of prime (perfect) laminations.

We say that a prime lamination $L \subset D_n$ is *small* if there exists a simple closed geodesic $C \subset \mathrm{int}(D_n)$ such that L and ∂D_n lie in distinct components of the two-component surface $D_n \setminus C$. Otherwise, we say that L is *large*.

2.9 Subtypes of simple geodesics and structure of Γ

Definition 2.13 (Subtypes of simple fundamental geodesics Σ_π). Let $V : \pi_1 \to \mathbb{Z}$ be a homomorphism defined by setting $V(u_i) = 1$ for $1 \leq i \leq n$. The number $V(v)$ is called the *volume* of an element $v \in \pi_1$.

Let $z \in \mathbb{Z}$. We denote the set of simple fundamental geodesics with volume z by $\Sigma_{\pi, z}$. Similarly, we denote by $\Sigma_{\pi, >z}$, $\Sigma_{\pi, \geq z}$, $\Sigma_{\pi, <z}$ and $\Sigma_{\pi, \leq z}$ the sets of all simple fundamental geodesics with volume $> z$, $\geq z$, $< z$ and $\leq z$, respectively.

Lemma 2.14. *For any simple fundamental geodesic γ we have*

$$1 \leq |V(\gamma)| \leq n - 1.$$

Equivalently, we have

$$\Sigma_\pi = \bigcup_{z \in \{\pm 1, \ldots, \pm(n-1)\}} \Sigma_{\pi, z}.$$

Next, we consider simple infinite geodesics. The following lemma easily follows from standard facts of lamination theory.

Lemma 2.15. *Suppose that $\gamma \in \Sigma_{\mathrm{inf}}$ is an infinite simple geodesic. Then one of the following conditions for the set $\mathrm{clos}(\gamma) \setminus \gamma \subset D_n$ holds:*

– *the set $\mathrm{clos}(\gamma) \setminus \gamma$ is empty;*
– *the set $\mathrm{clos}(\gamma) \setminus \gamma$ is a simple closed geodesic;*

– *the set* $\mathrm{clos}(\gamma) \smallsetminus \gamma$ *is a prime (perfect) lamination.*

Now, we introduce notation for subtypes of simple infinite geodesics.

Definition 2.16. Let γ be a simple infinite geodesic.

- If $\mathrm{clos}(\gamma) \smallsetminus \gamma = \varnothing$, then γ 'falls into a puncture'. In this case, we say that γ is a *puncture geodesic.*
- If $\mathrm{clos}(\gamma) \smallsetminus \gamma$ is a simple closed geodesic (say, C), then γ 'coils' to C. In this case, we say that γ is a *coil geodesic.* If γ coils to C clockwise (respectively, counterclockwise), we say that γ is a *right* (respectively, *left) coil geodesic.*
- If $\mathrm{clos}(\gamma) \smallsetminus \gamma$ is a perfect lamination, we say that γ is a *perfect geodesic.* We also say that γ is *large* or *small*, according to whether the corresponding perfect lamination is large or small.

We use the following notation:

- $\boldsymbol{\Sigma}_\wr$ denotes the set of all puncture geodesics;
- $\boldsymbol{\Sigma}_\rho$ denotes the set of all coil geodesics;
- $\boldsymbol{\Sigma}_\clubsuit$ denotes the set of all perfect geodesics.

Thus, we have

$$\boldsymbol{\Sigma}_{\mathrm{inf}} = \boldsymbol{\Sigma}_\wr \cup \boldsymbol{\Sigma}_\rho \cup \boldsymbol{\Sigma}_\clubsuit.$$

We also set

$$\boldsymbol{\Sigma}_* := \boldsymbol{\Sigma}_\rho \cup \boldsymbol{\Sigma}_\clubsuit.$$

Theorem 2.17. *Suppose $n \geq 3$, i.e. suppose that the disc D_n which we consider has $n \geq 3$ punctures. Then the set $\boldsymbol{\Sigma}_*$ is a $*$-Cantor subset of $\boldsymbol{\Gamma}$ (see Definition 2.12). Each perfect geodesic is a two-sided limit point of $\boldsymbol{\Sigma}_*$. Each coil geodesic is a one-sided limit point of $\boldsymbol{\Sigma}_*$ (each right coil geodesic is a right-hand limit point, while each left coil geodesic is a left-hand limit point). In particular, $\boldsymbol{\Sigma}_\rho$ is a countable set and $\boldsymbol{\Sigma}_\clubsuit$ is a continuum set.*

It follows that the space $\boldsymbol{\Gamma} \smallsetminus \boldsymbol{\Sigma}_\clubsuit$ is a countable collection of disjoint closed intervals, and $\boldsymbol{\Sigma}_\rho$ is the set of endpoints of these intervals (left coil geodesics are right endpoints, while right coil geodesics are left endpoints).

Suppose an interval $[\eta_1, \eta_2]$ (see Definition 2.9) is a connected component of $\boldsymbol{\Gamma} \smallsetminus \boldsymbol{\Sigma}_\clubsuit$. (Thus, η_1 is a left coil geodesic, while η_2 is a right coil geodesic.) Then one of the following two cases holds.

(i) *The interval $[\eta_1, \eta_2]$ contains exactly two simple fundamental geodesics (say, v_1 and v_2, where $v_1 \prec v_2$):*

$$\eta_1 \prec v_1 \prec v_2 \prec \eta_2.$$

In this case, each point of the open interval (v_1, v_2) is a finite nonfundamental simple geodesic and each point of the open intervals (η_1, v_1) and (v_2, η_2) is a nonsimple geodesic. Moreover, the following relations hold:

$$\eta_1 = v_1^{+\infty}, \quad v_1 \mathbf{u} = v_2, \quad \eta_2 = v_2^{+\infty};$$

$$2 - n \le V(v_1) \le -2, \quad 2 \le V(v_2) \le n - 2.$$

(ii) *The interval $[\eta_1, \eta_2]$ contains exactly four simple fundamental geodesics, v_1, v_2, v_3 and v_4 say, where $v_1 \prec v_2 \prec v_3 \prec v_4$. In this case, $[\eta_1, \eta_2]$ also contains exactly one puncture geodesic, ζ say, and we have*

$$\eta_1 \prec v_1 \prec v_2 \prec \zeta \prec v_3 \prec v_4 \prec \eta_2.$$

Furthermore, any point of the open intervals (v_1, v_2) and (v_3, v_4) is a finite nonfundamental simple geodesic and each point of the set $(\eta_1, v_1) \cup (v_2, \zeta) \cup (\zeta, v_3) \cup (v_4, \eta_2)$ is a nonsimple geodesic. Moreover, the following relations hold:

$$v_1 \in \boldsymbol{\Sigma}_{\pi, -(n-1)}, \quad v_2 \in \boldsymbol{\Sigma}_{\pi, 1}, \quad v_3 \in \boldsymbol{\Sigma}_{\pi, -1}, \quad v_4 \in \boldsymbol{\Sigma}_{\pi, n-1};$$

$$v_1 \mathbf{u} = v_2, \quad v_2 = v_3^{-1}, \quad v_3 \mathbf{u} = v_4;$$

and

$$\eta_1 = v_1^{+\infty}, \quad \zeta = v_2^{+\infty} = v_3^{+\infty}, \quad \eta_2 = v_4^{+\infty}.$$

Corollary 2.18. *We have*

$$\mathrm{clos}(\boldsymbol{\Sigma}_\pi) \setminus \boldsymbol{\Sigma}_\pi = \boldsymbol{\Sigma}_*.$$

Furthermore, each perfect geodesic is a two-sided limit point of $\boldsymbol{\Sigma}_\pi$, each right coil geodesic is a right-hand limit point of $\boldsymbol{\Sigma}_\pi$, and each left coil geodesic is a left-hand limit point of $\boldsymbol{\Sigma}_\pi$.

Corollary 2.19. *We have*

$$\{v^{+\infty} \mid v \in \boldsymbol{\Sigma}_\pi\} = \boldsymbol{\Sigma}_\lambda \cup \boldsymbol{\Sigma}_\rho.$$

Furthermore, for any $\eta \in \boldsymbol{\Sigma}_\rho$ there is a unique $v \in \boldsymbol{\Sigma}_\pi$ such that $\eta = v^{+\infty}$, while for any $\zeta \in \boldsymbol{\Sigma}_\lambda$ there are exactly two simple fundamental geodesics v_1 and v_2 such that $\zeta = v_1^{+\infty} = v_2^{+\infty}$. (In the latter case, we have $v_1 v_2 = e$ and $|V(v_1)| = |V(v_2)| = 1$).

Corollary 2.20. *We have*

$$\Gamma \smallsetminus \Sigma = \bigcup_{v \in \Sigma_{\pi, >0}} (v, v^{+\infty}) \cup \bigcup_{v \in \Sigma_{\pi, <0}} (v^{+\infty}, v);$$

moreover, the open intervals in the right part of this equality are disjoint.

2.10 Several auxiliary definitions

Definition 2.21 (simplification). Corollary 2.20 implies that for any nonsimple geodesic $\gamma \in \Gamma \smallsetminus \Sigma$ there is a unique simple fundamental geodesic $s(\gamma) \in \Sigma_\pi$ such that there is no simple geodesic between γ and $s(\gamma)$. We say that $s(\gamma)$ is a *simplification* of γ.

For a simple geodesic $\gamma \in \Sigma$, we set $s(\gamma) = \gamma$. We have thus defined a mapping

$$s : \Gamma \to \Sigma.$$

Clearly, s is $\prec$-monotone. Note that by this definition we have $s(\Gamma \smallsetminus \Sigma) = \Sigma_\pi$ and $s(\Gamma_\pi) = \Sigma_\pi$.

Definition 2.22 (maximal elements, right-hand Σ_π-neighbours). Let $v, w \in \Sigma_\pi$. We say that w is the *right-hand Σ_π-neighbour* of v and write

$$\mathrm{Next}_r(v) = w$$

if $v \prec w$ and the interval (v, w) does not contain Σ_π elements. We say that $v \in \Sigma_\pi$ is a *maximal* element if v has no right-hand Σ_π-neighbour.

Theorem 2.17 implies the following result.

Corollary 2.23.

(i) *The set of all maximal elements in Σ_π is the set $\Sigma_{\pi, \geq 2}$.*
(ii) *If $v \in \Sigma_{\pi, <0}$, then $\mathrm{Next}_r(v) = v\mathbf{u}$.*
(iii) *If $v \in \Sigma_{\pi, 1}$, then $\mathrm{Next}_r(v) = v^{-1}$.*

Suppose v is a simple fundamental geodesic. Then the image $v \subset D_n$ is homeomorphic to a circle. We denote by D^v the punctured disc in D_n bounded by this circle. Note that $D^v = D^{v^{-1}}$.
We also define

$$\Gamma^v := \{\gamma \in \Gamma \mid \gamma \subset D^v\}.$$

For a subset $A \subset \Gamma$ we put $A^v := A \cap \Gamma^v$. Obviously, for each $v \in \Sigma_\pi$ the set Γ^v is a closed bounded subset of Γ. We have $\Gamma^v \subset [v, v^{-1}]$ whenever $V(v) > 0$, and $\Gamma^v \subset [v^{-1}, v]$ otherwise.

Lemma 2.24 (Cf. Corollary 2.18). *For each $v \in \Sigma_\pi$, we have*

$$\mathrm{clos}(\Sigma_\pi{}^v) \smallsetminus \Sigma_\pi{}^v = \Sigma_*{}^v.$$

Each perfect geodesic in $\Sigma_*{}^v$ is a two-sided limit point of $\Sigma_\pi{}^v$, each right coil geodesic in $\Sigma_*{}^v$ is a right-hand limit point of $\Sigma_\pi{}^v$, and each left coil geodesic in $\Sigma_*{}^v$ is a left-hand limit point of $\Sigma_\pi{}^v$.

Definition 2.25. Let $\gamma \in \Gamma$ and $v \in \Sigma_{\pi, >0}$. We say that v is the *envelope* for γ and write $v = \mathrm{env}(\gamma)$ if $\gamma \subset D^v$ and for each $w \in \Sigma_\pi$ we have $w \subset D^v$ whenever $\gamma \subset D^w$. Obviously, the envelope of a geodesic in Γ is unique if it exists.

Remark. Clearly, each puncture geodesic and each coil geodesic has an envelope. A perfect geodesic γ has an envelope if and only if γ is a small perfect geodesic (see Definition 2.16).

Remark. If $v \in \Sigma_{\pi, >0}$, then $\mathrm{env}(v^{+\infty}) = \mathrm{env}(v^{-\infty}) = v$. (Recall that by Corollary 2.19 we have $v^{\pm\infty} \in \Sigma_l \cup \Sigma_\rho$.)

2.11 Intersections of geodesics, disjoint geodesics

Definition 2.26 (types of intersections and intersection sets.). Let (γ_1, γ_2) be an ordered pair of geodesics from Γ. Let $v \in \pi_1 \smallsetminus e$. We say that (γ_1, γ_2) has a [transverse] *intersection of type v* if $v\gamma_2 \neq e$ and we have either $v \prec \gamma_1 \prec v\gamma_2$ or $v\gamma_2 \prec \gamma_1 \prec v$.

Let

$$\gamma_1 \pitchfork \gamma_2 := \{v \in \pi_1 \smallsetminus e \mid (\gamma_1, \gamma_2) \text{ has an intersection of type } v\}.$$

The set $\gamma_1 \pitchfork \gamma_2 \subset \pi_1 \smallsetminus e$ is called the *intersection set* of γ_1 with γ_2.

Proposition 2.27. *Suppose γ_1 and γ_2 are geodesics from Γ. Then $\gamma_1 \pitchfork \gamma_2 \neq \varnothing$ if and only if γ_1 and γ_2 have transverse intersections. In particular, $\gamma \in \Gamma$ is a simple geodesic if and only if $\gamma \pitchfork \gamma = \varnothing$.*

Remark 2.28. Suppose $\gamma_1 \in \Gamma_\pi$ and $\gamma_2 \in \Gamma_\pi$ are fundamental geodesics. Then the set $\gamma_1 \pitchfork \gamma_2$ is finite. Furthermore, the length (see Section 2.1) of each intersection type is less then the sum of the lengths of γ_1 and γ_2.

The definition immediately implies that for any $\gamma_1, \gamma_2 \in \Gamma$ we have $v \in \gamma_1 \pitchfork \gamma_2 \iff v^{-1} \in \gamma_2 \pitchfork \gamma_1$.

Definition 2.29 (∗-disjoint geodesics). Any two geodesics in Γ have at least one common point in D_n (the base point x_0), in other words, they are not disjoint. We say that two geodesics $\gamma_1, \gamma_2 \in \Gamma$ are ∗-*disjoint* if

$$\gamma_1 \cap \gamma_2 \cap \operatorname{int}(D_n) = \varnothing.$$

Proposition 2.27 implies that geodesics γ_1 and γ_2 are ∗-disjoint if and only if $\gamma_1 \neq \gamma_2$ and $\gamma_1 \pitchfork \gamma_2 = \varnothing$.

Lemma 2.30. *Suppose* $\gamma_1 \in \Gamma$ *and* $\gamma_2 \in \Gamma$ *are* ∗-*disjoint. Then the simplifications* $s(\gamma_1)$ *and* $s(\gamma_2)$ *are also* ∗-*disjoint.*

Definition 2.31 (Strongly and weakly ∗-disjoint geodesics). We say that two geodesics $\gamma_1, \gamma_2 \in \Gamma$ are *strongly* ∗-*disjoint* if there exist a pair of simple fundamental geodesics v_1 and v_2 such that $\gamma_1 \subset D^{v_1}$, $\gamma_2 \subset D^{v_2}$, and

$$D^{v_1} \cap D^{v_2} \cap \operatorname{int}(D_n) = \varnothing.$$

We say that two geodesics $\gamma_1, \gamma_2 \in \Gamma$ are *weakly* ∗-*disjoint* if they are ∗-disjoint but are not strongly ∗-disjoint.

Remark 2.32. Note that if η_1 and η_2 are strongly ∗-disjoint, then their envelopes $\operatorname{env}(\eta_1)$ and $\operatorname{env}(\eta_2)$ exist and they are strongly ∗-disjoint.

Lemma 2.33. *Let* $\eta_1, \eta_2 \in \Sigma_*$. *Suppose* η_1 *and* η_2 *are* ∗-*disjoint. Then:*

- *if* η_1 *and* η_2 *are strongly* ∗-*disjoint, then the subsets* $\operatorname{clos}(\eta_1) \smallsetminus \eta_1$ *and* $\operatorname{clos}(\eta_2) \smallsetminus \eta_2$ *of* D_n *are disjoint;*
- *if* η_1 *and* η_2 *are weakly* ∗-*disjoint, then* $\operatorname{clos}(\eta_1) \smallsetminus \eta_1 = \operatorname{clos}(\eta_2) \smallsetminus \eta_2$.

Proposition 2.34. *Suppose* $\eta_1, \eta_2 \in \Gamma$. *Then one and only one of the following conditions holds:*

a) $\eta_1 = \eta_2$ *and they are simple;*
b) η_1 *and* η_2 *are weakly* ∗-*disjoint;*
c) η_1 *and* η_2 *are strongly* ∗-*disjoint;*
d) η_1 *and* η_2 *have transverse intersections.*

2.12 Algorithms

In this subsection we discuss computational procedures that deal with the above-defined relations in the free group.

By the words 'an algorithm is given a fundamental geodesic' we mean that an algorithm is given an element of the set $\pi_1^{(1)} \subset \pi_1$ (as a word in generators and their inverses).

Lemma 2.35. *There exists an algorithm that, given a pair v, w of nontrivial elements of the fundamental group π_1, detects whether $v \prec w$.*

Proof. Obviously, such an algorithm can be constructed with the help of the rules described in Section 2.6.2. □

Lemma 2.36. *There exists an algorithm that, given a pair v, w of fundamental geodesics and a nontrivial element $y \in \pi_1 \smallsetminus e$, detects whether the pair (v, w) has an intersection of type y.*

Proof. Follows from definition of an intersection type and Lemma 2.35.
□

Corollary 2.37. *There exists an algorithm that, given a pair v, w of fundamental geodesics, computes the intersection set $v \pitchfork w \subset \pi_1$.*

Outline of the proof. Formally, one can look through all the elements of π_1 with lengths less then the sum of the lengths of v and w (see Remark 2.28), each time applying Lemma 2.36. Certainly, there are much faster procedures for computing the intersection set. □

Corollary 2.38. *There exists an algorithm that detects whether two given fundamental geodesics are $*$-disjoint or not.*

Proof. Follows from Definition 2.29 and Corollary 2.37. □

Corollary 2.39. *There exists an algorithm that detects whether a given fundamental geodesic is simple.*

Proof. Follows from Proposition 2.27 and Corollary 2.37. □

Lemma 2.40. *There exists an algorithm that computes the simplification $s(v)$ of a given fundamental geodesic v.*

Outline of the proof. It can be proved that for each nonsimple geodesic $\gamma \in \Gamma$ the simplification $s(\gamma)$ and its inverse $(s(\gamma))^{-1}$ are the only simple elements of the intersection set $\gamma \pitchfork \gamma$. Thus, if a given element $v \in \pi_1^{(1)}$ is not simple (for by Corollary 2.39, we can detect simplicity) we can find the set $\{s(v), (s(v))^{-1}\}$ by Corollaries 2.37 and 2.39. The rest is easy.

Of course, we use a different procedure to compute $s(v)$ in practice.
□

Lemma 2.41. *There exists an algorithm that, given a simple fundamental geodesic v, detects whether v is maximal, and if v is not maximal,*

computes the element of the group π_1 that represents the Σ_π-geodesic
$\text{Next}_r(v)$.

Proof. Such an algorithm can easily be constructed with the help of
Corollary 2.23. □

3 Braids and disc automorphisms

In this section, we pass from the study of the punctured disc itself to
the study of its automorphisms. We use the constructions, definitions
and notation of the previous section.

3.1 Braids and disc automorphisms

Definition 3.1. The *mapping class group* of the punctured disc D_n is
the group

$$\text{MCG}(D_n) := \text{Homeo}(D_n, \partial D_n) / \text{Homeo}_0(D_n, \partial D_n),$$

where $\text{Homeo}(D_n, \partial D_n)$ is the group of all orientation-preserving homeomorphisms of D_n that fix ∂D_n pointwise, and $\text{Homeo}_0(D_n, \partial D_n)$ is the
normal subgroup consisting of all homeomorphisms isotopic to the iden-
tity rel ∂D_n.

It is well known that the braid group B_n is isomorphic to the mapping
class group $\text{MCG}(D_n)$ of the n-punctured disc D_n (see [2] for example).
A canonical isomorphism between these groups can be constructed as
follows: the generator $\sigma_j \in B_n$ is assigned the class of a homeomorphism
that exchanges the 'punctures' z_j and z_{j+1} (see Section 2.2) by 'rotating'
them clockwise in the 'simplest possible way' (this is a *Dehn half-twist*).

In what follows, we identify a braid $\beta \in B_n$ with the corresponding
class of homeomorphisms in $\text{MCG}(D_n)$.

Definition 3.2 (Artin's action). The group $B_n \cong \text{MCG}(D_n)$ acts in a
natural way on the free group $\pi_1 = \pi_1(D_n, x_0)$ by group automorphisms.
This action is called *Artin's action*. The following rules determine this
action completely:

$$\sigma_j(u_i) = \begin{cases} u_i, & \text{if } i \neq j, j+1 \\ u_{i+1}, & \text{if } i = j \\ u_i^{-1} u_{i-1} u_i, & \text{if } i = j+1. \end{cases}$$

Lemma 3.3. *For each braid $\beta \in B_n$ the following conditions are equivalent:*

- $\beta(u_1) = u_1$;
- *β can be represented by a word in the letters $\{\sigma_2, \sigma_3, \ldots, \sigma_{n-1}\}$ and their inverses.*

Definition 3.4 (The fundamental braid Δ). The braid

$$\Delta := (\sigma_1 \sigma_2 \ldots \sigma_{n-1})(\sigma_1 \sigma_2 \ldots \sigma_{n-2}) \ldots (\sigma_1 \sigma_2)(\sigma_1) \in B_n$$

is called the *fundamental braid*. The braid $\Delta^2 \in B_n$ generates the center of the group B_n whenever $n > 2$. (The center of B_n is an infinite cyclic group.)

Remark.

(i) The elements of $\pi_1^{(0)}$ (i.e. the powers of the distinguished element **u**) exhaust all B_n-invariant elements in π_1.

(ii) In the group B_n of automorphisms of the free group $\pi_1(D_n, x_0)$, the only inner automorphisms of $\pi_1(D_n, x_0)$ are the powers of $\Delta^2 \in B_n$. These inner automorphisms are conjugations by powers of **u**; conjugation by **u** corresponds to the action of Δ^{-2}.

Artin's action, being an action by group automorphisms, can be uniquely extended to the continuous action of B_n on the compact space $\pi_1 \cup \partial \pi_1$. We call the latter action the *extended Artin's action*.

3.2 Braid group action on the circle Θ

In this subsection, we describe a natural action of the group $B_n \cong \mathrm{MCG}(D_n)$ on the circle Θ. We call this action the *Nielsen-Thurston action*. The description of the Nielsen-Thurston action can be found in [16, 25].

Theorem 3.5. *The braid group B_n acts naturally on the marked circle $(\Theta, \mathbf{x})$ by orientation-preserving homeomorphisms.*

Proof. First, we observe that every homeomorphism $\phi : D_n \to D_n$ which is identical on the boundary, has a unique lift $\phi^\sim : D_n^\sim \to D_n^\sim$ such that $\phi^\sim(\mathbf{x}) = \mathbf{x}$. This yields a homomorphism

$$\mathrm{Homeo}(D_n, \partial D_n) \to \mathrm{Homeo}_+(D_n^\sim), \quad \phi \mapsto \phi^\sim.$$

The homeomorphism $\phi^\sim$ is uniquely extended to a homeomorphism $\phi^\approx$ of the disc $D_n^\approx$ (the proof is standard: see [12] for example.)

Next, we observe that the restricted homeomorphism $\phi^\sim|_{\partial D_n^\sim = \Theta}$ is completely determined by the class $[\phi] \in \mathrm{MCG}(D_n) \cong B_n$. Indeed, we observe that for each homeomorphism $\psi \in \mathrm{Homeo}_0(D_n, \partial D_n)$, the homeomorphism $\psi^\sim|_{\partial D_n^\sim}$ is identical, and since the set $\partial D_n^\sim$ is dense in Θ, the autohomeomorphism $\psi^\sim|_\Theta$ is also identical. Thus, we have a homomorphism

$$B_n = \mathrm{MCG}(D_n) \to \mathrm{Homeo}_+(\Theta), \quad [\phi] \mapsto \phi^\sim|_\Theta.$$

$\square$

We denote by $\beta(y)$ (or just by βy) the result of the action of an element $\beta \in B_n$ on a point $y \in \Theta$.

Theorem 3.6. *The natural inclusion $\pi_1 \subset \Theta$ is equivariant with respect to the Nielsen-Thurston action of B_n on the circle Θ and Artin's action of B_n on the free group π_1. Consequently, the continuous mapping $\pi_1 \cup \partial\pi_1 \to \Theta$ is equivariant with respect to the Nielsen-Thurston action and extended Artin's action.*

Remark. It can be shown that the 'braid automorphisms' of the free group π_1 are the only automorphisms of π_1 that can be extended to the orientation-preserving homeomorphism of the circle $\Theta \supset \pi_1$.

Lemma 3.7. *Let $\beta \in B_n$, $v \in \pi_1$, and $y \in \Theta$. Then*

$$\beta(vy) = (\beta v)(\beta y).$$

(using the notation of Definition 2.6).

3.3 Braid group action on Γ

The definition of the Nielsen-Thurston action directly implies that this action is identical on the arc $I \subset \Theta$ (see the notation at the start of Section 2.7). Then it is identical on the closure $\mathrm{clos}(I) \subset \Theta$. Therefore, an action of B_n on $\Gamma = \Theta \smallsetminus \mathrm{clos}(I)$ is defined.

Proposition 3.8. *The braid group B_n acts naturally on the space Γ by orientation-preserving homeomorphisms.*

We refer to this action also as the 'Nielsen-Thurston action'.

Note that the Nielsen-Thurston action preserves the 'inner structure' of Γ, which we studied in Section 2. In particular, this action preserves the sets of finite, infinite, and fundamental geodesics; it also preserves simplicity; volume is invariant under the action; the action preserves the sets of puncture, coil and perfect geodesics (and of course preserves the sets of right and left puncture geodesics, and sets of small and large perfect geodesics). Thus, the Nielsen-Thurston action preserves all types and subtypes of geodesics described above.

The relations which we defined on Γ are also invariant under the Nielsen-Thurston action. First of all, the action preserves, by definition, the ordering $\prec$. This implies that for any $\gamma_1, \gamma_2 \in \Gamma$ and any $\beta \in B_n$ we have

$$\beta\gamma_1 \pitchfork \beta\gamma_2 = \beta(\gamma_1 \pitchfork \gamma_2).$$

Therefore, each braid takes a pair of $*$-disjoint elements to a pair of $*$-disjoint elements. Furthermore, each braid takes a pair of strongly (respectively, weakly) $*$-disjoint elements to a pair of strongly (respectively, weakly) $*$-disjoint elements.

We state the following lemma for convenient reference.

Lemma 3.9. *Let* $v, v_1, v_2 \in \Sigma_\pi$; $\gamma, \gamma_1, \gamma_2 \in \Gamma$; *and* $\beta \in B_n$. *Then we have:*

$$\gamma_1 \prec \gamma_2 \iff \beta(\gamma_1) \prec \beta(\gamma_2)$$

$$\gamma \subset D^v \iff \beta(\gamma) \subset D^{\beta(v)}$$

$$v = \mathrm{env}(\gamma) \iff \beta(v) = \mathrm{env}(\beta(\gamma))$$

$$v_2 = \mathrm{Next}_r(v_1) \iff \beta(v_2) = \mathrm{Next}_r(\beta(v_1))$$

$$v = s(\gamma) \iff \beta(v) = s(\beta(\gamma)).$$

Lemma 3.10. *For each* $k \in \{1, \ldots, n-1\}$, *the set* $\Sigma_{\pi, k}$ *coincides with the orbit of the element* $u_1 u_2 \cdots u_k$ *under the action of the braid group* B_n:

$$\Sigma_{\pi, k} = B_n(u_1 u_2 \cdots u_k).$$

Lemma 3.11. *Suppose* $v, w \in \Sigma_{\pi, 1}$. *Then* v *and* w *are* $*$-disjoint if and only if there exists a braid $\beta \in B_n$ such that $\{\beta(v), \beta(w)\} = \{u_1, u_2\}$.

Lemma 3.12. *Consider the disc* D_n *with* $n \geq 3$ *punctures. Then for any point* $\gamma \in \Gamma$, *the set of limit points of the orbit* $B_n(\gamma)$ *is* Σ_*. *Each perfect geodesic is a two-sided limit point of* $B_n(\gamma)$ *and each coil geodesic is a one-sided limit point of* $B_n(\gamma)$. *Furthermore, each right coil geodesic* *is a right-hand limit point of $B_n(\gamma)$ and each left coil geodesic is a left-hand limit point of $B_n(\gamma)$.*

3.4 Braid group action on the circle Γ/Δ^2

The proofs of the statements presented in this subsection can be found in [20].

Lemma 3.13. *The homeomorphism $\Delta^2 : \Gamma \to \Gamma$ has no fixed points.*

Corollary 3.14. *The quotient space Γ/Δ^2 is homeomorphic to the circle. The braid group B_n therefore acts naturally on the circle Γ/Δ^2.*

3.5 The Nielsen-Thurston classification for braids

We recall the definitions of types of braids in the Nielsen-Thurston classification.

- A braid β is said to be *periodic* if $\beta^p = \Delta^{2q}$ for a pair of nonzero integers p, q.
- A braid β is said to be *reducible* if the corresponding class of homeomorphisms of D_n contains a homeomorphism preserving a nonempty family of pairwise disjoint simple closed geodesics $C \subset \text{int}(D_n)$ (we assume that D_n has been endowed with a hyperbolic metric). Such a family is called a *reduction system*.

  If a braid β is reducible and not periodic, it has a nonempty *canonical reduction system* $C \subset \text{int}(D_n)$. In this case, in each component† of the surface $D_n \smallsetminus C$ the induced homeomorphism is either periodic or pseudo-Anosov.
- A braid β is said to be *pseudo-Anosov* if it is neither periodic nor reducible. In this case, by the Nielsen-Thurston classification theorem, there exist two transversal geodesic laminations $L_s \subset D_n$ and $L_u \subset D_n$ (stable and unstable) such that $\psi(L_s) = L_s$ and $\psi(L_u) = L_u$ for a certain homeomorphism $\psi : D_n \to D_n$ representing β.

The Nielsen-Thurston type of a braid is a conjugacy invariant. Hence the type of a closed braid is defined in a natural way.

---

† By 'component' we mean a minimal collection of connected components that is invariant under the action (of the class) of our homeomorphism.

3.6 On pseudo-Anosov braids

Theorem 3.15. *Let $\beta \in B_n$ be a braid of pseudo-Anosov type. Then for a certain positive integer $m \in \mathbb{N}$, the set of fixed points of the homeomorphism $\beta^m : \Gamma/\Delta^2 \to \Gamma/\Delta^2$ (see Corollary 3.14) is nonempty and finite. Furthermore, each point of this set of fixed points is either an attracting or repelling point of the homeomorphism.*

Notation. We use the terms of the previous theorem. We denote by P_β^s and P_β^u, respectively, the sets of attracting and repelling points of the homeomorphism $\beta^m : \Gamma/\Delta^2 \to \Gamma/\Delta^2$. Note that these sets do not depend on the choice of m; in particular, the set $P_\beta := P_\beta^s \cup P_\beta^u$ is exactly the set of points with finite orbits for the homeomorphism $\beta : \Gamma/\Delta^2 \to \Gamma/\Delta^2$.

We denote by T_β^s and T_β^u the inverse images in Γ of the sets $P_\beta^s \subset \Gamma/\Delta^2$ and $P_\beta^u \subset \Gamma/\Delta^2$, respectively. The geodesics of T_β^s and T_β^u are said to be *stable* and *unstable T-geodesics*, respectively, for β (the 'T' is for 'Thurston'). We also denote the set $T_\beta^s \cup T_\beta^u$ by T_β.

Theorem 3.16. *Let $L_s(\beta)$ and $L_u(\beta)$ be the stable and unstable laminations, respectively, for β. Then*

$$T_\beta^s = \{\gamma \in \Gamma \mid \gamma \cap L_s(\beta) = \varnothing\} = \{\gamma \in \Gamma \mid \mathrm{clos}(\gamma) \smallsetminus \gamma = L_s(\beta)\};$$

$$T_\beta^u = \{\gamma \in \Gamma \mid \gamma \cap L_u(\beta) = \varnothing\} = \{\gamma \in \Gamma \mid \mathrm{clos}(\gamma) \smallsetminus \gamma = L_u(\beta)\}.$$

In particular, each geodesic in T_β is a large perfect geodesic.

Notation. For a pseudo-Anosov braid $\beta \in B_n$ we put

$$q(\beta) := |P_\beta^s| = |P_\beta^u|.$$

Also, we denote by $r(\beta)$ the smallest positive integer such that the homeomorphism $\beta^{r(\beta)} : \Gamma/\Delta^2 \to \Gamma/\Delta^2$ has fixed points. We denote by $s(\beta)$ an integer number such that the homeomorphism $\beta^{r(\beta)}\Delta^{2s(\beta)} : \Gamma \to \Gamma$ fixes the set T_β of T-geodesics pointwise.

The existence of $r(\beta)$ and $s(\beta)$ follows from Theorem 3.15. It can be shown that we have

$$r(\beta) \le q(\beta) \le n - 2.$$

Note also that $r(\beta)$ divides $q(\beta)$.

3.7 On reducible braids

Let $\beta \in B_n$ be a reducible braid. Let $C \subset D_n$ be a reduction system for β. We denote by M the component of $D_n \smallsetminus C$ that contains ∂D_n. Since M is obtained from the punctured disc D_n by cutting out several discs, the surface M is homeomorphic to a disc with m punctures. Furthermore, we have $m < n$ because each of the cut-out discs contains at least two punctures of the disc D_n. The definitions imply that our mapping class $\beta \in B_n \cong \mathrm{MCG}(D_n)$ induces a certain mapping class β_M in the group $\mathrm{MCG}(M) \cong \mathrm{MCG}(D_m)$.

Definition 3.17 (Satellites and companions). Let $\widehat{\alpha}$ be the closed m-braid that corresponds to the class β_M. We say that the closed braid $\widehat{\alpha}$ is a *companion* of $\widehat{\beta}$, and $\widehat{\beta}$ is a *satellite* of $\widehat{\alpha}$.

Let $\beta \in B_n$ be a reducible nonperiodic braid. Then β has canonical reduction system $\mathcal{C} \subset D_n$. The companion of $\widehat{\beta}$ that corresponds to $\mathcal{C}$ is called the *principal companion* of $\widehat{\beta}$.

Note that a principal companion is not reducible: it is either pseudo-Anosov or periodic.

Lemma 3.18.

(i) *For every reducible closed braid $\widehat{\beta}$ the set of companions for $\widehat{\beta}$ is finite.*

(ii) *Suppose $\widehat{\alpha}$ is a companion of a reducible closed braid $\widehat{\beta}$. Then for any $k \in \mathbb{Z}$ the closed braid $\widehat{\alpha}^k$ is a companion of $\widehat{\beta}^k$.*

(iii) *For every periodic reducible closed braid $\widehat{\beta}$, the set of companions for $\widehat{\beta}$ consists of periodic braids.*

Definition 3.19 (pure-reducible braids). We say that a braid is *pure reducible* if it fixes an isotopy class of essential simple closed curves in D_n; in other words, a pure reducible braid is a reducible braid that has a reduction system composed of an only one component.

Note that a pure reducible braid represents a multi-component link (i.e. not a knot).

Definition 3.20 (Split braids). We recall the notion of a split (closed) braid. A closed braid is *split* if it is a satellite of a trivial closed 2-braid.

3.8 On periodic braids

We use standard notation:

$$\delta := \sigma_1 \sigma_2 \cdots \sigma_{n-1} \in B_n, \qquad \delta_* := \sigma_1 \delta \in B_n.$$

The following well-known result follows from the Kérékjartò theorem (see [13, 27, 11]).

Proposition 3.21. *Each braid of periodic type in B_n is conjugate either to a power of δ or to a power of δ_*.*

3.9 Algorithms

There are several well-known algorithms that determine the type (in the Nielsen-Thurston classification) of a given braid: see, e.g. [1, 18]. Furthermore, for the 'reducible case' there are standard methods to compute the canonical reduction system of a reducible braid. Hence, the principal companion of a given reducible braid can be computed, and the pure-reducibility can be detected.

In the 'pseudo-Anosov case' there are standard procedures to compute various characteristics of a given pseudo-Anosov braid. In particular, for a given pseudo-Anosov braid, each T-geodesic can be described with any prescribed accuracy, and the numbers $q(\beta)$, $r(\beta)$, and $s(\beta)$ can be computed.

4 The Destabilization Algorithm

In this section, we describe an algorithm that detects whether the conjugacy class of a given braid admits a destabilization.

4.1 Geometric reformulation for the destabilization problem

The following result is a variation of a well-known folklore theorem.

Theorem 4.1. *Suppose $\beta \in B_n$. Then the following conditions are equivalent:*

a) *$\widehat{\beta}$ is a destabilizable closed braid;*
b) *there is an element $v \in \Sigma_{\pi, 1}$ such that v and $\beta(v)$ are $*$-disjoint. (Here we use the notation introduced in Definitions 2.13, 2.29 and 3.2.)*

It turns out that the restriction $v \in \Sigma_{\pi, 1}$ in this theorem is redundant.

Theorem 4.2. *Suppose $\beta \in B_n$. Then the following conditions are equivalent:*

a) *there exists a simple fundamental geodesic $v \in \Sigma_{\pi,1}$ with volume 1 such that v and $\beta(v)$ are $*$-disjoint;*
b) *there exists a simple fundamental geodesic $v \in \Sigma_\pi$ such that v and $\beta(v)$ are $*$-disjoint;*
c) *there exists a fundamental geodesic $v \in \Gamma_\pi$ such that v and $\beta(v)$ are $*$-disjoint.*

Proof that a)$\Rightarrow$b)$\Rightarrow$c). These implications are trivial.

Proof that c)$\Rightarrow$b). Let $v \in \Gamma_\pi = \pi_1^{(1)}$ be an element of Γ_π such that v and $\beta(v)$ are $*$-disjoint. Consider the simplification $s(v)$. By Lemma 3.9, $s(\beta(v)) = \beta(s(v))$; and by Lemma 2.30, $s(v)$ and $\beta(s(v)) = s(\beta(v))$ are $*$-disjoint. It remains to note that since v is a fundamental geodesic, its simplification $s(v)$ is a simple fundamental geodesic.

Proof that b)$\Rightarrow$a). Let $v \in \Sigma_\pi$ be an element of Σ_π such that v and $\beta(v)$ are $*$-disjoint. Since the volumes of simple fundamental geodesics v and $\beta(v)$ are equal (see the remarks after Proposition 3.8), it follows easily that v and $\beta(v)$ are strongly $*$-disjoint. Evidently, there exists an element $w \in \Sigma_{\pi,1}$ such that $w \subset D^v$ (such an element is unique if $|V(v)| = 1$). By Lemma 3.9 we have $\beta(w) \subset D^{\beta(v)}$. Therefore, w and $\beta(w)$ are $*$-disjoint. $\qquad\square$

We say that a nontrivial element $v \in \pi_1$ is a *solution* for a braid $\beta \in B_n$ if v and $\beta(v)$ are $*$-disjoint. If $v \in \Sigma_\pi$ is a solution for β, we say that v is a *simple solution*; and if $v \in \Sigma_{\pi,1}$ is a solution for β, we say that v is a *prime solution*.

Proposition 4.3. *The sets of solutions, simple solutions and prime solutions for a braid β are each invariant under the action of the centralizer of β. In particular, each of these sets is invariant under the action of β and Δ^2.*

Proof. Obviously, it is sufficient to show that for each braid α such that $\alpha\beta = \beta\alpha$, geodesics $\alpha\gamma$ and $\beta(\alpha\gamma)$ are $*$-disjoint whenever γ and $\beta(\gamma)$ are $*$-disjoint.

Suppose that γ and $\beta(\gamma)$ are $*$-disjoint. Then, since the action of a braid preserve the $*$-disjointness, $\alpha(\gamma)$ and $\alpha(\beta(\gamma))$ are $*$-disjoint. It remains to note that since $\alpha\beta = \beta\alpha$, we have $\alpha(\beta(\gamma)) = \beta(\alpha(\gamma))$. $\qquad\square$

Algorithms. Using the proofs of Theorems 4.1 and 4.2, and also the lemmas from Subsection 2.12, one can easily prove that:

- there exists an algorithm that, given a braid β and an element $v \in \pi_1$, detects whether v is a solution for β;
- there exists an algorithm that, given a braid $\beta \in B_n$ and a β-solution $v \in \pi_1^{(1)}$, computes a prime solution for β and a braid $\alpha \in B_{n-1}$ such that $\widehat{\alpha} \in \mathfrak{D}(\widehat{\beta})$.

4.2 On the destabilizability of periodic braids

The periodic case is the most simple one.

Notation: the exponent sum. We denote by $\exp(\beta)$ the *exponent sum* of a braid β with respect to Artin's generators. Note that exp is a conjugacy invariant.

Theorem 4.4. *An n-braid $\widehat{\beta}$ of periodic type is destabilizable if and only if*

$$0 < |\exp(\beta)| < n^2 - n.$$

Proof. Suppose $\beta \in B_n$ is a braid of periodic type.

Step 1. First we prove that $\widehat{\beta}$ is destabilizable whenever $0 < |\exp(\beta)| < n^2 - n$. We use Proposition 3.21. It is obvious that

$$\exp(\delta) = n - 1, \quad \exp(\delta^k) = (n-1)k; \quad \exp(\delta_*) = n, \quad \exp(\delta_*^\ell) = n\ell.$$

These equalities and Proposition 3.21 imply that if $0 < |\exp(\beta)| < n^2 - n$, then $\widehat{\beta}$ is in the following family of $4n - 6$ conjugacy classes:

$$E := \{\,\widehat{\delta}^k : 0 < |k| < n\,\} \cup \{\,\widehat{\delta}_*^\ell : 0 < |\ell| < n - 1\,\}.$$

First we consider the following case: $\widehat{\beta} = \widehat{\delta}^k$, where $0 < k < n$. Note that for each $i \in \{1, \ldots, n-1\}$ we have $\delta(u_i) = u_{i+1}$. Consequently, since $0 < k < n$, we have $\delta^k(u_1) = u_{1+k}$. Obviously, the elements u_1 and u_{1+k} are $*$-disjoint. This means that u_1 is a (prime) solution for δ^k, and $\widehat{\beta} = \widehat{\delta}^k$ is destabilizable by Theorem 4.1.

Since the inverse of a destabilizable braid is destabilizable (as remarked in the introduction to this article), it follows that $\widehat{\delta}^{-k}$ is destabilizable.

Now, let $\widehat{\beta} = \widehat{\delta}_*^\ell$, where $0 < \ell < n - 1$. We observe that $\delta_*(u_i) = u_{i+1}$ whenever $2 \leq i \leq n - 1$. Consequently, since $0 < \ell < n - 1$, we have

$\delta_*^\ell(u_2) = u_{2+\ell}$. Obviously, the elements u_2 and $u_{2+\ell}$ are $*$-disjoint. This means that u_2 is a (prime) solution for δ_*^ℓ, and $\widehat{\beta} = \widehat{\delta_*^\ell}$ is destabilizable by Theorem 4.1. It follows that $\widehat{\delta}_*^{-\ell}$ is also destabilizable.

Step 2. We prove that if the condition $0 < |\exp(\beta)| < n^2 - n$ does not hold, then $\widehat{\beta}$ is not destabilizable. Proposition 3.21 implies that if $\widehat{\beta}$ is not in E, then

$$\widehat{\beta} \in \{\widehat{e}, \widehat{\Delta}^2, \widehat{\Delta}^{-2}\} \cup \{\widehat{\delta}^k : |k| > n\} \cup \{\widehat{\delta}_*^\ell : |\ell| > n - 1\}.$$

The braids $\widehat{e}$, $\widehat{\Delta}^2$, and $\widehat{\Delta}^{-2}$ are not destabilizable since each of them represents an n-component link. Braids from the sets $\{\widehat{\delta}^k : |k| > n\}$ and $\{\widehat{\delta}_*^\ell : |\ell| > n - 1\}$ are not destabilizable since each of these braids has *twist number* with absolute value greater then 1 (see [20]). □

4.3 The 'functions' $\mathcal{J}_\beta : \Gamma \to \Gamma$

Let $\beta \in B_n$ be a braid and let $v \in \pi_1$ be an element of the fundamental group. We denote by $f_{\beta;\,v}$ (or by f if β and v are fixed) the composition of homeomorphisms $\beta : \Theta \to \Theta$ and $v : \Theta \to \Theta$ (see Theorem 3.5 and Definition 2.6):

$$f_{\beta;\,v} := v \circ \beta : \Theta \to \Theta, \qquad \theta \mapsto v(\beta(\theta)).$$

Thus $f_{\beta;\,v}$ is an orientation-preserving homeomorphism of the circle Θ. We introduce the following notation:

$$A := A_{\beta;\,v} := \{\theta \in \Theta \mid f(\theta) \prec \theta \prec f(e)\};$$
$$B := B_{\beta;\,v} := \{\theta \in \Theta \mid f(e) \prec \theta \prec f(\theta)\}.$$

Obviously, A and B are disjoint open subsets of Θ. Note that $e \notin A \cup B$.

Lemma 4.5. *Let $\beta \in B_n$, let $v \in \pi_1 \setminus e$, and let $\gamma \in \Gamma$. Then*

$$v \in \gamma \pitchfork \beta(\gamma) \iff \gamma \in A_{\beta;\,v} \cup B_{\beta;\,v}.$$

Proof. Follows directly from the above definitions. □

Lemma 4.6. *Let $f := f_{\beta;\,v}$, where $\beta \in B_n$ and $v \in \pi_1 \setminus e$. Then for each $\theta \in \Theta$ we have*

$$f(\theta) \prec \theta \prec f(e) \iff f^{-1}(e) \prec \theta \prec f^{-1}(\theta);$$

$$f(e) \prec \theta \prec f(\theta) \iff f^{-1}(\theta) \prec \theta \prec f^{-1}(e).$$

Proof. We prove the implication $f(\theta) \prec \theta \prec f(e) \implies f^{-1}(e) \prec \theta \prec f^{-1}(\theta)$ (proofs for other three implications are quite similar).

Using the notation introduced in Section 2.6 (page 94), we rewrite the condition $f(\theta) \prec \theta \prec f(e)$ in the form

$$e \curvearrowright f(\theta) \curvearrowright \theta \curvearrowright f(e).$$

This is equivalent to

$$f(e) \curvearrowright e \curvearrowright f(\theta) \curvearrowright \theta.$$

Since the homeomorphism f^{-1} preserves orientation, it follows that

$$e \curvearrowright f^{-1}(e) \curvearrowright \theta \curvearrowright f^{-1}(\theta).$$

This means exactly that the condition $f^{-1}(e) \prec \theta \prec f^{-1}(\theta)$ holds. $\square$

Lemma 4.7. *Let $f := f_{\beta;\,v}$, where $\beta \in B_n$ and $v \in \pi_1 \smallsetminus e$.*

a) *Suppose $A_{\beta;\,v}$ is not empty. Let $\theta \in A_{\beta;\,v}$. Then*

$$\Big(f(\theta), f(e)\Big) \cap \Big(f^{-1}(e), f^{-1}(\theta)\Big) \subset A_{\beta;\,v}$$

where $(f(\theta), f(e))$ and $(f^{-1}(e), f^{-1}(\theta))$ are subintervals of $\Theta \smallsetminus e$. (The definition of $A_{\beta;\,v}$ and Lemma 4.6 imply that these subintervals are nonempty).

b) *Suppose that $B_{\beta;\,v}$ is not empty and that $\theta \in B_{\beta;\,v}$. Then*

$$\Big(f(e), f(\theta)\Big) \cap \Big(f^{-1}(\theta), f^{-1}(e)\Big) \subset B_{\beta;\,v}.$$

Proof. We prove assertion a), the proof of the second assertion being similar. To prove assertion a), it suffices to show that for any θ' such that

$$\theta' \in \Big(f(\theta), f(e)\Big) \cap \Big(f^{-1}(e), f^{-1}(\theta)\Big), \tag{0}$$

we have $\theta' \in A_{\beta;\,v}$.

Note that condition (0) is equivalent to the following pair of conditions

$$f(\theta) \prec \theta' \prec f(e) \tag{1}$$

$$f^{-1}(e) \prec \theta' \prec f^{-1}(\theta). \tag{2}$$

We divide the proof into three cases:

(i) the case where $\theta' = \theta$,
(ii) the case where $\theta \prec \theta'$, and
(iii) the case where $\theta' \prec \theta$.

(i) If $\theta' = \theta$, then $\theta' \in A_{\beta;\,v}$ by the initial assumption.

(ii) We rewrite (2) in the form

$$f^{-1}(e) \curvearrowright \theta' \curvearrowright f^{-1}(\theta). \tag{2'}$$

Since the homeomorphism f preserves orientation, from (2') it follows that

$$e \curvearrowright f(\theta') \curvearrowright \theta. \tag{3}$$

We observe that (3) is equivalent to

$$f(\theta') \prec \theta. \tag{3'}$$

Combining condition (3'), the assumption $\theta \prec \theta'$, and condition (1), we obtain

$$f(\theta') \prec \theta \prec \theta' \prec f(e). \tag{4}$$

This implies

$$f(\theta') \prec \theta' \prec f(e). \tag{5}$$

By the definition of $A_{\beta;\,v}$, (5) means that $\theta' \in A_{\beta;\,v}$.

(iii) Now, let $\theta' \prec \theta$. This assumption and condition (2) yield

$$f^{-1}(e) \prec \theta' \prec \theta. \tag{6}$$

We rewrite (6) in the form

$$f^{-1}(e) \curvearrowright \theta' \curvearrowright \theta. \tag{6'}$$

From (6'), since f preserves the cyclic ordering $\curvearrowright$, we obtain

$$e \curvearrowright f(\theta') \curvearrowright f(\theta). \tag{7}$$

This is equivalent to

$$f(\theta') \prec f(\theta). \tag{7'}$$

Combining (7') with (1) we obtain (5) again. Therefore $\theta' \in A_{\beta;\,v}$.

$\square$

Let $\beta \in B_n$ be a braid and let $v \in \pi_1$ be an element of the fundamental group. We define the map $\mathcal{J}_{\beta;\,v} : \Theta \setminus e \to \Theta \setminus e$ as follows (we write f for $f_{\beta;\,v}$):

$$\mathcal{J}_{\beta;\,v}(\gamma) := \begin{cases} \min_{\prec}\,(f(e);\ f^{-1}(\gamma)) & \text{if } \gamma \in A_{\beta;\,v}; \\ \min_{\prec}\,(f^{-1}(e);\ f(\gamma)) & \text{if } \gamma \in B_{\beta;\,v}; \\ \gamma, & \text{otherwise.} \end{cases}$$

The definitions of $A_{\beta;\,v}$ and $B_{\beta;\,v}$ and Lemma 4.6 imply that

a) $f(e) \neq e$ and $f^{-1}(\gamma) \neq e$ whenever $\gamma \in A_{\beta;\,v}$,
b) $f^{-1}(e) \neq e$ and $f(\gamma) \neq e$ whenever $\gamma \in B_{\beta;\,v}$.

This implies that $\mathcal{J}_{\beta;\,v}$ is well-defined, i.e. for each $\gamma \in \Theta \smallsetminus e$ we have $\mathcal{J}_{\beta;\,v}(\gamma) \in \Theta \smallsetminus e$.

Proposition 4.8 (Properties of $\mathcal{J}_{\beta;\,v}$). *Let $\beta \in B_n$ be a braid and let $v \in \pi_1$ be an element of the fundamental group.*

  (i) *For each $\gamma \in \Theta \smallsetminus e$ we have $\gamma \preceq \mathcal{J}_{\beta;\,v}(\gamma)$.*
  (ii) *The map $\mathcal{J}_{\beta;\,v}$ is nondecreasing (with respect to $\prec$).*
  (iii) *The map $\mathcal{J}_{\beta;\,v}$ is left$_\prec$-continuous.*
  (iv) *For each $\gamma \in \Gamma$ we have $\mathcal{J}_{\beta;\,v}(\gamma) \in \Gamma$.*

Proof.

  (i) Follows from the definition of $\mathcal{J}_{\beta;\,v}$ and Lemma 4.6.
  (ii) Suppose the converse, i.e. suppose that there exist points γ and γ' from $\Theta \smallsetminus e$ such that

$$\gamma \prec \gamma', \quad \mathcal{J}_{\beta;\,v}(\gamma) \succ \mathcal{J}_{\beta;\,v}(\gamma').$$

By assertion (i) of the proposition, we have $\gamma' \preceq \mathcal{J}_{\beta;\,v}(\gamma')$. Combining this inequality with the initial assumption, we obtain

$$\gamma \prec \gamma' \preceq \mathcal{J}_{\beta;\,v}(\gamma') \prec \mathcal{J}_{\beta;\,v}(\gamma).$$

In particular, we have $\gamma \prec \mathcal{J}_{\beta;\,v}(\gamma)$. Then from the definition of $\mathcal{J}_{\beta;\,v}$ we conclude that $\gamma \in A \cup B$. Note also that γ' lies in the interval $(\gamma, \mathcal{J}_{\beta;\,v}(\gamma))$. Furthermore, the definition of $\mathcal{J}_{\beta;\,v}$ and Lemma 4.7 imply that

$$(\gamma, \mathcal{J}_{\beta;\,v}(\gamma)) \subset A \cup B.$$

Therefore, γ and γ' lie in one and the same connected component of the set $A \cup B$.

Moreover, the definition of $\mathcal{J}_{\beta;\,v}$ directly implies that $\mathcal{J}_{\beta;\,v}$ is continuous and nondecreasing on each connected component of the open subset $A \cup B \subset \Theta \smallsetminus e$. Then, since γ and γ' lie in one and the same subinterval of $A \cup B$, we have $\mathcal{J}_{\beta;\,v}(\gamma) \preceq \mathcal{J}_{\beta;\,v}(\gamma')$. This is a contradiction and so (ii) is proved.

(iii) As mentioned above, the restrictions

$$\mathcal{J}_{\beta;\,v}|_A : \gamma \to \min_{\prec}\left(f(e);\ f^{-1}(\gamma)\right)$$

and

$$\mathcal{J}_{\beta;\,v}|_B : \gamma \to \min_{\prec}\left(f^{-1}(e);\ f(\gamma)\right)$$

are continuous. In other words, $\mathcal{J}_{\beta;\,v}$ is continuous (in particular, left$_{\prec}$-continuous) at each point of the open set $A \cup B$.

We now prove that $\mathcal{J}_{\beta;\,v}$ is left$_{\prec}$-continuous at each point $z \in (\boldsymbol{\Theta} \smallsetminus e) \smallsetminus (A \cup B)$. Note that by definition we have $\mathcal{J}_{\beta;\,v}(z) = z$. Further, for any point $z' \prec z$, by assertion (i) of the proposition we have $z' \preceq \mathcal{J}_{\beta;\,v}(z')$, and by assertion (ii) of the proposition we have $\mathcal{J}_{\beta;\,v}(z') \preceq \mathcal{J}_{\beta;\,v}(z) = z$. Thus, for any $z' \prec z$ we have

$$z' \preceq \mathcal{J}_{\beta;\,v}(z') \preceq z.$$

It follows trivially that $\mathcal{J}_{\beta;\,v}$ is left$_{\prec}$-continuous in z, as required.

(iv) Let $\gamma \in \boldsymbol{\Gamma} \subset \boldsymbol{\Theta} \smallsetminus e$. If $\gamma \in \boldsymbol{\Gamma} \smallsetminus (A \cup B)$, then $\mathcal{J}_{\beta;\,v}(\gamma) = \gamma$ and the result follows trivially. Now, suppose $\gamma \in \boldsymbol{\Gamma} \cap (A \cup B)$. We consider the case where $\gamma \in A$ (the proof in the case where $\gamma \in B$ is similar).

If $\gamma \in A$, then by definition we have

$$\mathcal{J}_{\beta;\,v}(\gamma) = \min_{\prec}\left(f(e);\ f^{-1}(\gamma)\right).$$

Moreover, by the definition of A we have $f(\gamma) \prec \gamma \prec f(e)$, and by Lemma 4.6 we have $f^{-1}(e) \prec \gamma \prec f^{-1}(\gamma)$. In particular, $\gamma \prec f(e)$ and $\gamma \prec f^{-1}(\gamma)$.

Assume that $\mathcal{J}_{\beta;\,v}(\gamma) \notin \boldsymbol{\Gamma}$. Then, since

$$\mathcal{J}_{\beta;\,v}(\gamma) = \min_{\prec}\left(f(e);\ f^{-1}(\gamma)\right)$$

and since $\gamma \prec f(e)$, $\gamma \prec f^{-1}(\gamma)$, it follows that $f(e)$, $f^{-1}(\gamma) \notin \boldsymbol{\Gamma}$. Furthermore, we observe that $f(e) = v$ and that $f^{-1}(\gamma) = \beta^{-1}(v^{-1}(\gamma))$. Then since $\beta^{-1}(v^{-1}(\gamma)) = f^{-1}(\gamma) \notin \boldsymbol{\Gamma}$ it follows that $v^{-1}(\gamma) \notin \boldsymbol{\Gamma}$. Thus, we have $v = f(e) \notin \boldsymbol{\Gamma}$ and $v^{-1}(\gamma) \notin \boldsymbol{\Gamma}$. Consequently, v is a power of the distinguished element $\mathbf{u}$ and we have $\gamma = v(v^{-1}(\gamma)) \notin \boldsymbol{\Gamma}$, a contradiction. Hence $\mathcal{J}_{\beta;\,v}(\gamma) \in \boldsymbol{\Gamma}$.

$\square$

Proposition 4.9 (More properties of $\mathcal{J}_{\beta;v}$). *Suppose $\beta \in B_n$, $v \in \pi_1 \setminus e$, and $\gamma \in \Gamma$. Then we have*

$$v \in \gamma \pitchfork \beta(\gamma) \iff \gamma \prec \mathcal{J}_{\beta;v}(\gamma);$$
$$v \notin \gamma \pitchfork \beta(\gamma) \iff \gamma = \mathcal{J}_{\beta;v}(\gamma).$$

If $v \in \gamma \pitchfork \beta(\gamma)$, then for each $\gamma' \in (\gamma, \mathcal{J}_{\beta;v}(\gamma))$ we have $v \in \gamma' \pitchfork \beta(\gamma')$.

Proof. Follows directly from Lemmas 4.6 and 4.7, and the definitions of the intersection set, $\mathcal{J}_{\beta;v}$, $A_{\beta;v}$, and $B_{\beta;v}$. $\square$

Let $\beta \in B_n$ be a braid. We set

$$\mathcal{J}_{\beta}(\gamma) := \sup_{\prec} \{\mathcal{J}_{\beta;v}(\gamma) \mid v \in \pi_1\}.$$

To show that the map $\mathcal{J}_{\beta} : \Gamma \to \Gamma$ is well-defined, we must prove that for any $\gamma \in \Gamma$ we have

$$\sup_{\prec} \{\mathcal{J}_{\beta;v}(\gamma) \mid v \in \pi_1\} \in \Gamma.$$

Suppose first that $\gamma \in \Gamma_{\pi}$. Then the set $\gamma \pitchfork \beta(\gamma)$ is either finite or empty (see Remark 2.28). Thus, by Proposition 4.9, we have

$$\sup_{\prec} \{\mathcal{J}_{\beta;v}(\gamma) \mid v \in \pi_1\}$$
$$= \begin{cases} \max_{\prec} \{\mathcal{J}_{\beta;v}(\gamma) \mid v \in \{\gamma \pitchfork \beta(\gamma)\}\} & \text{if } \gamma \pitchfork \beta(\gamma) \neq \varnothing; \\ \gamma & \text{if } \gamma \pitchfork \beta(\gamma) = \varnothing. \end{cases}$$

Equivalently, we have

$$\sup_{\prec} \{\mathcal{J}_{\beta;v}(\gamma) \mid v \in \pi_1\} = \max_{\prec} \{\mathcal{J}_{\beta;v}(\gamma) \mid v \in \{e \cup (\gamma \pitchfork \beta(\gamma))\}\}.$$

Hence, by Proposition 4.8(iv), we have $\sup_{\prec} \{\mathcal{J}_{\beta;v}(\gamma) \mid v \in \pi_1\} \in \Gamma$ whenever $\gamma \in \Gamma_{\pi}$.

Now, let $\gamma \in \Gamma$. It can be easily seen that there exists a fundamental geodesic $\gamma' \in \Gamma_{\pi}$ such that $\gamma \prec \gamma'$. As established just above, $\sup_{\prec} \{\mathcal{J}_{\beta;v}(\gamma') \mid v \in \pi_1\} \in \Gamma$. Assertions (i) and (ii) of Proposition 4.8 trivially imply that

$$\gamma \preceq \sup_{\prec} \{\mathcal{J}_{\beta;v}(\gamma) \mid v \in \pi_1\} \preceq \sup_{\prec} \{\mathcal{J}_{\beta;v}(\gamma') \mid v \in \pi_1\}.$$

Therefore $\sup_{\prec} \{\mathcal{J}_{\beta;v}(\gamma) \mid v \in \pi_1\} \in \Gamma$, as required.

Proposition 4.10 (Properties of $\mathcal{J}_{\beta}$). *Let β be a braid. Then*

(i) *For each $\gamma \in \Gamma$, we have $\gamma \preceq \mathcal{J}_{\beta}(\gamma)$;*
(ii) *The map $\mathcal{J}_{\beta}$ is nondecreasing (with respect to $\prec$);*
(iii) *The map $\mathcal{J}_{\beta}$ is left-$\prec$-continuous.*

Proof. Follows from the definition of $\mathcal{J}_\beta$ and from Proposition 4.8. $\quad\square$

Proposition 4.11 (Properties of $\mathcal{J}_\beta$). *Let β be a braid. Then*

(i) $\gamma \prec \mathcal{J}_\beta(\gamma) \iff \gamma \pitchfork \beta(\gamma) \neq \varnothing$;

(ii) $\gamma = \mathcal{J}_\beta(\gamma) \iff \gamma \pitchfork \beta(\gamma) = \varnothing$.

(iii) *Suppose $\mathcal{J}_\beta(\gamma) \succ \gamma$. Suppose $\gamma' \in (\gamma, \mathcal{J}_\beta(\gamma))$. Then*

$$\gamma' \pitchfork \beta(\gamma') \neq \varnothing.$$

Proof. Follows from the definition of $\mathcal{J}_\beta$ and from Proposition 4.9. $\quad\square$

Lemma 4.12. *Let $\beta \in B_n$ and let $v \in \Sigma_{\pi, >0}$. Suppose $v^{+\infty} \pitchfork \beta(v^{+\infty}) \neq \varnothing$. Then $\mathcal{J}_\beta(v) \succ v^{+\infty}$.*

We omit the proof.

Proposition 4.13.

(i) *For any $\beta \in B_n$ and each $v \in \pi \smallsetminus e$, we have*

$$\mathcal{J}_{\beta;\,v}(\mathbf{\Gamma}_\pi) \subset \mathbf{\Gamma}_\pi.$$

(ii) *For any $\beta \in B_n$, we have*

$$\mathcal{J}_\beta(\mathbf{\Gamma}_\pi) \subset \mathbf{\Gamma}_\pi.$$

Proof. Follows easily from the definitions. $\quad\square$

Proposition 4.14.

(i) *There exists an algorithm that, given a braid β, an element $v \in \pi \smallsetminus e$, and an element $w \in \mathbf{\Gamma}_\pi$, computes the element $\mathcal{J}_{\beta;\,v}(w)$.*

(ii) *There exists an algorithm that, given a braid β and an element $v \in \mathbf{\Gamma}_\pi$, computes the element $\mathcal{J}_\beta(v)$.*

Proof. This follows from the lemmas of Section 2.12 and the definitions. $\quad\square$

4.4 Obstacles

Let β be a braid. A geodesic γ from the set $\Sigma_* = \Sigma_\rho \cup \Sigma_\clubsuit$ is called a β-*obstacle* if γ and $\beta(\gamma)$ have no transverse intersections but are not strongly $*$-disjoint. In other words, $\gamma \in \Sigma_*$ is a β-obstacle if either $\gamma = \beta(\gamma)$ or γ and $\beta(\gamma)$ are weakly $*$-disjoint.

Lemma 4.15. *The set of β-obstacles for a pseudo-Anosov braid β is either empty or equal to the set of T-geodesics for β.*

4.5 The Fundamental Algorithm

In this subsection, we present an algorithm that, given a braid β and an interval without β-obstacles $[\gamma, \eta) \subset \mathbf{\Gamma}$, where $\gamma, \eta \in \mathbf{\Sigma}_\pi$, detects whether $[\gamma, \eta)$ contains simple solutions for β, and outputs such a solution if one exists.

The Fundamental Algorithm.

  Input: $\beta \in B_n$; $\gamma, \eta \in \mathbf{\Sigma}_\pi$.

⊛0. Assign $t := 0$; $x_0 := \gamma$.

⊛1. Compare x_t and η; if $x_t \succeq \eta$, then stop:

the interval $[\gamma, \eta)$ does not contain simple solutions for β.

⊛2. Check whether x_t is a solution for β; if yes, then stop:

$x_t \in [\gamma, \eta)$ is a simple solution for β.

⊛3. If x_t has a right-hand $\mathbf{\Sigma}_\pi$-neighbour (see Definition 2.22), then:

assign $x_{t+1} := \mathrm{Next}_r(x_t)$.

⊛4. If x_t is maximal, then:

  − compute $\mathcal{J}_\beta(x_t)$;
  − compute the simplification $s(\mathcal{J}_\beta(x_t))$;
  − assign $x_{t+1} := s(\mathcal{J}_\beta(x_t))$.

⊛5. Reassign $t := t + 1$, and go to Mark ⊛1.

Proposition 4.16. *Let $\beta \in B_n$. Let $\gamma, \eta \in \mathbf{\Sigma}_\pi$. Suppose that the interval $[\gamma, \eta) \subset \mathbf{\Gamma}$ does not contain β-obstacles. Then the Fundamental Algorithm, given $(\beta; \gamma, \eta)$ as input, takes a finite number of steps.*

Furthermore, it stops on Mark ⊛1 if and only if $[\gamma, \eta)$ does not contain simple solutions for β; and it stops on Mark ⊛2 if and only if $[\gamma, \eta)$ contains a simple solution for β.

Proof. We analyse how the algorithm works. It takes as input a triple $(\beta; \gamma, \eta)$ and assigns $x_0 := \gamma$ on Mark ⊛0. After that, on Marks ⊛1 and ⊛2 the algorithm checks if the following condition holds

(C_0) \qquad\qquad $(x_0 \succ \eta)$ or (x_0 is a solution).

If (C_0) holds, then the algorithm stops and outputs an 'answer'. Assume now that (C_0) does not hold. Then algorithm proceeds to Marks ⊛3 and ⊛4 to compute x_1.

Claim 1.

(i) $x_1 \in \Sigma_\pi$;

(ii) $x_1 \succ x_0$;

(iii) The interval $[x_0, x_1)$ does not contain simple solutions for β.

Proof of the claim. Suppose first that x_0 is not a maximal element. In this case we have $x_1 = \text{Next}_r(x_0)$. Hence, $x_1 \in \Sigma_\pi$ and $x_1 \succ x_0$ by the definitions. Since $x_1 = \text{Next}_r(x_0)$, the interval (x_0, x_1) does not contain simple geodesics. Since we assume that (C_0) does not hold, x_0 is not a solution, and so the interval $[x_0, x_1)$ does not contain simple solutions for β.

Suppose, therefore, that x_0 is maximal.

(i) We see that $x_1 = s(\mathcal{J}_\beta(x_0)) \in \Sigma_\pi$, by Proposition 4.13 and the definition of the simplification (Definition 2.20).

(ii) Consider the geodesic $\zeta := x_0^{+\infty}$. By the results of Section 2, ζ is a right coil geodesic and $\text{env}(\zeta) = x_0$. From Theorem 2.17 it follows by construction that $\zeta \in [\gamma, \eta)$.

Let us show that ζ and $\beta(\zeta)$ have transverse intersections. Assume otherwise: then, since $\zeta \in [\gamma, \eta)$ and $[\gamma, \eta)$ does not contain β-obstacles, we conclude that ζ and $\beta(\zeta)$ are strongly $*$-disjoint (see the definition of an obstacle). The envelopes $x_0 = \text{env}(\zeta)$ and $\beta(x_0) = \text{env}(\beta(\zeta))$ are therefore (strongly) $*$-disjoint (see Remark 2.32). This means that x_0 is a solution for β, which contradicts the assumption that (C_0) does not hold. Thus, ζ and $\beta(\zeta)$ have transverse intersections. Then by Lemma 4.12, we have $\mathcal{J}_\beta(x_0) \succ \zeta$, whence by Theorem 2.17 it follows that $x_1 = s(\mathcal{J}_\beta(x_0)) \succ x_0$.

(iii) First we recall that x_0 is not a solution (since we assume that (C_0) does not hold). Furthermore, by Proposition 4.11(ii), the interval $(x_0, \mathcal{J}_\beta(x_0))$ does not contain solutions for β. If $\mathcal{J}_\beta(x_0)$ is simple, then $x_1 = s(\mathcal{J}_\beta(x_0)) = \mathcal{J}_\beta(x_0)$ and this completes the proof. If $\mathcal{J}_\beta(x_0)$ is not a simple geodesic, then by the definition of the simplification there are no simple geodesics between $\mathcal{J}_\beta(x_0)$ and $s(\mathcal{J}_\beta(x_0))$, whence we easily deduce that the interval $[x_0, x_1) = [x_0, s(\mathcal{J}_\beta(x_0)))$ does not contain simple solutions for β.

The claim is proved. $\qquad\qquad\qquad\qquad\qquad\qquad\qquad\qquad\qquad\square$

Now we return to the algorithm. After computing x_1 on Marks ⊛3 and ⊛4, the algorithm proceeds to Mark ⊛5; it reassigns t from 0 to 1 and goes to Mark ⊛1. Being on Mark ⊛1 again, we can see that the situation is similar to the one in the very beginning, except that now the input is the triple $(\beta; x_1, \eta)$ instead of the triple $(\beta; x_0, \eta)$. The algorithm repeats the procedure described above: on Marks ⊛1 and ⊛2 the algorithm checks whether the following condition holds:

$$(C_1) \qquad\qquad (x_1 \succ \eta) \text{ or } (x_1 \text{ is a solution}).$$

If (C_1) holds, then the algorithm stops. If (C_1) is not true, then the algorithm proceeds to Marks ⊛3 and ⊛4 where it computes x_2. (By analogy with Claim 1, we can prove that $x_2 \in \Sigma_\pi$; $x_2 \succ x_1$; the interval $[x_1, x_2)$ does not contain simple solutions for β.) After that, the algorithm goes to Mark ⊛5, then to Mark ⊛1, and so on.

Now we are able to describe the process of the algorithm's work as a whole. The Fundamental Algorithm, given a triple $(\beta; \gamma, \eta)$, begins to generate a sequence $x_0(= \gamma), x_1, x_2, \ldots$ of elements of $\mathbf{\Gamma}$. Each time after computing x_k, the algorithm checks the condition

$$(C_k) \qquad\qquad (x_k \succ \eta) \text{ or } (x_k \text{ is a solution}).$$

If (C_k) holds, then the algorithm stops. If (C_k) is not true, then the algorithm generates x_{k+1}.

By induction on k we prove the following (see the proof of Claim 1):

Claim 2.

    (i) $x_0, x_1, \ldots, x_k \in \Sigma_\pi$;

    (ii) $x_0 \prec x_1 \prec \ldots \prec x_k$;

    (iii) The interval $[x_0, x_k) = [x_0, x_1) \cup \cdots \cup [x_{k-1}, x_k)$ does not contain simple solutions for β.

Now, we are ready to prove the following part of the proposition:

Claim 3. If the algorithm at some moment stops on Mark ⊛1, then $[\gamma, \eta)$ does not contain simple solutions for β. If the algorithm stops on Mark ⊛2, then $[\gamma, \eta)$ contains a simple solution for β.

Proof of the claim. Suppose the algorithm stops on Mark ⊛1. This means that for a certain nonnegative integer k we have $x_k \succeq \eta$. By assertion (iii) of Claim 2, the interval $[x_0, x_k)$ does not contain simple solutions for β. Consequently, since $x_k \succeq \eta$, the interval $[\gamma, \eta) \subset [x_0, x_k)$ does not contain simple solutions for β.

Suppose the algorithm stops on Mark ⊛2. This means that for a certain nonnegative integer k the geodesic x_k is a simple solution for β. By assertion (ii) of Claim 2, we have $\gamma = x_0 \preceq x_k$. Since the algorithm does not stop on Mark ⊛1 while checking x_k, we have $x_k \prec \eta$. Thus, the simple solution x_k is in $[\gamma, \eta)$.

The claim is proved.   □

Claim 4. The Fundamental Algorithm, given a braid and an interval without obstacles for this braid, stops after a finite number of steps.

Proof of the claim. Assume that the algorithm takes an infinite number of steps. Then the considerations above imply that the algorithm computes an infinite sequence of simple fundamental elements $x_0, x_1, x_2, \ldots$ such that

$$x_0 \prec x_1 \prec x_2 \prec \ldots$$

where $x_t \in [\gamma, \eta)$. Since the sequence $\{x_t\}$ is increasing and bounded above by η, there exists a limit

$$\zeta := \lim_{t \to \infty} x_t.$$

Let us analyse what kind of geodesic ζ is. Since ζ is a left-hand limit point of a sequence consisting of simple fundamental elements, Theorem 2.17 implies that ζ is either a perfect geodesic or a left coil geodesic.

Since each element of the sequence $\{x_t\}$ is in $[\gamma, \eta)$, we have $\zeta \in [\gamma, \eta]$. Moreover, since η is a finite geodesic, while ζ is an infinite one, we have $\zeta \neq \eta$. This implies that $\zeta \in [\gamma, \eta)$. Thus, since $[\gamma, \eta)$ does not contain β-obstacles, ζ is not a β-obstacle. Then by the definition of an obstacle, ζ and $\beta(\zeta)$ either have transverse intersections or are strongly $*$-disjoint.

Assume first that ζ and $\beta(\zeta)$ are strongly $*$-disjoint. Then the envelopes $\mathrm{env}(\zeta)$ and $\mathrm{env}(\beta(\zeta)) = \beta(\mathrm{env}(\zeta))$ are strongly $*$-disjoint (see Remark 2.32). Therefore, for each simple fundamental geodesic $v \subset D^{\mathrm{env}(\zeta)}$ we have $\beta(v) \subset D^{\mathrm{env}(\beta(\zeta))}$, i.e. each simple fundamental geodesic $v \subset D^{\mathrm{env}(\zeta)}$ is a simple solution for β. Combining this fact with Lemma 2.24, one can easily conclude that ζ is a right-hand limit for simple solutions of β. On the other hand, since each interval $[\gamma, x_k)$ does not contain simple solutions for β (assertion (iii) of Claim 2), it follows that the interval $[\gamma, \zeta)$ does not contain simple solutions for β, giving a contradiction.

Now, assume that ζ and $\beta(\zeta)$ have transverse intersections. Then by Proposition 4.11(i), $\mathcal{J}_\beta(\zeta) \succ \zeta$. On the other hand, $\mathcal{J}_\beta$ is left-continuous, whence it follows that for a certain $\zeta' \prec \zeta$ we have $\mathcal{J}_\beta(\zeta') \succ \zeta$. Then by assertion (ii) of Proposition 4.10, for any $\zeta'' \succ \zeta'$ we have $\mathcal{J}_\beta(\zeta'') \succ \zeta$. Further, by definition of ζ there exists $N \in \mathbb{N}$ such that $x_t \succ \zeta'$ whenever $t \geq N$. Note that at least one of the elements x_N, x_{N+1}, x_{N+2}, x_{N+3} is maximal (this follows from Theorem 2.17). Say x_{N+r} is maximal, where $r \in \{0, 1, 2, 3\}$; then $x_{N+r+1} = s(\mathcal{J}_\beta(x_{N+r}))$. Since $N + r \geq N$, we have $x_{N+r} \succ \zeta'$ and hence $\mathcal{J}_\beta(x_{N+r}) \succ \zeta$. Therefore, since ζ is in Σ_*, we have $x_{N+r+1} = s(\mathcal{J}_\beta(x_{N+r})) \succ \zeta$, which again gives a contradiction.

Thus, we have established that ζ and $\beta(\zeta)$ neither coincide nor are weakly $*$-disjoint; also, ζ and $\beta(\zeta)$ are neither strongly $*$-disjoint nor have transverse intersections. But this is impossible by Proposition 2.34. $\square$

The theorem now follows from Claims 3 and 4. $\square$

4.6 Suitable sets of intervals for a pseudo-Anosov braid

Definition 4.17. Let β be a braid of pseudo-Anosov type. Let

$$E := \{[v_1, v_1'), \ldots, [v_k, v_k')\}, \quad v_i, v_i' \in \Sigma_\pi$$

be a finite set of subintervals of $\mathbf{\Gamma}$. We define

$$\widetilde{E} := [v_1, v_1') \cup \ldots \cup [v_k, v_k') \subset \mathbf{\Gamma}.$$

We say that the set E is *suitable* for β if

$$\bigcup_{r,s \in \mathbb{Z}} \Delta^{2r} \beta^s(\widetilde{E}) = \mathbf{\Gamma} \setminus T_\beta,$$

where $T_\beta \subset \mathbf{\Gamma}$ is the set of T-geodesics for β (see Section 3.6).

Proposition 4.18. *Suppose β is a braid of pseudo-Anosov type. Suppose E is a suitable set of intervals for β. Then the following conditions hold.*

(i) *If there is a simple solution for β, then there is a simple solution for β in the set $\widetilde{E}$.*

(ii) *The intervals of E do not contain β-obstacles.*

Proof. The first condition follows from the definition of a suitable set of intervals and Proposition 4.3. The second follows from the definition of a suitable set of intervals and Lemma 4.15. $\square$

An algorithm that computes a suitable set of intervals. Let β be a pseudo-Anosov braid. First, we compute the numbers

$$q := q(\beta), \quad r := r(\beta), \quad s := s(\beta)$$

(see Section 3.6 for notation). Furthermore, let $t_1 \in T_\beta$ be an arbitrary T-geodesic (we can 'choose' t_1 from the interval $[u_1, \Delta^2(u_1)] \subset \Gamma$, for example). The definitions imply that the interval $[t_1, \Delta^2(t_1)] \subset \Gamma$ contains exactly $2q + 1$ T-geodesics for β. Let

$$T_\beta \cap [t_1, \Delta^2(t_1)] = \{t_1, \ldots, t_{2q}, t_{2q+1} = \Delta^2(t_1)\}$$

where $t_1 \prec t_2 \prec \ldots \prec t_{2q+1}$. As observed in Section 3.9, we can compute the (large perfect) geodesics t_1, ..., t_{2q+1} to any prescribed accuracy. It follows that we can compute a set of fundamental geodesics $\{w_1, \ldots, w_{2q}\} \subset \Gamma_\pi$ such that $w_i \in (t_i, t_{i+1})$ for each $i \in \{1, \ldots, 2q\}$.

Then we compute the simplifications $y_i := s(w_i)$. By Theorem 2.17 it follows that $y_i \in (t_i, t_{i+1})$ for each $i \in \{1, \ldots, 2q\}$.

For each $i \in \{1, \ldots, 2m\}$ we then compute the element $y_i' := \beta^r \Delta^{2s}(y_i)$. Since the homeomorphism $\beta^r \Delta^{2s} : \Gamma \to \Gamma$ fixes T_β pointwise, we have $y_i' \in (t_i, t_{i+1})$ for each $i \in \{1, \ldots, 2q\}$.

Now, for $i \in \{1, \ldots, 2q\}$ we assign $v_i := y_i$ and $v_i' := y_i'$ if $y_i \prec y_i'$; while we assign $v_i := y_i'$ and $v_i' := y_i$ if $y_i' \prec y_i$.

We claim that the resulting set $\{[v_1, v_1'], \ldots, [v_{2q}, v_{2q}']\}$ of intervals is suitable for β. Indeed, we obviously have

$$\bigcup_{z \in \mathbb{Z}} (\beta^r \Delta^{2s})^z([v_i, v_i']) = (t_i, t_{i+1}).$$

It follows that

$$F := \bigcup_{z \in \mathbb{Z}} (\beta^r \Delta^{2s})^z([v_i, v_i') \cup \cdots \cup [v_{2q}, v_{2q}'))$$

$$= (t_1, t_2) \cup \cdots \cup (t_{2q}, t_{2q+1}) = (t_1, t_{2q+1}) \setminus T_\beta.$$

Then we have (recall that $t_{2q+1} = \Delta^2(t_1)$)

$$\bigcup_{z \in \mathbb{Z}} \Delta^{2z}(F) = \Gamma \setminus T_\beta.$$

Remark. Note that if $r = r(\beta) \neq 1$ then the suitable set of intervals $\{[v_1, v_1'], \ldots, [v_{2q}, v_{2q}']\}$ that we compute is 'excessive'. Its subset $\{[v_1, v_1'], \ldots, [v_{2q/r}, v_{2q/r}']\}$ consisting of $2q/r$ intervals is still a suitable set of intervals.

4.7 On the destabilizability of reducible braids

Theorem 4.19. *If the principal companion of a reducible braid* $\widehat{\beta}$ *is destabilizable, then* $\widehat{\beta}$ *is destabilizable.*

Proof. The proof is obvious from the 'knot point of view' (see remarks in the introduction)   □

Theorem 4.20. *Suppose* $\widehat{\beta}$ *is reducible (nonperiodic), but not pure-reducible braid. Then* $\widehat{\beta}$ *is destabilizable if and only if its principal companion is destabilizable.*

We omit the proof.

4.8 A destabilization algorithm

Let β be a braid. To detect whether the closed braid $\widehat{\beta}$ is destabilizable, we first of all determine the Thurston type of β (see Sections 3.5 and 3.9). Then we proceed to a corresponding subalgorithm (for periodic, pseudo-Anosov, and reducible (nonperiodic) braid types, respectively).

Periodic case

To recognize whether a closed n-braid $\widehat{\beta}$ of periodic type is destabilizable, it is sufficient to compute the exponent sum $\exp(\beta)$. (For by Theorem 4.4, $\widehat{\beta}$ is destabilizable if and only if $0 < |\exp(\beta)| < n^2 - n$.)

Pseudo-Anosov case

The procedure of checking whether a given braid $\widehat{\beta}$ of pseudo-Anosov type is destabilizable consists of two steps:

1. Find a 'suitable set of intervals' for β (the outline of an appropriate algorithm was given in the previous section).
2. Check whether intervals of the suitable set contain a 'simple solution' (apply the Fundamental Algorithm from Section 4.5 to each interval of the suitable set).

If intervals of the suitable set do not contain a simple solution, then $\widehat{\beta}$ is not destabilizable. This follows from Theorems 4.1 and 4.2 and from Proposition 4.18.

Conversely, if intervals of the suitable set contain a simple solution, then $\widehat{\beta}$ is destabilizable. This follows from Theorems 4.1 and 4.2.

The definition of a suitable set of intervals and Propositions 4.16 and 4.18 imply that the algorithm just described stops after a finite number of steps.

Reducible case

To recognize whether a given braid $\widehat{\beta}$ of reducible nonperiodic type is destabilizable, we begin by computing of the canonical reduction system and of the principal companion for $\widehat{\beta}$. We also detect whether $\widehat{\beta}$ is pure-reducible. After that, we consider two subcases: the 'generic' one, where $\widehat{\beta}$ is not pure-reducible; and the 'special' one, where $\widehat{\beta}$ is pure-reducible.

We do not describe the subalgorithm for the 'special' case in this work. This case is not difficult to analyse, but this needs a lot of additional constructions and definitions. Note that the 'special' subcase includes the *further* subcase of split braids. Clearly, a split braid is destabilizable if and only if one of its components is destabilizable.

In the 'generic' case, we check (using the procedures described above for the periodic and pseudo-Anosov cases) whether the principal companion $\widehat{\alpha}$ of $\widehat{\beta}$ is destabilizable (recall that $\widehat{\alpha}$ is either pseudo-Anosov or periodic). By Theorem 4.20, a reducible but not pure-reducible braid $\widehat{\beta}$ is destabilizable if and only if $\widehat{\alpha}$ is destabilizable.

Bibliography

[1] M. Bestvina and M. Handel. Train-tracks for surface homeomorphisms. *Topology (1)*, 34:109–140, 1995.

[2] J. S. Birman. *Braids, links, and mapping class groups*, volume 82 of *Ann. of Math. Stud.* Princeton Univ. Press, Princeton, NJ, 1974.

[3] J. S. Birman. On the conjugacy problem in the braid group. Errata: *Braids, links, and mapping class groups*, Princeton Univ. Press, 1974. *Canad. J. Math.*, 34(6):1396–1397, 1982.

[4] J. S. Birman and T. E. Brendle. Braids: a survey. Preprint, 2004.

[5] J. S. Birman and W. V. Menasco. Studying links via closed braids IV: Split links and composite links. *Inv. Math.*, 102:115–139, 1990.

[6] J. S. Birman and W. V. Menasco. Studying links via closed braids II: On a theorem of Bennequin. *Topology Appl.*, 40:71–82, 1991.

[7] J. S. Birman and W. V. Menasco. Studying links via closed braids I: A finiteness theorem. *Pacific J. Math.*, 154(1):17–36, 1992.

[8] J. S. Birman and W. V. Menasco. Studying links via closed braids V: Closed braid representations of the unlink. *Trans. Amer. Math. Soc.*, 329(2):585–606, 1992.

[9] J. S. Birman and W. V. Menasco. Studying links via closed braids VI: A non-finiteness theorem. *Pacific J. Math.*, 156(2):265–285, 1992.

[10] J. S. Birman and W. V. Menasco. Studying links via closed braids III: Classifying links which are closed 3-braids. *Pacific J. Math.*, 161(1):25–113, 1993.

[11] L. E. J. Brouwer. Über die periodischen Transformationen der Kugel. *Math. Ann.*, 80:39–41, 1919.

[12] A. Casson and S. Bleiler. *Automorphisms of surfaces after Nielsen and Thurston*, volume 9 of *London Math. Soc. Stud. Texts*. Cambridge Univ. Press, Cambridge, 1988.

[13] A. Constantin and B. Kolev. The theorem of Kerékjártó on periodic homeomorphisms of the disc and the sphere. *Enseign. Math. (2)*, 40:193–204, 1994.

[14] P. Dehornoy, I. Dynnikov, D. Rolfsen, and B. Wiest. *Why are braids orderable?*, volume 14 of *Panor. Synthèses*. Soc. Math. France, Paris, 2002.

[15] A. Fathi, F. Laudenbach, and V. Poenaru, editors. *Travaux de Thurston sur les surfaces*. Number 66–67 in Astérisque. Soc. Math. France, Paris, 1991. Séminaire Orsay, Reprint of *Travaux de Thurston sur les surfaces*, 1979.

[16] É. Ghys. Groups acting on the circle. *Enseign. Math. (2)*, 47:329–407, 2001.

[17] R. Kirby. Problems in low-dimensional topology. In R. Kirby, editor, *Geometric topology (Athens, GA, 1993)*, volume 2 of *AMS/IP Stud. Adv. Math.*, pages 35–473. Amer. Math. Soc., Providence, RI, 1997.

[18] J. E. Los. Pseudo-Anosov maps and invariant train tracks in the disc: a finite algorithm. *Proc. London Math. Soc. (3)*, 66(2):400–430, 1993.

[19] A. V. Malyutin. Orderings on braid groups, operations over closed braids, and confirmation of Menasco's conjecture. *Zap. Nauchn. Sem. S.-Peterburg. Otdel. Mat. Inst. Steklov. (POMI)*, 267:163–169, 2000. Translation in *J. Math. Sci.*, 113(6):822–826, 2003.

[20] A. V. Malyutin. Twist number of (closed) braids. *Algebra i Analiz*, 16(5):59–91, 2004.

[21] A. V. Malyutin and N. Yu. Netsvetaev. Dehornoy's ordering on the braid group and braid moves. *Algebra i Analiz*, 15(3):170–187, 2003. Translation in *St. Petersburg Math. J.*, 15(3):437–448, 2004.

[22] A. A. Markov. Über die freie Äquivalenz der geschlossenen Zöpfe. *Mat. Sb. (1)*, 43:73–78, 1936.

[23] J. McCool. On reducible braids. In *Word problems, II (Conf. on Decision Problems in Algebra, Oxford, 1976)*, volume 95 of *Stud. Logic Foundations Math.*, pages 261–295. North-Holland, Amsterdam, 1980.

[24] J. McCool. Erratum: "On reducible braids". *Canad. J. Math.*, 34(6):1398, 1982.

[25] H. Short and B. Wiest. Orderings of mapping class groups after Thurston. *Enseign. Math. (2)*, 46:279–312, 2000.

[26] W. P. Thurston. On the geometry and dynamics of diffeomorphisms of surfaces. *Bull. Amer. Math. Soc. (N.S.)*, 19:417–431, 1988.

[27] B. von Kerékjártó. Über die periodischen Transformationen der Kreis-scheibe und der Kugelfläche. *Math. Ann.*, 80:36–38, 1919.

Steklov Institute of Mathematics, Petersburg Department

Russian Academy of Sciences

27, Fontanka

St.Petersburg 191023

Russia

malyutin@pdmi.ras.ru

n-dimensional local fields and adeles on n-dimensional schemes

Denis V. Osipov

1 Introduction

The notion of an n-dimensional local field appeared in the works of
A. N. Parshin and K. Kato in the middle of the 1970s. These fields
generalize the usual local fields (which are 1-dimensional in this sense)
and help us to see higher-dimensional algebraic schemes from the local
point of view.

With every flag

$$X_0 \subset X_1 \ldots \subset X_{n-1} \qquad (\dim X_i = i)$$

of irreducible subvarieties on a scheme X ($\dim X = n$) one can canon-
ically associate a ring $K_{(X_0,\ldots,X_{n-1})}$. In the case where everything is
regularly embedded, the ring is an n-dimensional local field.

Originally, higher-dimensional local fields were used to develop the
generalization of class field theory to schemes of arbitrary dimension (see
the work of A. N. Parshin, K. Kato, S. V. Vostokov and others, [22, 11]).
However, many problems of algebraic varieties can be reformulated in
terms of higher-dimensional local fields and higher adelic theory.

For a scheme X there is an adelic object

$$\mathbb{A}_X = \prod{}' K_{(X_0,\ldots,X_{n-1})}$$

where the product is taken over all the flags with respect to certain
restrictions on components of adeles. In [19] A. N. Parshin defined adeles
on algebraic surfaces, which generalize usual adeles on curves. A. A.
Beilinson introduced a simplicial approach to adeles and generalized to
arbitrary dimensional Noetherian schemes in [2].

A. N. Parshin, A. A. Beilinson, A. Huber, A. Yekutiely, V. G. Lomadze and others have described connections of higher adelic groups with

131

cohomology of coherent sheaves [19, 2, 9, 29, 3, 4], intersection theory [21, 12, 15, 4], Chern classes [21, 10, 4], the theory of residues [19, 29, 2, 13, 4] and torus actions [5]. This paper is a review of the basic notions of higher-dimensional local fields and adeles on higher-dimensional schemes.

The paper is organized as follows. In Section 2 we give a general definition and formulate classification theorems for n-dimensional local fields. We describe how n-dimensional local fields appear from algebraic varieties and arithmetic schemes.

In Section 3 we define higher-dimensional adeles and adelic complexes. Starting from an example of adelic complexes on algebraic curves, we give a general simplicial definition for arbitrary Noetherian schemes, which is due to A. A. Beilinson. We formulate theorems about adelic resolutions of quasicoherent sheaves on Noetherian schemes. We apply these general constructions to algebraic sufaces to obtain adelic complexes on algebraic surfaces, which were introduced by A. N. Parshin.

In Section 4 we describe restricted adelic complexes. In constrast to the adelic complexes from Section 3, restricted adelic complexes are associated with a single flag of subvarieties. A. N. Parshin introduced restricted adeles for algebraic surfaces in [23, 24]; the author introduced restricted adelic complexes for arbitrary schemes in [17]. We also give the reconstruction theorem on restricted adelic complexes.

In the last section we briefly describe reciprocity laws on algebraic surfaces.

The author is very grateful to A. N. Parshin for many discussions on higher-dimensional local fields and adeles. He is also grateful to M. Taylor for interesting discussions and his hospitality during the visit to the University of Manchester, which was sponsored by an LMS grant. The author also acknowledges the support of the Russian Foundation for Basic Research, via grant no. 05-01-00455.

2 n-dimensional local fields

2.1 Classification theorems

We fix a perfect field k.

We say that K is a *local field of dimension* 1 with *residue field* k, if K is a fraction field of the complete discrete valuation ring $\mathcal{O}_K$ with residue field $\bar{K} = k$. We denote by ν_K the discrete valuation of K and by m_K the maximal ideal of the ring $\mathcal{O}_K$.

Such a field has the structure $K \supset \mathcal{O}_K \to \bar{K} = k$. As examples of such fields we have the field of power series

$$K = k((t)), \qquad \mathcal{O}_K = k[[t]], \qquad \bar{K} = k$$

and the field of p-adic numbers

$$K = \mathbb{Q}_p, \qquad \mathcal{O}_K = \mathbb{Z}_p, \qquad k = \mathbb{F}_p.$$

Moreover, we have only the following possibilities, see [27, Ch. II]:

Theorem 2.1. *Let K be a local field of dimension 1 with residue field k. Then K is one of the following:*

(i) $K = k((t))$ *is the power series field if* char $K =$ char k;
(ii) *if* char $K = 0$ *and* char $k = p$ *then either*

    (a) $K = \mathrm{Frac}(W(k))$ *where $\mathcal{O}_K = W(k)$ is the Witt ring of k (for example, $K = \mathbb{Q}_p$), or*
    (b) K *is a finite totally ramified extension of* $\mathrm{Frac}(W(k))$.

Now we give the following inductive definition.

Definition 2.2. *We say that a field K is a local field of dimension n with last residue field k if*

(i) $n = 0$ *and* $K = k$;
(ii) $n \geq 1$ *and K is the fraction field of a complete discrete valuation ring $\mathcal{O}_K$ whose residue field $\bar{K}$ is a local field of dimension $n-1$ with last residue field k.*

Thus, a local field of dimension $n \geq 1$ has the following inductive structure:

$$K = K^{(0)} \supset \mathcal{O}_K \to \bar{K} = K^{(1)} \supset \mathcal{O}_{\bar{K}} \to \bar{K}^{(1)}$$
$$= K^{(2)} \supset \mathcal{O}_{K^{(2)}} \to \ldots \bar{K}^{(n)} = k,$$

where for a discrete valuation field F we let $\mathcal{O}_F$ be the ring of integers in F and $\bar{F}$ be the residue field. The maximal ideals in $\mathcal{O}_{K^{(i)}}$ are denoted by $m_{K^{(i)}}$. Each field $K^{(i)}$ is a local field of dimension $n - i$ with last residue field k.

Definition 2.3. *A collection of elements $t_1, \ldots, t_n \in \mathcal{O}_K$ is called a system of local parameters for the n-dimensional local field K if t_1 is a generator of m_K and the images of $t_2, \ldots t_n$ in $\bar{K}$ form a system of local parameters for the $(n-1)$-dimensional local field $K^{(1)}$.*

An example of an n-dimensional local field is $K = k((t_n))\ldots((t_1))$. For this field we have

$$K^{(i)} = k((t_n))\ldots((t_{i+1})) \quad, \quad \mathcal{O}_{K^{(i)}} = k((t_n))\ldots((t_{i+2}))[[t_{i+1}]].$$

In the case $n = 2$ we have

$$K \supset \mathcal{O}_K \to \bar{K} \supset \mathcal{O}_{\bar{K}} \to k.$$

We can construct the following examples of 2-dimensional local fields with last residue field k. These examples depend on the characteristic of the 2-dimensional field K and the characteristics of its residue fields.

- $K = k((t_2))((t_1))$ and $\operatorname{char} K = \operatorname{char} \bar{K} = \operatorname{char} k$.
- $K = F((t))$, where F is a local field of dimension 1 with residue field k such that $\operatorname{char}(F) \neq \operatorname{char}(k)$, for example $F = \mathbb{Q}_p$.
- $K = F\{\{t\}\}$, where F is a local field of dimension 1 with residue field k.

The field $F\{\{t\}\}$ has the following description: $a \in F\{\{t\}\}$ if and only if

$$a = \sum_{i=-\infty}^{+\infty} a_i t^i, \qquad a_i \in F,$$

where $\lim_{i \to -\infty} \nu_F(a_i) = +\infty$ and $\nu_F(a_i) > c_a$ for some integer c_a.

We define the discrete valuation $\nu_{F\{\{t\}\}}$ by putting

$$\nu_{F\{\{t\}\}}(a) = \min \nu_F(a_i).$$

Then the ring $\mathcal{O}_{F\{\{t\}\}}$ consists of elements a such that all $a_i \in \mathcal{O}_F$, and the maximal ideal $m_{F\{\{t\}\}}$ consists of elements a such that all $a_i \in m_F$. Therefore

$$\overline{F\{\{t\}\}} = \bar{F}((t)).$$

We remark that for $F = k((u))$ the field $F\{\{t\}\}$ is isomorphic to the field $k((t))((u))$.

There exists the following classification theorem, see [4, 20, 30, 31].

Theorem 2.4. *Let K be an n-dimensional local field with finite last residue field.*

(i) *If $\operatorname{char}(K) = p$, then K is isomorphic to $\mathbb{F}_q((t_n))\ldots((t_1))$.*

(ii) *If $\operatorname{char}(K^{(n-1)}) = 0$, then K is isomorphic to $F((t_{n-1}))\ldots((t_1))$, where F is a 1-dimensional local field.*

(iii) *If* $\text{char}(K^{(m)}) = 0$ *and* $\text{char}(K^{(m+1)}) = p$, *then* K *is a finite extension of a field*

$$F\{\{t_n\}\}\ldots\{\{t_{m+2}\}\}((t_m))\ldots((t_1)) \qquad (*)$$

and there is a finite extension of K *which is of the form* $(*)$, *but possibly with different* F *and* t_i.

We remark that if π is a local parameter for a 1-dimensional local field F then $t_1, \ldots, t_m, \pi, t_{m+2}, \ldots, t_n$ are local parameters for a field

$$F\{\{t_n\}\}\ldots\{\{t_{m+2}\}\}((t_m))\ldots((t_1)).$$

2.2 Local fields which come from algebraic geometry

Consider an algebraic curve C over the field k. We fix a smooth point p on C and let

$$K_p = \text{Frac}(\hat{\mathcal{O}}_p),$$

where

$$\hat{\mathcal{O}}_p = \varprojlim_n \mathcal{O}_p/m_p^n$$

is a completion of the local ring $\mathcal{O}_p$ at the point p on the curve C. Then

$$K_p = k(p)((t)), \qquad (2.1)$$

where $k(p) = \mathcal{O}_p/m_p$ is the residue field of the point p on the curve C. This is the finite extension of the field k, and t is a local parameter of the point p on the curve C.

We see that the field K_p corresponds to case (i) of the classification theorem 2.1.

Now we consider a field of algebraic numbers K, i.e. a finite extension of the field $\mathbb{Q}$. Let A be the ring of integers of the field K; then let $X = \text{Spec } A$ (so that X is a 1-dimensional scheme). We fix a closed point $p \in X$, which corresponds to a maximal ideal in A. Then the completion of the field K at the point p is

$$K_p = \text{Frac}\,(\varprojlim_n A_p/m_p^n). \qquad (2.2)$$

We see that K_p is a 1-dimensional local field with residue field $\mathbb{F}_q$, and the field K_p corresponds to case (ii) of the classification theorem 2.1.

Now we give the definitions for the general situation. Let R be a ring,

p a prime ideal of R, M an R-module; let $S_p = R \setminus p$. We write $S_p^{-1}M$ for the localization of M at S_p. For an ideal a of R set

$$C_a M = \varprojlim_{n \in \mathbb{N}} M/a^n M.$$

Let X be a Noetherian scheme of dimension n. Let $\delta = (p_0, \ldots, p_n)$ be a chain of points of X (i.e. the chain of integral irreducible subschemes when considering the closures of the points p_i) such that $p_{i+1} \in \overline{\{p_i\}}$ for any i, where $\overline{\{p_i\}}$ is the closure of the point p_i in X. We suppose that for all i, we have $\dim p_i = i$.

Restrict δ to some affine open neighbourhood $\operatorname{Spec} B$ of the closed point p_n on X. Then δ determines a chain of prime divisors of the ring B, which we denote by the same letters $(p_0, \ldots, p_n)$. We define a ring as follows.

Definition 2.5.

$$K_\delta \overset{\text{def}}{=} C_{p_0} S_{p_0}^{-1} \ldots C_{p_n} S_{p_n}^{-1} B \tag{2.3}$$

This definition of K_δ does not depend on the choice of affine neighbourhood $\operatorname{Spec} B$ of the point p_n on the scheme X, see [9, Prop. 3.1.3, Prop. 3.2.1]. We remark that the ring $C_{p_n} S_{p_n}^{-1} B$ from (2.3) coincides with the completion $\hat{\mathcal{O}}_{p_n, X}$ of the local ring of the point p_n on the scheme X.

We now consider examples of (2.3) for small n.

Example 2.6. Let X be an irreducible 1-dimensional scheme (e.g. an irreducible curve over the field k, or the spectrum of the ring of algebraic integers). If p is a smooth point of X and η is a generic point of X, then for $\delta = (\eta, p)$ we see that K_δ is a 1-dimensional local field, which coincides with the field K_p from (2.1) or (2.2).

Example 2.7. Now let X be an irreducible algebraic surface over the field k. Let C be an irreducible divisor of X and p a point on C. We suppose that p is a smooth point on X and on C. Let η be a generic point of X, and consider $\delta = (\eta, C, p)$.

Fix a local parameter $t \in k(X)$ of the curve C on X at the point p ($t = 0$ is a local equation of the curve C at the point p on X). We also fix the local parameter $u \in k(X)$ at the point p on X which is transversal to the local parameter t (the divisor $u = 0$ is transversal to the divisor $t = 0$ at the point p). We fix any affine neighbourhood $\operatorname{Spec} B$ of p on X.

Then

$$C_p S_p^{-1} B = k(p)[[u, t]]$$
$$C_C S_C^{-1} C_p S_p^{-1} B = k(p)((u))[[t]]$$

and

$$K_\delta = C_\eta S_\eta^{-1} C_C S_C^{-1} C_p S_p^{-1} B = k(p)((u))((t)).$$

Hence K_δ is a 2-dimensional local field with last residue field $k(p)$. We see that the field K_δ corresponds to case (i) of the classification theorem 2.4.

Example 2.8. The previous example can be generalized. Let $p_0, \ldots, p_n$ be a flag of irreducible subvarieties on an n-dimensional algebraic variety X over the field k, such that $\dim p_i = n - i$, $p_{i+1} \subset p_i$ for all i and the point p_n is a smooth point on all subvarieties p_i. We can choose a system of local parameters $t_1, \ldots, t_n \in \mathcal{O}_{p_n, X}$ of the point p_n on X such that for every i, the equations $t_1 = 0, \ldots, t_i = 0$ define a subvariety p_i in some neighbourhood of the point p_n on X. Then according to (2.3) and similarly to the previous example we have for $\delta = (p_0, \ldots, p_n)$

$$K_\delta = k(p)((t_n)) \ldots ((t_1)).$$

Example 2.9. Now we suppose that the scheme X is an *arithmetic surface*, i.e. $\dim X = 2$ and we have a flat, projective morphism

$$f : X \to Y = \operatorname{Spec} A,$$

where A is the ring of integers of a number field K. We consider two kinds of integral irreducible 1-dimensional closed subscheme C on X.

(i) The subscheme C is *horizontal*, i.e. $f(C) = Y$. We consider a point $x \in C$ which is smooth on X and C. Let $\delta = (\eta, C, x)$, where η is a generic point of X. Then

$$K_\delta = L((t)),$$

where $t = 0$ is a local equation of C at the point x on X and $L \supset K_{f(x)} \supset \mathbb{Q}_p$ is a finite extension. Thus K_δ is a 2-dimensional local field with finite last residue field.

We see that this field K_δ corresponds to case (ii) of the classification theorem 2.4.

(ii) The subscheme C is *vertical*, i.e. it is a component of a fibre of f. This C is defined over some finite field $\mathbb{F}_q$. We consider a point $x \in C$ such that the morphism f is smooth at x and the point x is also defined over the field $\mathbb{F}_q$. Let $\delta = (\eta, C, x)$, where η is a generic point of X, and apply (2.3). For any affine neighbourhood $\operatorname{Spec} B$ of x on X the ring $C_x S_x^{-1} B$ coincides with the completion $\hat{O}_{x,X}$ of the local ring at the point x on X. But since f is a smooth map at p,

$$\hat{O}_{x,X} = O_{K_{f(x)}}[[u]],$$

and so

$$K_\delta = K_{f(x)}\{\{u\}\},$$

where $K_{f(x)} \supset \mathbb{Q}_p$ is a finite extension. Thus K_δ is a 2-dimensional local field with finite last residue field.

We see that this field K_δ corresponds to case (iii) of the classification theorem 2.4.

Note that in both these cases we have a canonical embedding f^* of the 1-dimensional local field $K_{f(x)}$ into the 2-dimensional local field K_δ.

Now we consider only *excellent Noetherian schemes* X (e.g. a scheme of finite type over a field, over $\mathbb{Z}$, or over a complete semi-local Noetherian ring; see [14, §34] and [6, §7.8]).

We introduce the following notation (see [21]). Let $\delta = (p_0, \ldots, p_n)$. Let a subscheme $X_i = \overline{\{p_i\}}$ be the closure of the point p_i in X. We introduce by induction the schemes X'_{i,α_i} in the following diagram

$$
\begin{array}{ccccccc}
X_0 & \supset & X_1 & \supset & X_2 & \supset & \cdots \\
\uparrow & & \uparrow & & \uparrow & & \\
X'_0 & \supset & X_{1,\alpha_1} & & & & \\
& & \uparrow & & & & \\
& & X'_{1,\alpha_1} & \supset & X_{2,\alpha_2} & & \\
& & & & \uparrow & & \\
& & & & \vdots & &
\end{array}
$$

Here X' denotes the normalization of a scheme X, and X_{i,α_i} is an integral irreducible subscheme in $X'_{i-1,\alpha_{i-1}}$ which is mapped onto X_i. By any such diagram we obtain the collection of indices $(\alpha_1, \ldots \alpha_n)$. The finite set of all such collections of indices is denoted by Λ_δ.

Such a collection of indices $(\alpha_1, \ldots \alpha_n) \in \Lambda_\delta$ determines a chain of discrete valuations in the following way. The integral irreducible subvariety X_{1,α_1} of the normal scheme X'_0 defines the discrete valuation of the field of functions on X_0. The residue field of this discrete valuation is the field of functions on the normal scheme X'_{1,α_1}, and the integer irreducible subscheme X_{2,α_2} defines the discrete valuation here. We proceed further for $\alpha_3, \ldots, \alpha_n$ in this way.

Moreover, there is the following theorem [21]. (See also [29, Theorem 3.3.2] for the proof).

Theorem 2.10. *Let X be an integral excellent n-dimensional Noetherian scheme. Then, for $\delta = (p_0, \ldots, p_n)$ the ring K_δ is an Artinian ring and*

$$K_\delta = \prod_{(\alpha_1,\ldots,\alpha_n)\in\Lambda_\delta} K_{(\alpha_1,\ldots,\alpha_n)}$$

where every $K_{(\alpha_1,\ldots,\alpha_n)}$ is an n-dimensional local field.

Example 2.11. To illustrate this theorem we now compute the ring K_δ in the following situation. Let p be a smooth point on an irreducible algebraic surface X over k. Suppose an irreducible curve $C \subset X$ contains the point p, but has a *node singularity* at p (i.e. the completed local ring of the point p on the curve C is $k[[t, u]]/tu$ for some local formal parameters u, t of the point p on X).

Let $\delta = (\eta, C, p)$, where η is a generic point of X, and fix any affine neighbourhood $\operatorname{Spec} B$ of p on X. Then according to (2.3)

$$C_p S_p^{-1} B = k(p)[[u, t]]$$
$$C_C S_C^{-1} C_p S_p^{-1} B = k(p)((u))[[t]] \oplus k(p)((t))[[u]]$$
$$K_\delta = C_\eta S_\eta^{-1} C_C S_C^{-1} C_p S_p^{-1} B = k(p)((u))((t)) \oplus k(p)((t))((u)).$$

3 Adeles and adelic complexes

3.1 Adeles on curves

Let C be a smooth connected algebraic curve over the field k. For any coherent sheaf $\mathcal{F}$ on C we consider an adelic space $\mathbb{A}_C(\mathcal{F})$, given by

$$\mathbb{A}_C(\mathcal{F}) = \left\{ \{f_p\} \in \prod_{p\in C} \mathcal{F} \otimes_{\mathcal{O}_C} K_p \quad : \quad \begin{array}{l} f_p \in \mathcal{F} \otimes_{\mathcal{O}_C} \mathcal{O}_{K_p} \\ \text{for almost all } p \end{array} \right\}$$

where the product is over all closed points p of the curve C.

We construct the following complex $\mathcal{A}_C(\mathcal{F})$:

$$
\begin{array}{ccccc}
\mathcal{F} \otimes_{\mathcal{O}_C} k(C) & \times & \displaystyle\prod_{p\in C} \mathcal{F} \otimes_{\mathcal{O}_C} \mathcal{O}_{K_p} & \longrightarrow & \mathbb{A}_C(\mathcal{F}) \\
a & \times & b & \longmapsto & a + b.
\end{array}
$$

Theorem 3.1 (see e.g. [26]). *The cohomology groups of the complex* $\mathcal{A}_C(\mathcal{F})$ *coincide with the cohomology groups* $H^*(C, \mathcal{F})$, *where* $\mathcal{F}$ *is any coherent sheaf on* C.

Proof. We give here a sketch of the proof. We construct an adelic complex $\mathcal{A}_U(\mathcal{F})$ of the sheaf $\mathcal{F}$ for any open subset $U \subset C$. Taking into account all U we obtain a complex of sheaves $\mathcal{A}(\mathcal{F})$ on the curve C.

For small affine U we find that the complex

$$0 \longrightarrow \mathcal{F}(U) \longrightarrow \mathcal{A}_U(\mathcal{F}) \longrightarrow 0$$

is exact, since we can apply the approximation theorem for Dedekind rings over fields. Therefore the complex $\mathcal{A}(\mathcal{F})$ is a resolution of the sheaf $\mathcal{F}$ on C. By construction this resolution is a *flasque* resolution of the sheaf $\mathcal{F}$ on C, and so calculates the cohomology of the sheaf $\mathcal{F}$ on the curve C. □

3.2 Adeles on higher-dimensional schemes

In this section we give a generalization of adelic complexes to schemes of arbitrary dimensions.

For algebraic surfaces adelic complexes were introduced by A. N. Parshin in [19]. We will later give a detailed exposition of adelic complexes on algebraic surfaces, as an application of general machinery constructed for arbitrary Noetherian schemes by A. Beilinson in [2]. For a good exposition and proofs of Beilinson's results see [9].

Definition of adelic spaces

We introduce the following notation. For any Noetherian scheme X let $P(X)$ be the set of points of the scheme X. Consider $p, q \in P(X)$ and define the relation $p \geq q$ if $q \in \overline{\{p\}}$, i.e. the point p is in the closure of the point q. Then $\geq$ is a partial order on $P(X)$. Let $S(X)$ be the simplicial set induced by $(P(X), \geq)$, i.e.

$$S(X)_m = \{(p_0, \ldots, p_m) \mid p_i \in P(X); p_i \geq p_{i+1}\}$$

is the set of m-simplices of $S(X)$ with the usual boundary δ_i^n and degeneracy maps σ_i^n for $n \in \mathbb{N}$, $0 \leq i \leq n$.

Let $K \subset S(X)_n$. For $p \in P(V)$ we let

$$_pK \stackrel{\text{def}}{=} \{(p_1, \ldots, p_n) \in S(V)_{n-1} \mid (p, p_1 \ldots, p_n) \in K\}.$$

Let $\mathbf{QS}(X)$ and $\mathbf{CS}(X)$ be the categories of quasicoherent and coherent sheaves, respectively, on the scheme X. Let $\mathbf{Ab}$ be the category of Abelian groups.

We have the following proposition, see [2, 9, 8], which is also a definition.

Proposition 3.2. *Let $S(X)$ be the simplicial set associated to the Noetherian scheme X. Then for integer $n \geq 0$, $K \in S(X)_n$ there exist functors*

$$\mathbb{A}(K, \cdot) \; : \; \mathbf{QS}(X) \longrightarrow \mathbf{Ab}$$

uniquely determined by the properties (i), (ii), (iii), *which are additive and exact.*

(i) $\mathbb{A}(K, \cdot)$ *commutes with direct limits.*

(ii) *For $n = 0$ and $\mathcal{F}$ a coherent sheaf on X*

$$\mathbb{A}(K, \mathcal{F}) = \prod_{p \in K} \varprojlim_{l} \mathcal{F}_p / m_p^l \mathcal{F}_p.$$

(iii) *For $n > 0$ and $\mathcal{F}$ a coherent sheaf on X*

$$\mathbb{A}(K, \mathcal{F}) = \prod_{p \in P(V)} \varprojlim_{l} \mathbb{A}(_\eta K, \mathcal{F}_p / m_p^l \mathcal{F}_p)$$

Remark 3.3. Since any quasicoherent sheaf on an excellent Noetherian scheme is a direct limit of coherent sheaves, we can apply property (i) of this proposition to define $\mathbb{A}(K, \mathcal{F})$ on quasicoherent sheaves.

Local factors

By induction on the definition of $\mathbb{A}(K, \mathcal{F})$ we get the following proposition [9, Prop. 2.1.4.].

Proposition 3.4. *For any integer $n > 0$, $K \subset S(X)_n$, and any quasicoherent sheaf $\mathcal{F}$ on X*

$$\mathbb{A}(K, \mathcal{F}) \subset \prod_{\delta \in K} \mathbb{A}(\delta, \mathcal{F}).$$

The inclusion is a natural transformation of functors.

From this proposition we see that $\mathbb{A}(K, \mathcal{F})$ is a kind of complicated adelic product inside $\prod_{\delta \in K} \mathbb{A}(\delta, \mathcal{F})$. It is therefore important to study the local factors $\mathbb{A}(\delta, \mathcal{F})$ for $\delta \in S(X)_n$. We have the following two propositions from [9] about these local factors.

Proposition 3.5. *Let* $\delta = (p_0, \ldots, p_n) \in S(X)_n$. *Let* U *be an open affine subscheme which contains the point* p_n *and therefore all of* δ. *Let* $M = \mathcal{F}(U)$. *Then for a quasicoherent sheaf* $\mathcal{F}$

$$\mathbb{A}(\delta, \mathcal{F}) = \mathbb{A}(\delta, \tilde{M}),$$

where $\tilde{M}$ *is a quasicoherent sheaf on* U *which corresponds to* M.

In the following proposition, local factors $\mathbb{A}(\delta, \mathcal{F})$ are computed for affine schemes.

Proposition 3.6. *Let* $X = \operatorname{Spec} R$ *and* $\mathcal{F} = \tilde{M}$ *for some* R-*module* M. *Further let* $\delta = (p_0, \ldots, p_n) \in S(X)_n$. *Then*

$$\mathbb{A}(\delta, \mathcal{F}) = C_{p_0} S_{p_0}^{-1} \ldots C_{p_n} S_{p_n}^{-1} R \otimes_R M. \tag{3.1}$$

$C_{p_0} S_{p_0}^{-1} \ldots C_{p_n} S_{p_n}^{-1} R$ *is a flat Noetherian* R-*algebra, and for finitely generated* R-*modules*

$$C_{p_0} S_{p_0}^{-1} \ldots C_{p_n} S_{p_n}^{-1} R \otimes_R M = C_{p_0} S_{p_0}^{-1} \ldots C_{p_n} S_{p_n}^{-1} M.$$

We now compare (3.1) and (2.3) for K_δ. We see that for an n-dimensional Noetherian scheme X, for $\delta \in S(X)_n$ and a quasicoherent sheaf $\mathcal{F}$

$$\mathbb{A}(\delta, \mathcal{F}) = K_\delta \otimes_{\mathcal{O}_X} \mathcal{F}.$$

Remark 3.7. Due to Theorem 2.10, for $\delta \in S(X)_n$ the local factors $\mathbb{A}(\delta, \mathcal{O}_X)$ on an excellent Noetherian integral n-dimensional scheme X are finite products of n-dimensional local fields.

Adelic complexes

Now we want to define adelic complexes on the scheme X. Consider the simplicial set $S(X)$ with the usual boundary maps δ_i^n and degeneracy maps σ_i^n for $n \in \mathbb{N}$, $0 \leq i \leq n$.

We note the following property, see [9, Prop. 2.1.5.].

Proposition 3.8. *Let* $K, L, M \subset S(X)_n$ *such that* $K \cup M = L$ *and* $K \cap M = \emptyset$. *Then there are natural transformations* i *and* π *of functors*

$$i(\cdot) : \mathbb{A}(K, \cdot) \longrightarrow \mathbb{A}(L, \cdot)$$
$$\pi(\cdot) : \mathbb{A}(L, \cdot) \longrightarrow \mathbb{A}(M, \cdot)$$

such that the following diagram is commutative and has split-exact rows for all quasicoherent sheaves $\mathcal{F}$ *on* X:

$$0 \longrightarrow \mathbb{A}(K,\mathcal{F}) \overset{i(\mathcal{F})}{\longrightarrow} \mathbb{A}(L,\mathcal{F}) \overset{\pi(\mathcal{F})}{\longrightarrow} \mathbb{A}(M,\mathcal{F}) \longrightarrow 0$$

$$0 \longrightarrow \prod_{\delta \in K} \mathbb{A}(\delta,\mathcal{F}) \longrightarrow \prod_{\delta \in L} \mathbb{A}(\delta,\mathcal{F}) \longrightarrow \prod_{\delta \in M} \mathbb{A}(\delta,\mathcal{F}) \longrightarrow 0 .$$

The proof is by induction on Proposition 3.2.

Definition 3.9. *Let $K \subset S(X)_0$, $\mathcal{F}$ a quasicoherent sheaf. We write*

$$d^0(K,\mathcal{F}) \; : \; \Gamma(X,\mathcal{F}) \longrightarrow \mathbb{A}(K,\mathcal{F})$$

for the canonical map, which is a natural transformation of functors.

Definition 3.10. *Let $K \subset S(X)_{n+1}$, $L \subset S(X)_n$, $\delta_i^{n+1}K \subset L$ for some $i \in \{0,\ldots,n+1\}$. We define natural transformations of functors*

$$d_i^{n+1}(K,L,\cdot) \; : \; \mathbb{A}(L,\cdot) \longrightarrow \mathbb{A}(K,\cdot)$$

by the following properties.

(i) *If $i = 0$ and $\mathcal{F}$ is a coherent sheaf on X, then we apply the functor $\mathbb{A}(_pK,\cdot)$ to $\mathcal{F} \to \mathcal{F}_p/m_p^l\mathcal{F}_p$ and compose this map with the projection of Proposition 3.8 for $L \supset {}_pK$. We use the universal property of $\prod_{p \in P(X)} \varprojlim$.*

(ii) *If $i = 1$, $n = 0$ and $\mathcal{F}$ is a coherent sheaf on X, then the projection of Proposition 3.8 for $L \supset {}_pK$ is composed with the following map: the maps $d^0(_pK, \mathcal{F}_p/m_p^l\mathcal{F}_p)$ form a projective system for $l \in \mathbb{N}$ and we apply $\prod_{p \in P(X)} \varprojlim$ to them.*

(iii) *If $i > 0$, $n > 0$, $\mathcal{F}$ is a coherent sheaf, then the hypothesis $\delta_i^{n+1}K \subset L$ implies $\delta_{i-1}^n(_pK) \subset {}_pL$ for all $p \in P(X)$. Set*

$$d_i^{n+1}(K,L,\mathcal{F}) = \prod_{p \in P(X)} \varprojlim_{l \in \mathbb{N}} d_{i-1}^n(_pK, {}_pL, \mathcal{F}_p/m_p^l\mathcal{F}_p).$$

(iv) *$d_i^{n+1}(K,L,\cdot)$ commutes with direct limits.*

For $\delta \in S(X)_{n+1}$ and $\delta' = \delta_i^{n+1}(\delta) \in S(X)_n$, by Definition 3.10 we have local boundary maps

$$d_i^{n+1} \quad : \quad \mathbb{A}(\delta',\mathcal{F}) \longrightarrow \mathbb{A}(\delta,\mathcal{F}).$$

For $K \subset S(X)_{n+1}$, $L \subset S(X)_n$ with $\delta_i^{n+1} K \subset L$ we define

$$D_i^{n+1}(\mathcal{F}) \quad : \quad \prod_{\delta \in L} \mathbb{A}(\delta, \mathcal{F}) \longrightarrow \prod_{\delta \in K} \mathbb{A}(\delta, \mathcal{F}),$$

where $(x_\delta)_{\delta \in L} \mapsto (y_\delta)_{\delta \in K}$ is given by $y_\delta = \prod d_i^{n+1}(x_{\delta'})$.

The following proposition from [9] is useful for the computation of boundary maps. It describes the boundary maps d_i^{n+1} by means of the boundary maps D_i^{n+1} on the product of local factors.

Proposition 3.11. *Let $K \subset S(X)_{n+1}$, $L \subset S(X)_n$ with $\delta_i^{n+1} K \subset L$. The following diagram commutes:*

$$
\begin{array}{ccc}
\mathbb{A}(L, \mathcal{F}) & \xrightarrow{\;d_i^{n+1}\;} & \mathbb{A}(K, \mathcal{F}) \\
\downarrow & & \downarrow \\
\displaystyle\prod_{\delta \in L} \mathbb{A}(\delta, \mathcal{F}) & \xrightarrow{\;D_i^{n+1}\;} & \displaystyle\prod_{\delta \in K} \mathbb{A}(\delta, \mathcal{F}).
\end{array}
$$

The proof is by induction on the definitions.

For the scheme X we consider the set $S(X)_n^{(red)}$ of nondegenerate n-dimensional simplices. (A simplex $(p_0, \ldots p_n)$ is *nondegenerate* if $p_i \neq p_{i+1}$ for all i.) For any $n \geq 0$ and any quasicoherent sheaf $\mathcal{F}$ on X we write

$$\mathbb{A}_X^n(\mathcal{F}) = \mathbb{A}(S(X)_n^{(red)}, \mathcal{F}).$$

Consider the boundary maps $d_i^{n+1} : \mathbb{A}_X^n(\mathcal{F}) \longrightarrow \mathbb{A}_X^{n+1}(\mathcal{F})$. These satisfy the following equalities:

$$d_j^n d_i^n = d_i^n d_{j-1}^n \qquad\qquad i < j. \tag{3.2}$$

For $n \geq 1$ we define $d_n : \mathbb{A}_X^{n-1}(\mathcal{F}) \longrightarrow \mathbb{A}_X^n(\mathcal{F})$ by

$$d^n = \sum_{j=0}^n (-1)^j d_j^n. \tag{3.3}$$

We then have the following proposition, which is also a definition.

Proposition 3.12. *The differentials d^n make $\mathbb{A}_X^*(\mathcal{F})$ into a cohomological complex of Abelian groups $A_X(\mathcal{F})$, which we call the adelic complex of the sheaf $\mathcal{F}$ on X.*

The proof follows by direct calculations with (3.2) and (3.3).

Theorem 3.13. *If $\mathcal{F}$ is a quasicoherent sheaf on a Noetherian scheme X, then*

$$H^i(\mathcal{A}_X(\mathcal{F})) = H^i(X, \mathcal{F}) \qquad \text{for all } i.$$

Proof. The proof of this theorem is a generalization of the proof of Theorem 3.1. For any open subscheme $U \subset X$ we consider the complex

$$0 \longrightarrow \mathcal{F}(U) \xrightarrow{d^0} \mathbb{A}_U^0(\mathcal{F}) \xrightarrow{d^1} \mathbb{A}_U^1(\mathcal{F}) \xrightarrow{d^2} \dots \xrightarrow{d^n} \mathbb{A}_U^n(\mathcal{F}) \xrightarrow{d^{n+1}} \dots \quad (3.4)$$

Taking into account all U, we see that the complex (3.4) is a complex of sheaves on X. Moreover, by Proposition 3.8 the sheaves in this complex are *flasque* sheaves, since $S(U)_n^{(red)} \subset S(X)_n^{(red)}$ for any n.

By [9, Th. 4.1.1], for any affine scheme U the complex (3.4) is an exact complex. Therefore we have constructed a flasque resolution of the sheaf $\mathcal{F}$ on X. This resolution calculates the cohomology of the sheaf $\mathcal{F}$ on X. $\qquad \square$

Remark 3.14. Here we have constructed *reduced adeles*, as we used only nondegenerate simplices in $S(X)$. These reduced adeles really carry information and they are part of the full complex, see [9].

3.3 Adeles on algebraic surfaces

In this section we verify that the general adelic complex constructed in the previous section coincides with the adelic complex for curves from Section 3.1. We also give an application of this general construction of adelic complexes to algebraic surfaces.

Consider a smooth connected algebraic curve C over a field k. The set $S(C)_0^{(red)}$ consists of the generic point η and all closed points p of the curve C; the set $S(C)_1^{(red)}$ consists of all pairs (η, p). For any coherent sheaf $\mathcal{F}$ on C we can compute by definition

$$\mathbb{A}_C^0(\mathcal{F}) = \mathcal{F} \otimes_{\mathcal{O}_C} k(C) \times \prod_{p \in C} \mathcal{F} \otimes_{\mathcal{O}_C} \mathcal{O}_{K_p}.$$

Let $K \subset S(C)_0^{(red)}$ be a subset consisting of all the closed points of the curve C. We have, by definition

$$\mathbb{A}^1_C(\mathcal{F}) = \mathbb{A}(K, \mathcal{F}_\eta)$$
$$= \mathbb{A}(K, \mathcal{F} \otimes_{\mathcal{O}_C} k(C))$$
$$= \mathbb{A}\left(K, \varinjlim_{D \in \mathrm{Div}(C)} \mathcal{F} \otimes_{\mathcal{O}_C} \mathcal{O}_C(D)\right)$$
$$= \varinjlim_{D \in \mathrm{Div}(C)} \mathbb{A}(K, \mathcal{F} \otimes_{\mathcal{O}_C} \mathcal{O}_C(D))$$
$$= \varinjlim_{D \in \mathrm{Div}(C)} \prod_{p \in C} \mathcal{F} \otimes_{\mathcal{O}_C} \mathcal{O}_{K_p}(D).$$

Therefore the adelic complex constructed in Section 3.2 coincides with the adelic complex for curves from Section 3.1.

Now consider a smooth connected algebraic surface X over a field k:

- The set $S(X)_0^{(red)}$ consists of the generic point η of X, generic points of all irreducible curves $C \subset X$ and all closed points $p \in X$.
- The set $S(X)_1^{(red)}$ consists of all pairs (η, C), (η, p) and (C, p). (In our notation we identify the generic point of a curve $C \subset X$ with the curve C.)
- The set $S(X)_2^{(red)}$ consists of all triples (η, C, p).

Consider $\delta = (\eta, C, p)$, and let f be a natural map from the local ring $\mathcal{O}_{p,X}$ to the completion $\hat{\mathcal{O}}_{p,X}$. The curve C defines a prime ideal $\mathbf{C}'$ in the ring $\mathcal{O}_{p,X}$. Let $\mathbf{C}_1, \ldots, \mathbf{C}_n$ be all prime ideals of height 1 in the ring $\hat{\mathcal{O}}_{p,X}$ such that for any i we have $f^{-1}(\mathbf{C}_i) = \mathbf{C}'$. Any such $\mathbf{C}_i$ we will call a *germ of C at p*.

For any such germ $\mathbf{C}_i$ we define a 2-dimensional local field

$$K_{p,\mathbf{C}_i} = \mathrm{Frac} \varprojlim_l \left((\hat{\mathcal{O}}_{p,X})_{(\mathbf{C}_i)} / \mathbf{C}_i^l (\hat{\mathcal{O}}_{p,X})_{(\mathbf{C}_i)} \right)$$

where the ring $(\hat{\mathcal{O}}_{x,X})_{(\mathbf{C}_i)}$ is a localization of the ring $\hat{\mathcal{O}}_{p,X}$ along the prime ideal $\mathbf{C}_i$. Then according to the formula (2.3), we have (see [4, 19])

$$\mathbb{A}(\delta, \mathcal{O}_X) = K_\delta = \bigoplus_{i=1}^{i=n} K_{p,\mathbf{C}_i}$$

and similarly

$$\mathbb{A}((C, p), \mathcal{O}_X) = \bigoplus_{i=1}^{i=n} \mathcal{O}_{K_{p,\mathbf{C}_i}}.$$

We compute by definition

$$\mathbb{A}((\eta, C), \mathcal{O}_X) = K_C,$$

where the field K_C is the completion of the field $k(X)$ along the discrete valuation given by an irreducible curve C on X.

From the definition we see that $\mathbb{A}((\eta, p), \mathcal{O}_X)$ is a subring of $\mathrm{Frac}(\hat{\mathcal{O}}_{p,X})$ generated by subrings $k(X)$ and $\hat{\mathcal{O}}_{p,X}$. We denote this subring by K_p.

By definition we compute

$$\mathbb{A}((\eta), \mathcal{O}_X) = k(X),$$
$$\mathbb{A}((C), \mathcal{O}_X) = \mathcal{O}_{K_C},$$
$$\mathbb{A}((p), \mathcal{O}_X) = \hat{\mathcal{O}}_{p,X}.$$

Remark 3.15. The local boundary maps d_i^n give natural embeddings of the rings

$$\mathbb{A}((\eta), \mathcal{O}_X), \ \mathbb{A}((C), \mathcal{O}_X), \ \mathbb{A}((p), \mathcal{O}_X),$$
$$\mathbb{A}((\eta, p), \mathcal{O}_X), \ \mathbb{A}((\eta, C), \mathcal{O}_X) \ \text{and} \ \mathbb{A}((C, p), \mathcal{O}_X)$$

into the ring K_δ.

By definition,

$$\mathbb{A}_X^0(\mathcal{O}_X) = k(X) \times \prod_{C \subset X} \mathcal{O}_{K_C} \times \prod_{p \in X} \hat{\mathcal{O}}_{p,X}.$$

From Proposition 3.2 and similarly to the case of algebraic curves, we can compute the ring $\mathbb{A}_X^2(\mathcal{O}_X)$, see [4, 21] for details. For any prime ideal $\mathbf{C} \subset \hat{\mathcal{O}}_{p,X}$ of height 1 we define the subring $\hat{\mathcal{O}}_{p,X}(\infty\mathbf{C})$ of $K_{p,\mathbf{C}}$ by

$$\hat{\mathcal{O}}_{p,X}(\infty\mathbf{C}) = \varprojlim_l t_{\mathbf{C}}^{-l} \hat{\mathcal{O}}_{p,X},$$

where $t_{\mathbf{C}}$ is a generator of the ideal $\mathbf{C}$ in $\hat{\mathcal{O}}_{p,X}$. The ring $\hat{\mathcal{O}}_{p,X}(\infty\mathbf{C})$ does not depend on the choice of $t_{\mathbf{C}}$.

By $p \in \mathbf{C} \subset X$ we denote a germ at p of an irreducible curve C in X. Now we have

$$\mathbb{A}_X^2(\mathcal{O}_X) = \left\{ \{f_{p,\mathbf{C}}\} \in \prod_{p \in \mathbf{C} \subset X} K_{p,\mathbf{C}} : \ \text{C1 and C2 hold} \right\}$$

where the conditions C1 and C2 are as follows.

C1. There exists a divisor D on X such that for any $p \in \mathbf{C} \subset X$

$$\nu_{K_{p,\mathbf{C}}}(f_{p,\mathbf{C}}) \geq \nu_{\mathbf{C}}(D).$$

C2. For any irreducible curve $C \subset X$, any integer k and all except a finite number of points $p \in C$ we have that inside the group $(K_{p,\mathbf{C}} \bmod \mathbf{C}^k \mathcal{O}_{K_{p,\mathbf{C}}})$

$$f_{p,\mathbf{C}} \bmod \mathbf{C}^k \mathcal{O}_{K_{p,\mathbf{C}}} \in \hat{\mathcal{O}}_{p,X}(\infty \mathbf{C}) \bmod \mathbf{C}^k \mathcal{O}_{K_{p,\mathbf{C}}}.$$

Here we supposed that the curve C has at p a germ $\mathbf{C}$.

We have

$$\mathbb{A}_X^1(\mathcal{O}_X) = \begin{cases} \left(\prod_{C \subset X} K_C \right) \cap \mathbb{A}_X^2(\mathcal{O}_X) \times \left(\prod_{p \in X} K_p \right) \cap \mathbb{A}_X^2(\mathcal{O}_X) \\ \times \left(\prod_{p \in C \subset X} \mathcal{O}_{K_{p,\mathbf{C}}} \right) \cap \mathbb{A}_X^2(\mathcal{O}_X), \end{cases}$$

where the intersection is taken inside $\prod_{p \in C \subset X} K_{p,\mathbf{C}}$ with respect to Remark 3.15 and the diagonal embeddings

$$\prod_{C \subset X} K_C \longrightarrow \prod_{p \in C \subset X} K_{p,\mathbf{C}} \quad, \quad \prod_{p \in X} K_p \longrightarrow \prod_{p \in C \subset X} K_{p,\mathbf{C}}.$$

From the defining formula (3.3) and the explicit description of the rings $\mathbb{A}_X^*(\mathcal{O}_X)$ it is easy to see the differentials d^n in the complex $\mathcal{A}_X(\mathcal{O}_X)$ ([4, 21]). Indeed, let

$$A_0 = k(X) \quad, \quad A_1 = \prod_{C \subset X} \mathcal{O}_{K_C} \quad, \quad A_2 = \prod_{p \in X} \hat{\mathcal{O}}_{p,X},$$

and

$$A_{01} = \left(\prod_{C \subset X} K_C \right) \cap \mathbb{A}_X^2(\mathcal{O}_X), \qquad A_{02} = \left(\prod_{p \in X} K_p \right) \cap \mathbb{A}_X^2(\mathcal{O}_X),$$

$$A_{12} = \left(\prod_{p \in C \subset X} \mathcal{O}_{K_{p,\mathbf{C}}} \right) \cap \mathbb{A}_X^2(\mathcal{O}_X), \quad A_{012} = \mathbb{A}_X^2(\mathcal{O}_X).$$

Then the adelic complex $\mathcal{A}_X(\mathcal{O}_X)$ is

$$
\begin{array}{ccccc}
A_0 \oplus A_1 \oplus A_2 & \longrightarrow & A_{01} \oplus A_{02} \oplus A_{12} & \longrightarrow & A_{012} \\
(a_0, a_1, a_2) & \mapsto & (a_1 - a_0, a_2 - a_0, a_2 - a_1) & & \\
& & (a_{01}, a_{02}, a_{12}) & \mapsto & a_{01} - a_{02} + a_{12}.
\end{array}
$$

Remark 3.16. We note the following interesting property (see [4] and

[24, Remark 5]). Suppose that X is a projective surface. For any subset $I \subset [0, 1, 2]$ there is an embedding $A_I \hookrightarrow A_{012}$. Now for any subsets $I, J \subset [0, 1, 2]$ we have that inside the group A_{012}

$$A_I \cap A_J = A_{I \cap J}.$$

This property is also true for the corresponding components of the adelic complex of any locally free sheaf on X.

4 Restricted adelic complexes

In this section we describe restricted adelic complexes. The main difference from the adelic complexes constructed in Section 3 is that restricted adelic complexes are associated with one fixed chain (or flag) of irreducible subvarieties of a scheme X.

Restricted adelic complexes come from the so-called *Krichever correspondence* [17, 24]; see also [23] for connections with the theory of ζ-functions of algebraic curves. For algebraic curves, restricted adelic complexes originally come from the theory of integrable systems, see [25]: they were constructed for algebraic surfaces by A. N. Parshin in [24], and for higher-dimensional schemes by the author in [17].

4.1 Restricted adelic complexes on algebraic curves and surfaces

Consider an irreducible algebraic curve C over k. We fix a smooth closed point $p \in C$. For any coherent sheaf $\mathcal{F}$ of rank r on C we consider the following complex:

$$
\begin{array}{ccc}
\Gamma(C \setminus p, \mathcal{F}) \oplus (\mathcal{F} \otimes_{\mathcal{O}_C} \mathcal{O}_{K_p}) & \longrightarrow & \mathcal{F} \otimes_{\mathcal{O}_C} K_p \\
(a_0 \oplus a_1) & \longmapsto & a_1 - a_0.
\end{array}
\tag{4.1}
$$

Note that for the torsion-free sheaf $\mathcal{F}$ we have natural embeddings

$$\mathcal{F} \otimes_{\mathcal{O}_C} \mathcal{O}_{K_p} \longrightarrow \mathcal{F} \otimes_{\mathcal{O}_C} K_p$$
$$\Gamma(C \setminus p, \mathcal{F}) \longrightarrow \mathcal{F} \otimes_{\mathcal{O}_C} K_p,$$

where the last embedding is given by

$$\Gamma(C \setminus p, \mathcal{F}) \longrightarrow \Gamma(\operatorname{Spec} \mathcal{O}_p \setminus p, \mathcal{F}) \longrightarrow \Gamma(\operatorname{Spec} \mathcal{O}_{K_p} \setminus p, \mathcal{F}) = \mathcal{F} \otimes_{\mathcal{O}_C} K_p.$$

After the choice of basis of module $\mathcal{F}_p$ over the ring $\mathcal{O}_p$ we have

$$\mathcal{F} \otimes_{\mathcal{O}_C} K_p = K_p^{\oplus r}.$$

Therefore in this case the complex (4.1) is a complex of subgroups inside $K_p^{\oplus r}$, where K_p is a 1-dimensional local field.

For the following theorem see, for example, [23, 24].

Theorem 4.1. *The cohomology groups of the complex* (4.1) *coincide with the cohomology groups* $H^*(C, \mathcal{F})$.

The chain of quasi-isomorphisms between the complex (4.1) and the adelic complex $\mathcal{A}_C(\mathcal{F})$ was constructed in [24]. This proves Theorem 4.1. We remark that it is important for the proof that $C \setminus p$ is an affine curve; see also Remark 4.13 below.

The complex (4.1) is called the *restricted adelic complex* on C associated with the point p.

Now let X be an algebraic surface over k. We fix an irreducible curve $C \subset X$ and a point $p \in C$ which is a smooth point on both C and X. Let $\mathcal{F}$ be a torsion-free coherent sheaf on X.

The following notation is from [23, 24]. Let $x \in C$ and let $\hat{\mathcal{F}}_x$, $\hat{\mathcal{F}}_C$, $\hat{\mathcal{F}}_\eta$ be completions of stalks of the sheaf $\mathcal{F}$ at scheme points given by x, the irreducible curve C and the generic point η of X respectively.

Let

$$B_x(\mathcal{F}) = \bigcap_{\mathbf{D} \neq C} \left((\hat{\mathcal{F}}_x \otimes K_x) \cap (\hat{\mathcal{F}}_x \otimes \mathcal{O}_{K_x, \mathbf{D}}) \right)$$

where $\mathbf{D}$ runs over all germs at x of irreducible curves on X which are not equal to C, and the intersection is done inside the group $\hat{\mathcal{F}}_x \otimes K_x$. Let

$$B_C(\mathcal{F}) = (\hat{\mathcal{F}}_C \otimes K_C) \cap \left(\bigcap_{x \neq p} B_x \right)$$

where the intersection is done inside $\hat{\mathcal{F}}_p \otimes K_{x,\mathbf{C}}$. For all closed points $x \neq p$ of C and all germs $\mathbf{C}$ at x of C,

$$A_C(\mathcal{F}) = B_C(\mathcal{F}) \cap \hat{\mathcal{F}}_C,$$

$$A(\mathcal{F}) = \hat{\mathcal{F}}_\eta \cap \left(\bigcap_{x \in X - C} \hat{\mathcal{F}}_p \right).$$

We note that $A(\mathcal{F}) = \Gamma(X - C, \mathcal{F})$, and for the smooth point $x \in C$ the space $B_x(\mathcal{O}_X)$ coincides with the space $\hat{\mathcal{O}}_{p,X}(\infty \mathbf{C})$ from Section 3.3.

Theorem 4.2 ([24, Th. 3]). *Let X be an irreducible algebraic surface over a field k, $C \subset X$ be an irreducible curve, and $p \in C$ be a smooth* point on both C and X. Let $\mathcal{F}$ be a torsion-free coherent sheaf on X. Assume that the surface $X - C$ is affine. Then there exists a chain of quasi-isomorphisms between the adelic complex $\mathcal{A}_X(\mathcal{F})$ and the complex

$$A(\mathcal{F}) \oplus A_C(\mathcal{F}) \oplus \hat{\mathcal{F}}_p \longrightarrow \left\{ \begin{array}{c} B_C(\mathcal{F}) \oplus B_p(\mathcal{F}) \\ \oplus(\hat{\mathcal{F}}_p \otimes \mathcal{O}_{K_{p,\mathbf{C}}}) \end{array} \right\} \longrightarrow \hat{\mathcal{F}}_p \otimes K_{p,\mathbf{C}}$$

$$(4.2)$$

where the first map is given by

$$(a_0, a_1, a_2) \mapsto (a_1 - a_0, a_2 - a_0, a_2 - a_1)$$

and the second by

$$(a_{01}, a_{02}, a_{12}) \mapsto a_{01} - a_{02} + a_{12}.$$

Under the conditions of this theorem the cohomology groups of the complex (4.2) coincide with the cohomology groups of the adelic complex $\mathcal{A}_X(\mathcal{F})$, and therefore they are equal to $H^*(X, \mathcal{F})$.

Definition 4.3. *The complex* (4.2) *is called the* restricted adelic complex *on X associated with the curve C and the point $p \in C$.*

For the proof of the following proposition, see [24, Prop. 4].

Proposition 4.4. *Under the conditions of Theorem 4.2 we suppose also that $\mathcal{F}$ is a locally free sheaf, X is a projective variety, the local rings of X are Cohen-Macaulay and the curve C is a locally complete intersection. Then inside the field $K_{x,\mathbf{C}}$ we have*

$$B_C(\mathcal{F}) \cap B_p(\mathcal{F}) = A(\mathcal{F}).$$

Let the rank of $\mathcal{F}$ be r. Then after the choice of basis for the 1-dimensional free $\hat{\mathcal{O}}_{p,X}$-module $\hat{\mathcal{F}}_p$ we have

$$\hat{\mathcal{F}}_p = \hat{\mathcal{O}}_{p,X}^{\oplus r},$$

$$\hat{\mathcal{F}}_p \otimes K_{p,\mathbf{C}} = K_{p,\mathbf{C}}^{\oplus r},$$

$$\hat{\mathcal{F}}_p \otimes \hat{\mathcal{O}}_{K_{p,\mathbf{C}}} = \hat{\mathcal{O}}_{K_{p,\mathbf{C}}}^{\oplus r},$$

and

$$B_p(\mathcal{F}) = B_p^{\oplus r} = \hat{\mathcal{O}}_{p,X}(\infty\mathbf{C})^{\oplus r} \quad , \quad A_C(\mathcal{F}) = A(\mathcal{F}) \cap \hat{\mathcal{O}}_{K_{p,\mathbf{C}}}^{\oplus r},$$

where the last intersection is done inside $K_{p,\mathbf{C}}^{\oplus r}$.

Now due to Proposition 4.4, the complex (4.2) is a complex of subgroups of $K_{p,\mathbf{C}}^{\oplus r}$ and is uniquely determined by one subgroup $B_C(\mathcal{F})$ of

$K_{p,\mathbf{C}}^{\oplus r}$. In fact, all the other components of (4.2) can be defined by intersections of $B_C(\mathcal{F})$ with subgroups of $K_{p,\mathbf{C}}^{\oplus r}$, which do not really depend on the sheaf $\mathcal{F}$.

4.2 Restricted adelic complexes on higher-dimensional schemes

In this section we construct restricted adelic complexes for arbitrary schemes. These complexes will generalize the corresponding complexes from Section 4.1.

General definitions

Let X be a Noetherian separated scheme. Consider a flag of closed subschemes

$$X \supset Y_0 \supset Y_1 \supset \ldots \supset Y_n$$

in X. Let J_j be the ideal sheaf of Y_j in X, $0 \le j \le n$. Let i_j be the embedding $Y_j \hookrightarrow X$. Let U_i be an open subscheme of Y_i complementing Y_{i+1}, $0 \le i \le n-1$. Let $j_i : U_i \hookrightarrow Y_i$ be the open embedding of U_i in Y_i, $0 \le i \le n-1$. Put $U_n = Y_n$ and let j_n be the identity morphism from U_n to Y_n.

Assume that every point $x \in X$ has an open affine neighbourhood $U \ni x$ such that $U \cap U_i$ is an affine scheme for any $0 \le i \le n$. In what follows, a flag of subschemes $\{Y_i,\ 0 \le i \le n\}$ with this condition is called a flag with *locally affine complements*.

Remark 4.5. As an example, the last condition (existence of locally affine complements) holds in the following cases:

- if Y_{i+1} is the Cartier divisor on Y_i for $0 \le i \le n-1$;
- if U_i is an affine scheme for any $0 \le i \le n-1$ (the intersection of two open affine subschemes on a separated scheme is an affine subscheme).

Consider the n-dimensional simplex and its standard simplicial set (without degeneracies). To be precise, consider the set

$$(\{0\}, \{1\}, \ldots, \{n\}).$$

Then the simplicial set $S = \{S_k\}$ is given by

- $S_0 \stackrel{\text{def}}{=} \{\eta \in \{0\}, \{1\}, \ldots, \{n\}\}$.

- for $k \geq 1$,

$$S_k \stackrel{\text{def}}{=} \{(\eta_0, \ldots, \eta_k), \quad \text{where } \eta_l \in S_0 \text{ and } \eta_{l-1} < \eta_l \}.$$

The boundary map ∂_i $(0 < i < k)$ is given by eliminating the ith component of the k-tuple $(\eta_0, \ldots, \eta_k)$ to give the ith face of $(\eta_0, \ldots, \eta_k)$.

Let $\mathbf{QS}(X)$ be the category of quasicoherent sheaves on X; let $\mathbf{Sh}(X)$ be the category of sheaves of Abelian groups on X. Let $f : Y \longrightarrow X$ be a morphism of schemes. From now on f^* always denotes the pull-back functor in the category of sheaves of Abelian groups, and f_* is the direct image functor in the category of sheaves of Abelian groups.

We give the following definition from [17].

Definition 4.6. *For any* $(\eta_0, \ldots, \eta_k) \in S_k$ *we define a functor*

$$V_{(\eta_0, \ldots, \eta_k)} : \mathbf{QS}(X) \longrightarrow \mathbf{Sh}(X),$$

which is uniquely determined by the following inductive conditions:

(i) $V_{(\eta_0, \ldots, \eta_k)}$ *commutes with direct limits.*

(ii) *If* $\mathcal{F}$ *is a coherent sheaf and* $\eta \in S_0$, *then*

$$V_\eta(\mathcal{F}) \stackrel{\text{def}}{=} \varprojlim_{m \in \mathbb{N}} (i_\eta)_* (j_\eta)_* (j_\eta)^* (\mathcal{F}/J_\eta^m \mathcal{F}).$$

(iii) *If* $\mathcal{F}$ *is a coherent sheaf and* $(\eta_0, \ldots, \eta_k) \in S_k$, $k \geq 1$, *then*

$$V_{(\eta_0, \eta_1, \ldots, \eta_k)}(\mathcal{F}) \stackrel{\text{def}}{=} \varprojlim_{m \in \mathbb{N}} V_{(\eta_1, \ldots, \eta_k)} \left((i_{\eta_0})_* (j_{\eta_0})_* (j_{\eta_0})^* (\mathcal{F}/J_{\eta_0}^m \mathcal{F}) \right).$$

We will sometimes use the equivalent notation for $V_{(\eta_0, \ldots, \eta_k)}(\mathcal{F})$, in which the closed subschemes are indicated explicitly:

$$V_{(\eta_0, \ldots, \eta_k)}(\mathcal{F}) = V_{(Y_{\eta_0}, \ldots, Y_{\eta_k})}(X, \mathcal{F}).$$

The following proposition [17, Prop. 1] is proved by induction.

Proposition 4.7. *Let* $\sigma = (\eta_0, \ldots, \eta_k) \in S_k$. *Then the following assertions hold.*

(i) *The functor* $V_\sigma : \mathbf{QS}(X) \longrightarrow \mathbf{Sh}(X)$ *is well defined.*

(ii) *The functor* V_σ *is exact and additive.*

(iii) *The functor* V_σ *is local on* X, *that is, for any open* $U \subset X$ *and any quasicoherent sheaf* $\mathcal{F}$ *on* X *we have*

$$V_{(Y_{\eta_0}, \ldots, Y_{\eta_k})}(X, \mathcal{F}) \mid_U = V_{(Y_{\eta_0} \cap U, \ldots, Y_{\eta_k} \cap U)}(U, \mathcal{F} \mid_U).$$

If $Y_j \cap U = \emptyset$, *then* $Y_i \cap U$ *is the empty subscheme of* U *defined by the ideal sheaf* $\mathcal{O}_U$.

(iv) *For any quasicoherent sheaf $\mathcal{F}$ on X, the sheaf $V_{(\eta_0,\ldots,\eta_k)}(\mathcal{F})$ is a sheaf of $\mathcal{O}_X$-modules supported on the subscheme Y_{η_k}. (In general, this sheaf is not quasicoherent.)*

(v) *For any quasicoherent sheaf $\mathcal{F}$ on X we have*

$$V_\sigma(\mathcal{F}) = V_\sigma(\mathcal{O}_X) \otimes_{\mathcal{O}_X} \mathcal{F}.$$

(vi) *If each U_i is affine, $0 \leq i \leq n$, then for any quasicoherent sheaf $\mathcal{F}$ on X and any $m \geq 1$ we have*

$$H^m(X, V_\sigma(\mathcal{F})) = 0.$$

Construction of the restricted adelic complex

We consider the standard n-simplex $S = \{S_k,\ 0 \leq k \leq n\}$ without degeneracies. If $\sigma = (\eta_0, \ldots, \eta_k) \in S_k$, then $\partial_i(\sigma)$ is the ith face of σ, $0 \leq i \leq k$. We define a morphism of functors $d_i(\sigma) : V_{\partial_i(\sigma)} \longrightarrow V_\sigma$, as the morphism that commutes with direct limits and coincides on coherent sheaves with the map

$$V_{\partial_i(\sigma)}(\mathcal{F}) \longrightarrow V_\sigma(\mathcal{F}) \tag{4.3}$$

defined by the following rules.

a) If $i = 0$, then (4.3) is obtained by applying the functor $V_{\partial_0(\sigma)}$ to the map

$$\mathcal{F} \longrightarrow (i_{\eta_0})_*(j_{\eta_0})_*(j_{\eta_0})^*(\mathcal{F}/J_{\eta_0}^m\mathcal{F})$$

and passing to the projective limit with respect to m;

b) If $i = 1$ and $k = 1$, then we have the natural map

$$(i_{\eta_0})_*(j_{\eta_0})_*(j_{\eta_0})^*(\mathcal{F}/J_{\eta_0}^m\mathcal{F}) \longrightarrow V_{(\eta_1)}((i_{\eta_0})_*(j_{\eta_0})_*(j_{\eta_0})^*(\mathcal{F}/J_{\eta_0}^m\mathcal{F})).$$

Passing to the projective limit with respect to m we get the map (4.3) in this case.

c) If $i \neq 0$ and $k > 1$, then we use induction on k to get the map

$$V_{\partial_{i-1}\cdot(\partial_0(\sigma))}((i_{\eta_0})_*(j_{\eta_0})_*(j_{\eta_0})^*(\mathcal{F}/J_{\eta_0}^m\mathcal{F}))$$
$$\longrightarrow V_{\partial_0(\sigma)}((i_{\eta_0})_*(j_{\eta_0})_*(j_{\eta_0})^*(\mathcal{F}/J_{\eta_0}^m\mathcal{F})).$$

Passing to the projective limit with respect to m we get the map (4.3) in this case.

Proposition 4.8 ([17, Prop. 3]). *For any $1 \leq k \leq n$, $0 \leq i \leq k$ let*

$$d_i^k \stackrel{\text{def}}{=} \sum_{\sigma \in S_k} d_i(\sigma) \quad : \quad \bigoplus_{\sigma \subset S_{k-1}} V_\sigma \longrightarrow \bigoplus_{\sigma \in S_k} V_\sigma.$$

We define the map

$$d_0^0 \quad : \quad id \longrightarrow \bigoplus_{\sigma \in S_0} V_\sigma$$

as the direct sum of the natural maps $\mathcal{F} \longrightarrow V_\sigma(\mathcal{F})$. *(Here id is the functor of the natural embedding of* $\mathbf{QS}(X)$ *into* $\mathbf{Sh}(X)$, $\mathcal{F}$ *is a quasicoherent sheaf on* X *and* $\sigma \in S_0$.)

Then for all $0 \leq i < j \leq k \leq n - 1$ *we have*

$$d_j^{k+1} d_i^k = d_i^{k+1} d_{j-1}^k. \tag{4.4}$$

Let

$$d^m \overset{\text{def}}{=} \sum_{0 \leq i \leq m} (-1)^i d_i^m.$$

Then, given any quasicoherent sheaf $\mathcal{F}$ on X, Proposition 4.8 enables us to construct the complex of sheaves $V(\mathcal{F})$ in the standard way:

$$\cdots \longrightarrow \bigoplus_{\sigma \in S_{m-1}} V_\sigma(\mathcal{F}) \overset{d^m}{\longrightarrow} \bigoplus_{\sigma \in S_m} V_\sigma(\mathcal{F}) \longrightarrow \cdots.$$

The property $d^{m+1} d^m = 0$ follows from (4.4) by an easy direct calculation.

Theorem 4.9 ([17, Th. 1]). *Let* X *be a Noetherian separated scheme and let* $Y_0 \supset Y_1 \supset \ldots \supset Y_n$ *be a flag of closed subschemes with locally affine complements. Assume that* $Y_0 = X$. *Then the following complex is exact:*

$$0 \longrightarrow \mathcal{F} \overset{d^0}{\longrightarrow} V(\mathcal{F}) \longrightarrow 0. \tag{4.5}$$

Proof. We give a sketch of the proof. It suffices to consider the case when the sheaf $\mathcal{F}$ is coherent.

Consider the exact sequence of sheaves

$$0 \longrightarrow \mathcal{H} \longrightarrow \mathcal{F} \longrightarrow (j_0)_*(j_0)^*\mathcal{F} \longrightarrow \mathcal{G} \longrightarrow 0.$$

Since the functors V_σ are exact for all σ, we obtain the following exact sequence of complexes of sheaves:

$$0 \longrightarrow V(\mathcal{H}) \longrightarrow V(\mathcal{F}) \longrightarrow V((j_0)_*(j_0)^*\mathcal{F}) \longrightarrow V(\mathcal{G}) \longrightarrow 0. \tag{4.6}$$

The sheaves $\mathcal{H}$ and $\mathcal{G}$ are supported on Y_1. Therefore by induction we may assume that the complexes

$$0 \longrightarrow \mathcal{H} \overset{d^0}{\longrightarrow} V(\mathcal{H}) \longrightarrow 0$$

$$0 \longrightarrow \mathcal{G} \xrightarrow{d^0} V(\mathcal{G}) \longrightarrow 0$$

are already exact. The complex

$$0 \longrightarrow (j_0)_*(j_0)^*\mathcal{F} \xrightarrow{d^0} V((j_0)_*(j_0)^*\mathcal{F}) \longrightarrow 0$$

is exact, because $V(j_0)_*(j_0)^*\mathcal{F}$ has the same components as $V_{\sigma'}(\mathcal{F})$ for degrees k and $k+1$, where $\sigma' = (0, \eta_0, \ldots, \eta_k)$ for $\sigma = (\eta_0, \ldots, \eta_k) \in S_k$. Now the theorem follows from (4.6).   □

For any $\sigma \in S_k$ we define

$$A_\sigma(\mathcal{F}) \overset{\text{def}}{=} H^0(X, V_\sigma(\mathcal{F})).$$

Proposition 4.10 ([17, Prop. 4]). *Let X be a Noetherian separated scheme, and let $Y_0 \supset Y_1 \supset \ldots \supset Y_n$ be a flag of closed subschemes such that U_i is affine, $0 \le i \le n$. Let $\sigma \in S_k$ be arbitrary. Then:*

  (i) *A_σ is an exact and additive functor:* $\mathbf{QS}(X) \longrightarrow \mathbf{Ab}$;
 (ii) *if $X = \operatorname{Spec} A$ and M is some A-module, then*

$$A_\sigma(\tilde{M}) = A_\sigma(\mathcal{O}_X) \otimes_A M.$$

Let $\mathcal{F}$ be any quasicoherent sheaf on X. Applying the functor $H^0(X, \cdot)$ to the complex $V(\mathcal{F})$ we obtain the complex $A(\mathcal{F})$ of Abelian groups:

$$\cdots \longrightarrow \bigoplus_{\sigma \in S_{m-1}} A_\sigma(\mathcal{F}) \longrightarrow \bigoplus_{\sigma \in S_m} A_\sigma(\mathcal{F}) \longrightarrow \cdots.$$

Theorem 4.11 ([17, Th. 2]). *Let X be a Noetherian separated scheme. Let $Y_0 \supset Y_1 \supset \ldots \supset Y_n$ be a flag of closed subschemes such that $Y_0 = X$ and U_i is affine, $0 \le i \le n$. Then the cohomology of the complex $A(\mathcal{F})$ coincides with that of the sheaf $\mathcal{F}$ on X: that is, for any i*

$$H^i(X, \mathcal{F}) = H^i(A(\mathcal{F})).$$

Proof. It follows from Theorem 4.9 and assertion (vi) of Proposition 4.7 that $V(\mathcal{F})$ is an acyclic resolution for the sheaf $\mathcal{F}$. Hence the cohomology of $\mathcal{F}$ may be calculated by means of global sections of this resolution. This proves Theorem 4.11.   □

This theorem immediately yields the following geometric corollary, see [17, Th. 3].

Theorem 4.12. *Let X be a projective algebraic scheme of dimension n over a field. Let $Y_0 \supset Y_1 \supset \ldots \supset Y_n$ be a flag of closed subschemes such that $Y_0 = X$ and Y_i is an ample divisor on Y_{i-1} for $1 \le i \le n$. Then for any quasicoherent sheaf $\mathcal{F}$ on X and any i we have*

$$H^i(X, \mathcal{F}) = H^i(A(\mathcal{F})).$$

Proof. Since Y_i is an ample divisor on Y_{i-1} for $1 \le i \le n$, we see that U_i is an affine scheme for all $0 \le i \le n-1$. Since $\dim Y_n = 0$, $U_n = Y_n$ is also affine. Applying Theorem 4.11 we complete the proof. □

Remark 4.13. For any quasicoherent sheaf $\mathcal{F}$ and any $\sigma = (\eta_0) \in S_0$, $A_\sigma(\mathcal{F})$ is the group of sections over U_{η_0} of the sheaf $\mathcal{F}$ lifted to the formal neighbourhood of the subscheme Y_{η_0} in X. The complex $A(\mathcal{F})$ can be interpreted as the Čech complex for this 'acyclic covering' of the scheme X.

Definition 4.14. *The complex $A(\mathcal{F})$ is called the* restricted adelic complex *on X associated with the flag $Y_0 \supset Y_1 \supset \ldots \supset Y_n$.*

Remark 4.15. Some remarks for curves and surfaces:

- if C is an algebraic curve, $Y_0 = C$ and $Y_1 = p$ is a smooth point, then $A(\mathcal{F})$ coincides with the complex (4.1); indeed,

$$A_0(\mathcal{F}) = \Gamma(C \backslash p, \mathcal{F}), \quad A_1(\mathcal{F}) = \mathcal{F} \otimes_{\mathcal{O}_C} \mathcal{O}_{K_p} \text{ and } A_{12}(\mathcal{F}) = \mathcal{F} \otimes_{\mathcal{O}_C} K_p;$$

- if X is an algebraic surface, $Y_0 = C$ and $Y_1 = p$ is a smooth point on both C and X, then $A(\mathcal{F})$ coincides with the complex (4.2); indeed,

$$A_0(\mathcal{F}) = A(\mathcal{F}), \quad A_1(\mathcal{F}) = A_C(\mathcal{F}), \quad A_2(\mathcal{F}) = \hat{\mathcal{F}}_p,$$
$$A_{01}(\mathcal{F}) = B_C(\mathcal{F}), \quad A_{02}(\mathcal{F}) = B_p(\mathcal{F}), \quad A_{12}(\mathcal{F}) = \hat{\mathcal{F}}_p \otimes \mathcal{O}_{K_{p,C}}$$

and $A_{012}(\mathcal{F}) = \hat{\mathcal{F}}_p \otimes K_{p,\mathbf{C}}$.

Reconstruction of the restricted adelic complex

From Proposition 4.8 we have the natural map

$$d_i(\sigma) : A_{\partial_i(\sigma)}(\mathcal{F}) \longrightarrow A_\sigma(\mathcal{F})$$

for any $\sigma \in S_k$, $1 \le k \le n$, and any i, $0 \le i \le k$.

Using (4.4), we obtain the natural map

$$A_{\sigma_1}(\mathcal{F}) \longrightarrow A_{\sigma_2}(\mathcal{F})$$

for any locally free sheaf $\mathcal{F}$ on X and any $\sigma_1, \sigma_2 \in S$, $\sigma_1 \subset \sigma_2$.

Proposition 4.16 ([17, Th. 4]). *Let X be a projective equidimensional Cohen-Macaulay scheme of dimension n over a field. Let*

$$Y_0 \supset Y_1 \supset \ldots \supset Y_n$$

be a flag of closed subschemes such that $Y_0 = X$ and Y_i is an ample Cartier divisor on Y_{i-1} for $1 \le i \le n$. Then the following assertions hold for any locally free sheaf $\mathcal{F}$ on X.

(i) *The natural map $H^0(X, \mathcal{F}) \longrightarrow A_\sigma(\mathcal{F})$ is an embedding for any $\sigma \in S_k$, $0 \le k \le n$.*

(ii) *The natural map $A_{\sigma_1}(\mathcal{F}) \longrightarrow A_{\sigma_2}(\mathcal{F})$ is an embedding for any locally free sheaf $\mathcal{F}$ on X and any $\sigma_1, \sigma_2 \in S$, $\sigma_1 \subset \sigma_2$.*

Remark 4.17. We note that any integral Noetherian scheme of dimension 1 is a Cohen-Macaulay scheme. Any normal Noetherian scheme of dimension 2 is a Cohen-Macaulay scheme, see [7, Ch. II, Th. 8.22A].

We denote the unique face of dimension n in S by $(0, 1, \ldots, n) \in S$. By Proposition 4.16 we can embed $A_{\sigma_1}(\mathcal{F})$ and $A_{\sigma_2}(\mathcal{F})$ in $A_{(0,1\ldots,n)}(\mathcal{F})$ for any $\sigma_1, \sigma_2 \in S$ and any locally free sheaf $\mathcal{F}$.

Now we may formulate the following theorem [17, Th. 5].

Theorem 4.18. *Let all the hypotheses of Proposition 4.16 be satisfied. Then the following assertions hold for any locally free sheaf $\mathcal{F}$ and any $\sigma_1, \sigma_2 \in S$.*

(i) *If $\sigma_1 \cap \sigma_2 = \emptyset$ then, inside $A_{(0,1,\ldots,n)}$,*

$$A_{\sigma_1}(\mathcal{F}) \cap A_{\sigma_2}(\mathcal{F}) = H^0(X, \mathcal{F}).$$

(ii) *If $\sigma_1 \cap \sigma_2 \ne \emptyset$ then, inside $A_{(0,1,\ldots,n)}$,*

$$A_{\sigma_1}(\mathcal{F}) \cap A_{\sigma_2}(\mathcal{F}) = A_{\sigma_1 \cap \sigma_2}(\mathcal{F}).$$

Remark 4.19. Theorem 4.18 is similar to the property of adelic complexes $\mathcal{A}_X(\mathcal{F})$ which was observed in Remark 3.16.

We assume that the hypotheses of Proposition 4.16 hold and that the field of definition of the scheme X is k. We also assume that $Y_n = p$, where p is a smooth point on each Y_i, $0 \le i \le n$.

Let us choose and fix local parameters $t_1, \ldots, t_n \in \widehat{\mathcal{O}}_{p,X}$ such that $t_i|_{Y_{i-1}} = 0$ is a local equation of the divisor Y_i in the formal neighbourhood of the point p on the scheme Y_{i-1}, $1 \le i \le n$.

Let $\mathcal{F}$ be a rank 1 locally free sheaf on X. We fix a trivialization e_p of $\mathcal{F}$ in a formal neighbourhood of the point p on X, that is, an isomorphism $e_p : \hat{\mathcal{F}}_p \longrightarrow \hat{\mathcal{O}}_{p,X}$. By our choice of the local parameters and the trivialization we can identify $A_{(0,1,\dots,n)}(\mathcal{F})$ with the n-dimensional local field $k(p)((t_n))\dots((t_1))$. Moreover, we fix a set of integers $0 \le j_1 \le \dots \le j_k \le n-1$. Define $\sigma_{(j_1,\dots,j_k)} \in S_{n-k}$ to be the set

$$\{i \,:\, 0 \le i \le n, \, i \ne j_1, \, \dots, \, i \ne j_k\}.$$

By Proposition 4.16 there is a natural embedding

$$A_{\sigma_{(j_1,\dots,j_k)}}(\mathcal{F}) \longrightarrow A_{(0,1,\dots,n)}(\mathcal{F}).$$

When we identify $A_{(0,1,\dots,n)}(\mathcal{F})$ with the field $k(p)((t_n))\dots((t_1))$, the subspace $A_{\sigma_{(j_1,\dots,j_k)}}(\mathcal{F})$ corresponds to the following k-subspace:

$$\left\{ \sum a_{i_1,\dots,i_n} t_n^{i_n} \dots t_1^{i_1} \,:\, a_{i_1,\dots,i_n} \in k(p), \atop i_{j_1+1} \ge 0, \, i_{j_2+1} \ge 0, \, \dots, \, i_{j_k+1} \ge 0 \right\}. \quad (4.7)$$

Thus, by Theorem 4.18, to determine the images of $A_\sigma(\mathcal{F})$ inside $k(p)((t_n))\dots((t_1))$ (for any $\sigma \in S$), it suffices to know only one image of $A_{(0,1,\dots,n-1)}$ in $k(p)((t_n))\dots((t_1))$. (All the others are obtained by taking the intersection of this image with the standard subspaces (4.7) of $k(p)((t_n))\dots((t_1))$.)

It is clear that these arguments generalize immediately to locally free sheaves $\mathcal{F}$ of rank r and to the spaces $k(p)((t_n))\dots((t_1))^{\oplus r}$.

These arguments lead to the following theorem, which enables one to reconstruct the restricted adelic complex $A(\mathcal{F})$; see also [17, Th. 6].

Theorem 4.20. *Let all the hypotheses of Proposition 4.16 be satisfied. We also assume that $Y_n = p$, where p is a smooth point on each Y_i, $0 \le i \le n$. Let $\mathcal{F}$ be a locally free sheaf on X. Then the subspace*

$$A_{(0,1,\dots,n-1)}(\mathcal{F}) \subset A_{(0,1,\dots,n)}(\mathcal{F})$$

uniquely determines the restricted adelic complex $A(\mathcal{F})$.

5 Reciprocity laws

Let L be a field with discrete valuation ν_L, valuation ring $\mathcal{O}_L$ and maximal ideal m_L. The tame symbol is

$$(f,g)_L = (-1)^{\nu_L(f)\nu_L(g)} \frac{f^{\nu_L(g)}}{g^{\nu_L(f)}} \quad \bmod \, m_L, \quad (5.1)$$

where f, g are elements of L^*.

There exists the following reciprocity law (see [26], for example).

Proposition 5.1. *Let C be a complete smooth algebraic curve over k. For any $f, g \in k(C)^*$ we have*

$$\prod_{p \in C} \mathrm{Nm}_{k(p)/k} \, (f, g)_{K_p} = 1, \tag{5.2}$$

where only finitely many terms in this product are not equal to 1. Here Nm is the norm map, and the product is taken over all closed points $p \in C$.

Let K be a 2-dimensional local field with last residue field k. There is a discrete valuation of rank 2 on K:

$$(\nu_1, \nu_2) : K^* \to \mathbb{Z} \oplus \mathbb{Z}.$$

Here $\nu_1 = \nu_K$ is the discrete valuation of the field K, and

$$\nu_2(b) \stackrel{\mathrm{def}}{=} \nu_{\bar{K}}(\overline{bt_1^{-\nu_1(b)}}),$$

where $\nu_1(t_1) = 1$. We note that ν_2 depends on the choice of local parameter t_1.

Let m_K be the maximal ideal of $\mathcal{O}_K$, and $m_{\bar{K}}$ be the maximal ideal of $\mathcal{O}_{\bar{K}}$. Define a map

$$\nu_K(\quad, \quad) \quad : \quad K^* \times K^* \longrightarrow \mathbb{Z}$$

as the composition of maps

$$K^* \times K^* \longrightarrow K_2(K) \stackrel{\partial_2}{\longrightarrow} \bar{K}^* \stackrel{\partial_1}{\longrightarrow} \mathbb{Z},$$

where ∂_i is the boundary map in algebraic K-theory. The map ∂_2 coincides with the tame symbol (5.1) with respect to the discrete valuation ν_1i, while the map ∂_1 coincides with the discrete valuation $\nu_{\bar{K}}$.

We may also define a map

$$(\quad, \quad, \quad)_K \quad : \quad K^* \times K^* \times K^* \longrightarrow k^*$$

as the composition of maps

$$K^* \times K^* \times K^* \longrightarrow K_3^M(K) \stackrel{\partial_3}{\longrightarrow} K_2(\bar{K}) \stackrel{\partial_2}{\longrightarrow} k^*,$$

where K_3^M is the Milnor K-group.

These maps have the following explicit expressions (see [4]):

$$\nu_K(f,g) = \nu_1(f)\nu_2(g) - \nu_2(f)\nu_1(g)$$

$$(f,g,h)_K = \text{sign}_K(f,g,h) f^{\nu_K(g,h)} g^{\nu_K(h,f)} h^{\nu_K(f,g)} \quad \text{mod } m_K \quad \text{mod } m_{\bar{K}}.$$

Here

$$\text{sign}_K(f,g,h) = (-1)^B,$$

where

$$B = \begin{cases} \nu_1(f)\nu_2(g)\nu_2(h) + \nu_1(g)\nu_2(f)\nu_2(h) & +\nu_1(h)\nu_2(g)\nu_2(f) \\ +\nu_2(f)\nu_1(g)\nu_1(h) + \nu_2(g)\nu_1(f)\nu_1(h) & +\nu_2(h)\nu_1(f)\nu_1(g). \end{cases}$$

Proposition 5.2. *For any $f,g,h \in K^*$*

$$\text{sign}_K(f,g,h) = (-1)^A$$

where

$$A = \nu_K(f,g)\nu_K(f,h) + \nu_K(f,g)\nu_K(f,h)$$
$$+ \nu_K(f,g)\nu_K(f,h) + \nu_K(f,g)\nu_K(f,h)\nu_K(g,h).$$

The proof follows from direct calculations modulo 2 with A and B, using the explicit expressions above.

Let X be a smooth algebraic surface over k. We recall the following notation from Section 3.3: if C is a curve on X, then K_C is the completion of the field $k(X)$ along the discrete valuation given by the irreducible curve C on X; if p is a point on X, then K_p is a subring in $\text{Frac}(\hat{\mathcal{O}}_{p,X})$ generated by subrings $k(X)$ and $\hat{\mathcal{O}}_{p,X}$.

There are the following reciprocity laws, see [4].

Theorem 5.3.

(i) *We fix a point $p \in X$ and take any $f,g,h \in K_p^*$. Then*

$$\sum_{X \supset C \ni p} \nu_{K_{p,C}}(f,g) = 0$$

where only finitely many terms in this sum are nonzero, and

$$\prod_{X \supset C \ni p} (f,g,h)_{K_{p,C}} = 1$$

where only finitely many terms in this product are not equal to 1; both the sum and the product are taken over all germs of irreducible curves on X at p.

(ii) *We fix an irreducible projective curve C on X and any $f, g, h \in$*
K_C^*. Then

$$\sum_{p \in C \subset X} [k(p) : k] \cdot \nu_{K_{p,\mathbf{C}}}(f, g) = 0$$

where only finitely many terms in this sum are nonzero, and

$$\prod_{p \in C \subset X} \mathrm{Nm}_{k(p)/k} (f, g, h)_{K_{p,\mathbf{C}}} = 1$$

where only finitely many terms in this product are not equal to 1;
both the sum and the product are taken over all points $p \in C$ and
all germs of irreducible curve C at p.

Remark 5.4. The relative reciprocity laws were constructed in [15]
(see also [16] for a short exposition) for a smooth projective morphism
f of a smooth algebraic surface X to a smooth algebraic curve S when
char $k = 0$. If $p \in \mathbf{C} \subset X$ then explicit formulas were constructed in [15]
for maps

$$K_2(K_{p,\mathbf{C}}) \longrightarrow K_{f(p)}^*.$$

Remark 5.5. For a 2-dimensional local field K, the map $\nu_K(\ ,\ )$ was
interpreted in [18] as the commutator of liftings of elements $f, g \in K^*$
in a central extension of the group K^* by $\mathbb{Z}$. From this interpretation
the reciprocity laws for $\nu_K(\ ,\ )$ were proved. The proof in [18] uses
adelic rings on an algebraic surface X. This is an abstract version of the
reciprocity law for $\nu_K(\ ,\ )$, like the abstract version of the reciprocity
law (5.2) for a projective curve in [1] (and in [28] for the residues of
differentials on a projective curve).

Remark 5.6. We do not describe the reciprocity laws for residues of
differentials of 2-dimensional local fields, which were formulated and
proved in [19] (see also [29]).

Remark 5.7. The symbols $\nu_K(\ ,\ )$ and $(\ ,\ ,\ )_K$ correspond to the
unramified and tamely ramified extensions of 2-dimensional local fields
when the last residue field is finite [22].

Bibliography

[1] E. Arbarello, C. De Concini, and V. G. Kac. The infinite wedge representation and the reciprocity law for algebraic curves. In *Theta functions—
Bowdoin 1987, Part 1 (Brunswick, ME, 1987)*, volume 49 of *Proc. Sympos. Pure Math.*, pages 171–190. Amer. Math. Soc., Providence, RI, 1989.

[2] A. A. Beĭlinson. Residues and adèles. *Funktsional. Anal. i Prilozhen*, 14(1):44–45, 1980. Translation in *Func. Anal. Appl.*, 14(1):34–35, 1980.

[3] T. Fimmel. Verdier duality for systems of coefficients over simplicial sets. *Math. Nachr.*, 190:51–122, 1998.

[4] T. Fimmel and A. N. Parshin. Introduction to higher adelic theory. preprint, 1999.

[5] S. O. Gorchinskiy and A. N. Parshin. Adelic Lefschetz formula for the action of a one-dimensional torus. *Tr. St-Peterbg. Mat. Obshch.*, 11:37–57, 2005. In Russian; see also arXiv math.AG/0408058.

[6] A. Grothendieck and J. A. Dieudonné. *Éléments de géométrie algébrique. I.*, volume 166 of *Die Grundlehren der mathematischen Wissenschaften*. Springer-Verlag, Berlin-Heidelberg-New York, 1971.

[7] R. Hartshorne. *Algebraic geometry.* Number 52 in Graduate Texts in Mathematics. Springer-Verlag, New York, 1977.

[8] A. Huber. Adele für Schemata und Zariski-Kohomologie. Schriftenreihe des Math. Inst. der Univ. Muenster, 3. Serie, Heft3, 1991.

[9] A. Huber. On the Parshin-Beilinson adeles for schemes. *Abh. Math. Semin. Univ. Hamb.*, 61:249–273, 1991.

[10] R. Hübl and A. Yekutieli. Adelic Chern forms and applications. *Amer. J. Math.*, 121(4):797–839, 1999.

[11] K. Kato and S. Saito. Global class field theory of arithmetic schemes. In *Applications of algebraic K-theory to algebraic geometry and number theory, Part I, II (Boulder, Colo., 1983)*, volume 55 of *Contemp. Math.*, pages 255–331. Amer. Math. Soc., Providence, RI, 1986.

[12] V. G. Lomadze. Index of intersection of divisors. *Izv. Akad. Nauk SSSR Ser. Mat.*, 44(5):1120–1130, 1980. Translation in *Math. USSR Izv.*, 17:343–352, 1981.

[13] V. G. Lomadze. On residues in algebraic geometry. *Izv. Akad. Nauk SSSR Ser. Mat.*, 45(6):1258–1287, 1981. Translation in *Math. USSR Izv.*, 19:495–520, 1982.

[14] H. Matsumura. *Commutative algebra*, volume 56 of *Math. Lecture Note Series*. Benjamin/Cummings, Reading, Mass., second edition, 1980.

[15] D. V. Osipov. Adelic constructions of direct images of differentials and symbols. *Mat. Sb.*, 188(5):59–84, 1997. Translation in *Sb. Math.*, 188(5):697–723, 1997; see also arXiv math.AG/9802112.

[16] D. V. Osipov. Adelic constructions for direct images of differentials and symbols. In *Invitation to higher local fields (Münster, 1999)*, volume 3 of *Geom. Topol. Monogr.*, pages 215–221 (electronic). Geom. Topol. Publ., Coventry, 2000. See also arXiv math.NT/0012152.

[17] D. V. Osipov. The Krichever correspondence for algebraic varieties. *Izv. Ross. Akad. Nauk Ser. Mat.*, 65(5):91–128, 2001. Translation in *Izv. Math.*, 65(5):941–975, 2001; see also arXiv math.AG/0003188.

[18] D. V. Osipov. Central extensions and reciprocity laws on algebraic surfaces. *Mat. Sb.*, 196(10):111–136, 2005. Translation in *Sb. Math.*, 196(10):1503–1527, 2005; see also arXiv math.NT/0501155.

[19] A. N. Parshin. On the arithmetic of two-dimensional schemes. I. Distributions and residues. *Izv. Akad. Nauk SSSR Ser. Mat.*, 40(4):736–773, 949, 1976. Translation in *Math. USSR Izv.*, 10(4):695–729, 1976.

[20] A. N. Parshin. Abelian coverings of arithmetic schemes. *Dokl. Akad. Nauk SSSR*, 243(4):855–858, 1978. Translation in *Soviet Math. Doklady*, 19:1438–1442, 1978.

[21] A. N. Parshin. Chern classes, adèles and *L*-functions. *J. Reine Angew. Math.*, 341:174–192, 1983.

[22] A. N. Parshin. Local class field theory. *Trudy Mat. Inst. Steklov.*, 165:143–170, 1984. Translation in *Proc. Steklov Inst. Math.*, 165(3):157–185, 1985.

[23] A. N. Parshin. Higher dimensional local fields and *L*-functions. In *Invitation to higher local fields (Münster, 1999)*, volume 3 of *Geom. Topol. Monogr.*, pages 199–213 (electronic). Geom. Topol. Publ., Coventry, 2000. See also arXiv math.NT/0012151.

[24] A. N. Parshin. Integrable systems and local fields. *Comm. Algebra*, 29(9):4157–4181, 2001. Short version at arXiv math.AG/9911097.

[25] G. Segal and G. Wilson. Loop groups and equations of KdV type. *Inst. Hautes Études Sci. Publ. Math.*, (61):5–65, 1985.

[26] J.-P. Serre. *Groupes algébriques et corps de classes*. Hermann, Paris, 1959.

[27] J.-P. Serre. *Local fields*, volume 67 of *Graduate Texts in Mathematics*. Springer-Verlag, New York, 1979. Translated from the French by Marvin Jay Greenberg.

[28] J. Tate. Residues of differentials on curves. *Ann. Sci. École Norm. Sup. (4)*, 1:149–159, 1968.

[29] A. Yekutieli. An explicit construction of the Grothendieck residue complex. *Astérisque*, (208):127, 1992. With an appendix by Pramathanath Sastry.

[30] I. B. Zhukov. Structure theorems for complete fields. In *Proceedings of the St. Petersburg Mathematical Society, Vol. III*, volume 166 of *Amer. Math. Soc. Transl. Ser. 2*, pages 175–192, Providence, RI, 1995. Amer. Math. Soc.

[31] I. B. Zhukov. Higher dimensional local fields. In *Invitation to higher local fields (Münster, 1999)*, volume 3 of *Geom. Topol. Monogr.*, pages 5–18 (electronic). Geom. Topol. Publ., Coventry, 2000. Cf. arXiv math.NT/0012132.

Steklov Mathematical Institute,

Gubkina str. 8,

119991, Moscow, Russia

d_osipov@mi.ras.ru

Cohomology of face rings, and torus actions

Taras E. Panov

1 Introduction

This article centres on the cohomological aspects of 'toric topology', a new and actively developing field on the borders of equivariant topology, combinatorial geometry and commutative algebra. The algebro-geometric counterpart of toric topology, known as 'toric geometry' or algebraic geometry of *toric varieties*, is now a well established field in algebraic geometry which is characterized by strong links with combinatorial and convex geometry (see the classical survey paper [10] or the more modern exposition [13]). Since the appearance of Davis and Januszkiewicz's work [11], where the concept of a (*quasi*)*toric manifold* was introduced as a topological generalization of a smooth compact toric variety, there has grown an understanding that most phenomena of smooth toric geometry may be modelled in the purely topological situation of smooth manifolds with a nicely behaved torus action.

One of the main results of [11] is that the equivariant cohomology of a toric manifold can be identified with the *face ring* of the quotient simple polytope; or, for more general classes of torus actions, with the face ring of a certain simplicial complex K. The ordinary cohomology of a quasi-toric manifold can also be effectively identified as the quotient of the face ring by a *regular sequence* of degree-two elements, which provides a generalization of the well-known Danilov-Jurkiewicz theorem of toric geometry. The notion of the face ring of a simplicial complex sits at the heart of Stanley's 'combinatorial commutative algebra' [24], linking geometrical and combinatorial problems concerning simplicial complexes with commutative and homological algebra. Our concept of toric topology aims at extending these links and developing new applications by

165

applying the full strength of the apparatus of equivariant topology of torus actions.

This article surveys certain new developments of toric topology related to the cohomology of face rings. Introductory remarks can be found at the beginning of each section and most subsections. A more detailed description of the history of the subject, together with an extensive bibliography, can be found in [8] and its extended Russian version [9].

The current article represents the work of the algebraic topology and combinatorics group at the Department of Geometry and Topology, Moscow State University, and the author thanks all its members for the collaboration and insight gained from numerous discussions, particularly mentioning Victor Buchstaber, Ilia Baskakov and Arseny Gadzhikurbanov. The author is also grateful to Nigel Ray for several valuable comments and suggestions that greatly improved this text, and for his hospitality during the visit to Manchester which was sponsored by an LMS grant. He also acknowledges the support of the Russian Foundation for Basic Research, via grant no. 04-01-00702.

2 Simplicial complexes and face rings

The notion of the face ring $\mathbf{k}[K]$ of a simplicial complex K is central to the algebraic study of triangulations. In this section we review its main properties, emphasising functoriality with respect to simplicial maps. Then we introduce the bigraded Tor-algebra $\mathrm{Tor}_{\mathbf{k}[v_1,\ldots,v_m]}(\mathbf{k}[K], \mathbf{k})$ using a finite free resolution of $\mathbf{k}[K]$ as a module over the polynomial ring. The corresponding bigraded Betti numbers are important combinatorial invariants of K.

2.1 Definition and main properties

Let $K = K^{n-1}$ be an arbitrary $(n-1)$-dimensional simplicial complex on an m-element vertex set V, which we usually identify with the set of ordinals $[m] = \{1, \ldots, m\}$. Those subsets $\sigma \subseteq V$ belonging to K are referred to as *simplices*; we also use the notation $\sigma \in K$. We count the empty set $\varnothing$ as a simplex of K. When it is necessary to distinguish between combinatorial and geometrical objects, we denote by $|K|$ the *geometrical realization* of K, which is a triangulated topological space.

Choose a ground commutative ring $\mathbf{k}$ with unit (we are mostly interested in the cases $\mathbf{k} = \mathbb{Z}, \mathbb{Q}$ or finite field). Let $\mathbf{k}[v_1, \ldots, v_m]$ be the graded polynomial algebra over $\mathbf{k}$ with $\deg v_i = 2$. For an arbitrary subset $\omega = \{i_1, \ldots, i_k\} \subseteq [m]$ denote by v_ω the square-free monomial $v_{i_1} \ldots v_{i_k}$. The *face ring* (or *Stanley-Reisner algebra*) of K is the quotient ring

$$\mathbf{k}[K] = \mathbf{k}[v_1, \ldots, v_m]/\mathcal{I}_K,$$

where $\mathcal{I}_K$ is the homogeneous ideal generated by all monomials v_σ such that σ is not a simplex of K. The ideal $\mathcal{I}_K$ is called the *Stanley-Reisner ideal* of K.

Example 2.1. Let K be a 2-dimensional simplicial complex shown on Figure 1. Then

$$\mathbf{k}[K] = \mathbf{k}[v_1, \ldots, v_5]/(v_1 v_5, v_3 v_4, v_1 v_2 v_3, v_2 v_4 v_5).$$

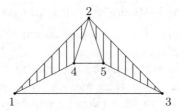

Fig. 1.

Despite its simple construction, the face ring appears to be a very powerful tool allowing us to translate the combinatorial properties of different particular classes of simplicial complexes into the language of commutative algebra. The resulting field of 'combinatorial commutative algebra', whose foundations were laid by Stanley in his monograph [24], has attracted a lot of interest from both combinatorialists and commutative algebraists.

Let K_1 and K_2 be two simplicial complexes on the vertex sets $[m_1]$ and $[m_2]$ respectively. A set map $\varphi \colon [m_1] \to [m_2]$ is called a *simplicial map* between K_1 and K_2 if $\varphi(\sigma) \in K_2$ for any $\sigma \in K_1$; we often identify such a φ with its restriction to K_1 (regarded as a collection of subsets of $[m_1]$), and use the notation $\varphi \colon K_1 \to K_2$.

Proposition 2.2. *Let $\varphi \colon K_1 \to K_2$ be a simplicial map. Define a map $\varphi^* \colon \mathbf{k}[w_1, \ldots, w_{m_2}] \to \mathbf{k}[v_1, \ldots, v_{m_1}]$ by*

$$\varphi^*(w_j) := \sum_{i \in \varphi^{-1}(j)} v_i.$$

Then φ^* induces a homomorphism $\mathbf{k}[K_2] \to \mathbf{k}[K_1]$, which we will also denote by φ^*.

Proof. We have to check that $\varphi^*(\mathcal{I}_{K_2}) \subseteq \mathcal{I}_{K_1}$. Suppose $\tau = \{j_1, \ldots, j_s\} \subseteq [m_2]$ is not a simplex of K_2. Then

$$\varphi^*(w_{j_1} \cdots w_{j_s}) = \sum_{i_1 \in \varphi^{-1}(j_1), \ldots, i_s \in \varphi^{-1}(j_s)} v_{i_1} \cdots v_{i_s}. \qquad (2.1)$$

We claim that $\sigma = \{i_1, \ldots, i_s\}$ is not a simplex of K_1 for any monomial $v_{i_1} \cdots v_{i_s}$ in the right hand side of the above identity. Indeed, if $\sigma \in K_1$ then $\varphi(\sigma) = \tau \in K_2$ by the definition of simplicial map, which leads to a contradiction. Hence the right hand side of (2.1) is in $\mathcal{I}_{K_1}$. $\qquad \square$

2.2 Cohen-Macaulay rings and complexes

Cohen-Macaulay rings and modules play an important role in homological commutative algebra and algebraic geometry. A standard reference for the subject is [6], where the reader may find proofs of the basic facts about Cohen-Macaulay rings and regular sequences mentioned in this subsection. In the case of simplicial complexes the Cohen-Macaulay property of the corresponding face rings leads to important combinatorial and topological consequences.

Let $A = \oplus_{i \geq 0} A^i$ be a finitely generated commutative graded algebra over $\mathbf{k}$. We assume that A is connected ($A^0 = \mathbf{k}$) and has only even-degree graded components, so that we do not need to distinguish between graded and non-graded commutativity. We denote by A_+ the positive-degree part of A and by $\mathcal{H}(A_+)$ the set of homogeneous elements in A_+.

A sequence $t_1, \ldots, t_n$ of algebraically independent homogeneous elements of A is called an *hsop* (homogeneous system of parameters) if A is a finitely generated $\mathbf{k}[t_1, \ldots, t_n]$-module (equivalently, if $A/(t_1, \ldots, t_n)$ has finite dimension as a $\mathbf{k}$-vector space).

Lemma 2.3 (Nöther normalization lemma). *Any finitely generated graded algebra A over a field $\mathbf{k}$ admits an hsop. If $\mathbf{k}$ has characteristic zero and A is generated by degree-two elements, then a degree-two hsop can be chosen.*

A degree-two hsop is called an *lsop* (linear system of parameters).

A sequence $\boldsymbol{t} = t_1, \ldots, t_k$ of elements of $\mathcal{H}(A_+)$ is called a *regular sequence* if t_{i+1} is not a zero divisor in $A/(t_1, \ldots, t_i)$ for $0 \leq i < k$. A regular sequence consists of algebraically independent elements, so it generates a polynomial subring in A. It can be shown that t is a regular sequence if and only if A is a *free* $\mathbf{k}[t_1, \ldots, t_k]$-module.

An algebra A is called *Cohen-Macaulay* if it admits a regular hsop t. It follows that A is Cohen-Macaulay if and only if it is a free and finitely generated module over its polynomial subring. If $\mathbf{k}$ is a field of characteristic zero and A is generated by degree-two elements, then one can choose t to be an lsop. A simplicial complex K is called *Cohen-Macaulay* (over $\mathbf{k}$) if its face ring $\mathbf{k}[K]$ is Cohen-Macaulay.

Example 2.4. Let $K = \partial \Delta^2$ be the boundary of a 2-simplex. Then

$$\mathbf{k}[K] = \mathbf{k}[v_1, v_2, v_3]/(v_1 v_2 v_3) \,.$$

The elements $v_1, v_2 \in \mathbf{k}[K]$ are algebraically independent, but do not form an hsop, since $\mathbf{k}[K]/(v_1, v_2) \cong \mathbf{k}[v_3]$ is not finite-dimensional as a $\mathbf{k}$-space. On the other hand, the elements $t_1 = v_1 - v_3$, $t_2 = v_2 - v_3$ of $\mathbf{k}[K]$ form an hsop, since $\mathbf{k}[K]/(t_1, t_2) \cong \mathbf{k}[t]/t^3$. It is easy to see that $\mathbf{k}[K]$ is a free $\mathbf{k}[t_1, t_2]$-module with one 0-dimensional generator 1, one 1-dimensional generator v_1, and one 2-dimensional generator v_1^2. Thus $\mathbf{k}[K]$ is Cohen-Macaulay and (t_1, t_2) is a regular sequence.

For an arbitrary simplex $\sigma \in K$ define its *link* and *star* as the subcomplexes

$$\mathrm{link}_K \sigma = \{\tau \in K : \sigma \cup \tau \in K, \ \sigma \cap \tau = \varnothing\};$$
$$\mathrm{star}_K \sigma = \{\tau \in K : \sigma \cup \tau \in K\}.$$

If $v \in K$ is a vertex, then $\mathrm{star}_K v$ is the subcomplex consisting of all simplices of K containing v, and all their subsimplices. Note also that $\mathrm{star}_K v$ is the cone over $\mathrm{link}_K v$.

The following fundamental theorem characterizes Cohen-Macaulay complexes combinatorially.

Theorem 2.5 (Reisner). *A simplicial complex K is Cohen-Macaulay over $\mathbf{k}$ if and only if for any simplex $\sigma \in K$ (including $\sigma = \varnothing$) and $i < \dim(\mathrm{link}_K \sigma)$, it holds that $\widetilde{H}_i(\mathrm{link}_K \sigma; \mathbf{k}) = 0$.*

Using standard techniques of PL topology the previous theorem may be reformulated in purely topological terms.

Proposition 2.6 (Munkres). *K^{n-1} is Cohen-Macaulay over $\mathbf{k}$ if and only if for an arbitrary point $x \in |K|$, it holds that*

$$\widetilde{H}_i(|K|; \mathbf{k}) = H_i(|K|, |K| \setminus x; \mathbf{k}) = 0 \qquad \text{for } i < n - 1.$$

Thus any triangulation of a sphere is a Cohen-Macaulay complex.

2.3 Resolutions and Tor-*algebras*

Let M be a finitely generated graded $\mathbf{k}[v_1, \ldots, v_m]$-module. A *free resolution* of M is an exact sequence

$$\ldots \xrightarrow{d} R^{-i} \xrightarrow{d} \ldots \xrightarrow{d} R^{-1} \xrightarrow{d} R^0 \longrightarrow M \longrightarrow 0 \quad (2.2)$$

where the R^{-i} are finitely generated graded free $\mathbf{k}[v_1, \ldots, v_m]$-modules and the maps d are degree-preserving. By the Hilbert syzygy theorem, there is a free resolution of M with $R^{-i} = 0$ for $i > m$. A resolution of the form (2.2) determines a bigraded differential $\mathbf{k}$-module $[R, d]$, where $R = \bigoplus R^{-i,j}$, $R^{-i,j} := (R^{-i})^j$ and $d \colon R^{-i,j} \to R^{-i+1,j}$. The bigraded cohomology module $H[R, d]$ has $H^{-i,k}[R, d] = 0$ for $i > 0$ and $H^{0,k}[R, d] = M^k$. Let $[M, 0]$ be the bigraded module with $M^{-i,k} = 0$ for $i > 0$, $M^{0,k} = M^k$, and zero differential. Then the resolution (2.2) determines a bigraded map $[R, d] \to [M, 0]$ inducing an isomorphism in cohomology.

Let N be another module; then applying the functor $\underline{} \otimes_{\mathbf{k}[v_1,\ldots,v_m]} N$ to a resolution $[R, d]$ we get a homomorphism of differential modules

$$[R \otimes_{\mathbf{k}[v_1,\ldots,v_m]} N, d] \to [M \otimes_{\mathbf{k}[v_1,\ldots,v_m]} N, 0],$$

which in general does not induce an isomorphism in cohomology. The $(-i)$th cohomology module of the cochain complex

$$\ldots \longrightarrow R^{-i} \underset{\mathbf{k}[v_1,\ldots,v_m]}{\otimes} N \longrightarrow \ldots \longrightarrow R^0 \underset{\mathbf{k}[v_1,\ldots,v_m]}{\otimes} N \longrightarrow 0$$

is denoted by $\mathrm{Tor}^{-i}_{\mathbf{k}[v_1,\ldots,v_m]}(M, N)$. Thus,

$$\mathrm{Tor}^{-i}_{\mathbf{k}[v_1,\ldots,v_m]}(M, N) := \frac{\mathrm{Ker}\left[d \colon R^{-i} \underset{\mathbf{k}[v_1,\ldots,v_m]}{\otimes} N \to R^{-i+1} \underset{\mathbf{k}[v_1,\ldots,v_m]}{\otimes} N\right]}{d(R^{-i-1} \underset{\mathbf{k}[v_1,\ldots,v_m]}{\otimes} N)}.$$

Since all the R^{-i} and N are graded modules, we actually have a *bigraded* $\mathbf{k}$-module

$$\mathrm{Tor}_{\mathbf{k}[v_1,\ldots,v_m]}(M, N) = \bigoplus_{i,j} \mathrm{Tor}^{-i,j}_{\mathbf{k}[v_1,\ldots,v_m]}(M, N).$$

The following properties of $\mathrm{Tor}^{-i}_{\mathbf{k}[v_1,\ldots,v_m]}(M, N)$ are well known.

Proposition 2.7.

(a) *The module* $\mathrm{Tor}^{-i}_{\mathbf{k}[v_1,\ldots,v_m]}(M, N)$ *does not depend on the choice of resolution in* (2.2);

(b) $\mathrm{Tor}^{-i}_{\mathbf{k}[v_1,\ldots,v_m]}(\,\cdot\,, N)$ *and* $\mathrm{Tor}^{-i}_{\mathbf{k}[v_1,\ldots,v_m]}(M, \cdot\,)$ *are covariant functors;*

(c) $\mathrm{Tor}^0_{\mathbf{k}[v_1,\ldots,v_m]}(M,N) \cong M \otimes_{\mathbf{k}[v_1,\ldots,v_m]} N$;

(d) $\mathrm{Tor}^{-i}_{\mathbf{k}[v_1,\ldots,v_m]}(M,N) \cong \mathrm{Tor}^{-i}_{\mathbf{k}[v_1,\ldots,v_m]}(N,M)$.

Now put $M = \mathbf{k}[K]$ and $N = \mathbf{k}$. Since $\deg v_i = 2$, we have

$$\mathrm{Tor}_{\mathbf{k}[v_1,\ldots,v_m]}\big(\mathbf{k}[K],\mathbf{k}\big) = \bigoplus_{i,j=0}^{m} \mathrm{Tor}^{-i,2j}_{\mathbf{k}[v_1,\ldots,v_m]}\big(\mathbf{k}[K],\mathbf{k}\big)$$

Define the *bigraded Betti numbers* of $\mathbf{k}[K]$ by

$$\beta^{-i,2j}\big(\mathbf{k}[K]\big) := \dim_{\mathbf{k}} \mathrm{Tor}^{-i,2j}_{\mathbf{k}[v_1,\ldots,v_m]}\big(\mathbf{k}[K],\mathbf{k}\big), \qquad 0 \le i,j \le m. \quad (2.3)$$

We also set

$$\beta^{-i}(\mathbf{k}[K]) = \dim_{\mathbf{k}} \mathrm{Tor}^{-i}_{\mathbf{k}[v_1,\ldots,v_m]}(\mathbf{k}[K],\mathbf{k}) = \sum_j \beta^{-i,2j}(\mathbf{k}[K]).$$

Example 2.8. Let K be the boundary of a square. Then

$$\mathbf{k}[K] \cong \mathbf{k}[v_1,\ldots,v_4]/(v_1v_3, v_2v_4).$$

Let us construct a resolution of $\mathbf{k}[K]$ and calculate the corresponding bigraded Betti numbers. The module R^0 has one generator 1 (of degree 0), and the map $R^0 \to \mathbf{k}[K]$ is the quotient projection. Its kernel is the ideal $\mathcal{I}_K$, generated by two monomials v_1v_3 and v_2v_4. Take R^{-1} to be a free module on two 4-dimensional generators, denoted v_{13} and v_{24}, and define $d\colon R^{-1} \to R^0$ by sending v_{13} to v_1v_3 and v_{24} to v_2v_4. Its kernel is generated by one element $v_2v_4v_{13} - v_1v_3v_{24}$. Hence, R^{-2} has one generator of degree 8, say a, and the map $d\colon R^{-2} \to R^{-1}$ is injective and sends a to $v_2v_4v_{13} - v_1v_3v_{24}$. Thus, we have a resolution

$$0 \longrightarrow R^{-2} \longrightarrow R^{-1} \longrightarrow R^0 \longrightarrow M \longrightarrow 0$$

where rank $R^0 = \beta^{0,0}(\mathbf{k}[K]) = 1$, rank $R^{-1} = \beta^{-1,4} = 2$ and rank $R^{-2} = \beta^{-2,8} = 1$.

The Betti numbers $\beta^{-i,2j}(\mathbf{k}[K])$ are important combinatorial invariants of the simplicial complex K. The following result expresses them in terms of homology groups of subcomplexes of K.

Given a subset $\omega \subseteq [m]$, we may restrict K to ω and consider the *full subcomplex* $K_\omega = \{\sigma \in K\colon \sigma \subseteq \omega\}$.

Theorem 2.9 (Hochster). *We have*

$$\beta^{-i,2j}\big(\mathbf{k}[K]\big) = \sum_{\omega \subseteq [m]\colon |\omega|=j} \dim_{\mathbf{k}} \widetilde{H}^{j-i-1}(K_\omega; \mathbf{k}),$$

where $\widetilde{H}^*(\cdot)$ denotes the reduced cohomology groups and we assume that $\widetilde{H}^{-1}(\varnothing) = \mathbf{k}$.

Hochster's original proof of this theorem uses rather complicated combinatorial and commutative algebra techniques. Later in Subsection 5.1 we give a topological interpretation of the numbers $\beta^{-i,2j}(\mathbf{k}[K])$ as the bigraded Betti numbers of a topological space, and prove a generalization of Hochster's theorem.

Example 2.10 (Koszul resolution). Let $M = \mathbf{k}$ with the $\mathbf{k}[v_1, \ldots, v_m]$-module structure defined via the map $\mathbf{k}[v_1, \ldots, v_m] \to \mathbf{k}$ sending each v_i to 0. Let $\Lambda[u_1, \ldots, u_m]$ denote the exterior $\mathbf{k}$-algebra on m generators. The tensor product $R = \Lambda[u_1, \ldots, u_m] \otimes \mathbf{k}[v_1, \ldots, v_m]$ (here and below we use $\otimes$ for $\otimes_{\mathbf{k}}$) may be turned into a differential bigraded algebra by setting

$$\text{bideg}\, u_i = (-1, 2), \quad \text{bideg}\, v_i = (0, 2),$$
$$du_i = v_i, \quad dv_i = 0, \tag{2.4}$$

and requiring d to be a derivation of algebras. An explicit construction of a cochain homotopy shows that $H^0[R, d] = \mathbf{k}$ and $H^{-i}[R, d] = 0$ for $i > 0$. Since $\Lambda[u_1, \ldots, u_m] \otimes \mathbf{k}[v_1, \ldots, v_m]$ is a free $\mathbf{k}[v_1, \ldots, v_m]$-module, it determines a free resolution of $\mathbf{k}$. It is known as the *Koszul resolution* and its expanded form (2.2) is as follows:

$$0 \to \Lambda^m[u_1, \ldots, u_m] \otimes \mathbf{k}[v_1, \ldots, v_m] \longrightarrow \cdots$$
$$\longrightarrow \Lambda^1[u_1, \ldots, u_m] \otimes \mathbf{k}[v_1, \ldots, v_m] \longrightarrow \mathbf{k}[v_1, \ldots, v_m] \longrightarrow \mathbf{k} \to 0$$

where $\Lambda^i[u_1, \ldots, u_m]$ is the subspace of $\Lambda[u_1, \ldots, u_m]$ spanned by monomials of length i.

Now let us consider the differential bigraded algebra $[\Lambda[u_1, \ldots, u_m] \otimes \mathbf{k}[K], d]$ with d defined as in (2.4).

Lemma 2.11. *There is an isomorphism of bigraded modules:*

$$\text{Tor}_{\mathbf{k}[v_1, \ldots, v_m]}(\mathbf{k}[K], \mathbf{k}) \cong H\big[\Lambda[u_1, \ldots, u_m] \otimes \mathbf{k}[K], d\big]$$

which endows $\text{Tor}_{\mathbf{k}[v_1, \ldots, v_m]}(\mathbf{k}[K], \mathbf{k})$ *with a bigraded algebra structure in a canonical way.*

Proof. Using the Koszul resolution in the definition of Tor, we calculate

$$\mathrm{Tor}_{\mathbf{k}[v_1,\ldots,v_m]}(\mathbf{k}[K], \mathbf{k})$$
$$\cong \mathrm{Tor}_{\mathbf{k}[v_1,\ldots,v_m]}(\mathbf{k}, \mathbf{k}[K])$$
$$= H\big[\Lambda[u_1,\ldots,u_m] \otimes \mathbf{k}[v_1,\ldots,v_m] \otimes_{\mathbf{k}[v_1,\ldots,v_m]} \mathbf{k}[K]\big]$$
$$\cong H\big[\Lambda[u_1,\ldots,u_m] \otimes \mathbf{k}[K]\big].$$

The cohomology in the right hand side is a bigraded algebra, providing an algebra structure for $\mathrm{Tor}_{\mathbf{k}[v_1,\ldots,v_m]}(\mathbf{k}[K], \mathbf{k})$. □

The bigraded algebra $\mathrm{Tor}_{\mathbf{k}[v_1,\ldots,v_m]}(\mathbf{k}[K], \mathbf{k})$ is called the Tor-*algebra* of the simplicial complex K.

Lemma 2.12. *A simplicial map* $\varphi\colon K_1 \to K_2$ *between two simplicial complexes on the vertex sets* $[m_1]$ *and* $[m_2]$ *respectively induces a homomorphism*

$$\varphi_t^*\colon \mathrm{Tor}_{\mathbf{k}[w_1,\ldots,w_{m_2}]}(\mathbf{k}[K_2], \mathbf{k}) \to \mathrm{Tor}_{\mathbf{k}[v_1,\ldots,v_{m_1}]}(\mathbf{k}[K_1], \mathbf{k}) \qquad (2.5)$$

of the corresponding Tor-*algebras.*

Proof. This follows directly from Propositions 2.2 and 2.7 (b). □

3 Toric spaces

Moment-angle complexes provide a functor $K \mapsto \mathcal{Z}_K$ from the category of simplicial complexes and simplicial maps to the category of spaces with torus action and equivariant maps. This functor allows us to use the techniques of equivariant topology in the study of combinatorics of simplicial complexes and commutative algebra of their face rings; in a way, it breathes geometrical life into Stanley's 'combinatorial commutative algebra'. In particular, the calculation of the cohomology of $\mathcal{Z}_K$ opens a way to a topological treatment of homological invariants of face rings.

The space $\mathcal{Z}_K$ was introduced for an arbitrary finite simplicial complex K by Davis and Januszkiewicz [11] as a technical tool in their study of (quasi)toric manifolds, a topological generalization of smooth algebraic toric varieties. Later this space turned out to be of great independent interest. For the subsequent study of $\mathcal{Z}_K$, its place within 'toric topology', and connections with combinatorial problems we refer to [8] and its extended Russian version [9]. Here we review the most important aspects of this study related to the cohomology of face rings.

3.1 Moment-angle complexes

The m-torus T^m is a product of m circles; we usually regard it as embedded in $\mathbb{C}^m$ in the standard way:

$$T^m = \{(z_1, \ldots, z_m) \in \mathbb{C}^m : |z_i| = 1, \quad i = 1, \ldots, m\}.$$

It is contained in the *unit polydisc*

$$(D^2)^m = \{(z_1, \ldots, z_m) \in \mathbb{C}^m : |z_i| \leq 1, \quad i = 1, \ldots, m\}.$$

For an arbitrary subset $\omega \subseteq V$, define

$$B_\omega := \{(z_1, \ldots, z_m) \in (D^2)^m : |z_i| = 1 \text{ for } i \notin \omega\}.$$

The subspace B_ω is homeomorphic to $(D^2)^{|\omega|} \times T^{m-|\omega|}$.

Given a simplicial complex K on $[m] = \{1, \ldots, m\}$, we define the *moment-angle complex* $\mathcal{Z}_K$ by

$$\mathcal{Z}_K := \bigcup_{\sigma \in K} B_\sigma \subseteq (D^2)^m. \tag{3.1}$$

The torus T^m acts on $(D^2)^m$ coordinatewise and each subspace B_ω is invariant under this action. Therefore, the space $\mathcal{Z}_K$ inherits a torus action. The quotient $(D^2)^m/T^m$ can be identified with the *unit m-cube*

$$I^m := \{(y_1, \ldots, y_m) \in \mathbb{R}^m : 0 \leq y_i \leq 1, \quad i = 1, \ldots, m\}.$$

The quotient B_ω/T^m is then the following $|\omega|$-dimensional face of I^m:

$$C_\omega := \{(y_1, \ldots, y_m) \in I^m : y_i = 1 \text{ if } i \notin \omega\}.$$

Thus the whole quotient $\mathcal{Z}_K/T^m$ is identified with a certain cubical subcomplex in I^m, which we denote by $\mathrm{cc}(K)$.

Lemma 3.1. *The cubical complex $\mathrm{cc}(K)$ is PL-homeomorphic to cone K.*

Proof. Let K' denote the barycentric subdivision of K (the vertices of K' correspond to nonempty simplices σ of K). We define a *PL* embedding $i_c \colon \mathrm{cone}\, K' \hookrightarrow I^m$ by mapping each vertex σ to the vertex $(\varepsilon_1, \ldots, \varepsilon_m)$, where $\varepsilon_i = 0$ if $i \in \sigma$ and $\varepsilon_i = 1$ otherwise, mapping the cone vertex to $(1, \ldots, 1) \in I^m$, and then extending linearly on the simplices of cone K'. The barycentric subdivision of a face $\sigma \in K$ is a subcomplex in K', which we denote $K'|_\sigma$. Under the map i_c the subcomplex cone $K'|_\sigma$ maps onto the face $C_\sigma \subset I^m$. Thus the whole complex cone K' maps homeomorphically onto $\mathrm{cc}(K)$, which concludes the proof. $\square$

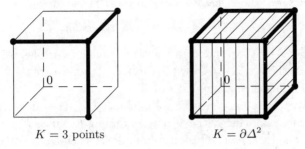

$K = 3$ points $\qquad\qquad\qquad K = \partial\Delta^2$

Fig. 2. Embedding $i_c\colon$ cone $K' \hookrightarrow I^m$.

It follows that the moment-angle complex $\mathcal{Z}_K$ can be defined by the pullback diagram

$$
\begin{array}{ccc}
\mathcal{Z}_K & \longrightarrow & (D^2)^m \\
\downarrow & & \downarrow{\scriptstyle\rho} \\
\text{cone } K' & \xrightarrow{\ i_c\ } & I^m
\end{array}
$$

where ρ is the projection onto the orbit space.

Example 3.2. The embedding i_c for two simple cases when K is a three point complex and the boundary of a triangle is shown in Figure 2. If $K = \Delta^{m-1}$ is the whole simplex on m vertices, then cc(K) is the whole cube I^m, and the above constructed PL-homeomorphism between cone$(\Delta^{m-1})'$ and I^m defines the *standard triangulation* of I^m.

The next lemma shows that the space $\mathcal{Z}_K$ is particularly nice for certain geometrically important classes of triangulations.

Lemma 3.3. *Suppose that K is a triangulation of an $(n-1)$-dimensional sphere. Then $\mathcal{Z}_K$ is a closed $(m+n)$-dimensional manifold.*

In general, if K is a triangulated manifold then $\mathcal{Z}_K \setminus \rho^{-1}(1,\ldots,1)$ is a noncompact manifold, where $(1,\ldots,1) \in I^m$ is the cone vertex and $\rho^{-1}(1,\ldots,1) \cong T^m$.

Proof. We only prove the first statement here; the proof of the second is similar and can be found in [9].

Each vertex v_i of K corresponds to a vertex of the barycentric subdivision K', which we continue to denote v_i. Let star$_{K'}\, v_i$ be the star of v_i in K', that is, the subcomplex consisting of all simplices of K' containing v_i, and all their subsimplices. The space cone K' has a canonical *face structure*, whose *facets* (codimension-1 faces) are

$$F_i := \operatorname{star}_{K'} v_i, \quad i = 1, \dots, m, \tag{3.2}$$

and whose i-faces are nonempty intersections of i-tuples of facets. In particular, the vertices (0-faces) in this face structure are the barycentres of $(n - 1)$-dimensional simplices of K.

For every such barycentre b we denote by U_b the subset of cone K' obtained by removing all faces not containing b. Since K is a triangulation of a sphere, cone K' is an n-ball; hence each U_b is homeomorphic to an open subset in I^n via a homeomorphism preserving the dimension of faces. Since each point of cone K' is contained in some U_b, this displays cone K' as a *manifold with corners*. Having identified cone K' with $\operatorname{cc}(K)$ and further $\operatorname{cc}(K)$ with $\mathcal{Z}_K/T^m$, we see that every point in $\mathcal{Z}_K$ lies in a neighbourhood homeomorphic to an open subset in $(D^2)^n \times T^{m-n}$ and thus in $\mathbb{R}^{m+n}$. $\qquad\square$

A particularly important class of examples of sphere triangulations arise from boundary triangulations of convex polytopes. Suppose P is a *simple* n-dimensional convex polytope, i.e. one where each vertex is contained in exactly n facets. Then the dual (or *polar*) polytope is *simplicial*, and we denote its boundary complex by K_P. K_P is then a triangulation of an $(n - 1)$-sphere. The faces of cone K'_P introduced in the previous proof coincide with those of P.

Example 3.4. Let $K = \partial \Delta^{m-1}$. Then $\mathcal{Z}_K = \partial((D^2)^m) \cong S^{2m-1}$. In particular, for $m = 2$ from (3.1) we get the familiar decomposition

$$S^3 = D^2 \times S^1 \cup S^1 \times D^2 \subset D^2 \times D^2$$

of a 3-sphere into a union of two solid tori.

Using faces (3.2) we can identify the isotropy subgroups of the T^m-action on $\mathcal{Z}_K$. Namely, the isotropy subgroup of a point x in the orbit space cone K' is the coordinate subtorus

$$T(x) = \{(z_1, \dots, z_m) \in T^m \colon z_i = 1 \text{ if } x \notin F_i\}.$$

In particular, the action is free over the interior (that is, near the cone point) of cone K'.

It follows that the moment-angle complex can be identified with the quotient

$$\mathcal{Z}_K = (T^m \times |\operatorname{cone} K'|)/\sim,$$

where $(t_1, x) \sim (t_2, y)$ if and only if $x = y$ and $t_1 t_2^{-1} \in T(x)$. In the case when K is the dual triangulation of a simple polytope P^n we may write $(T^m \times P^n)/\sim$ instead. The latter T^m-manifold is the one introduced by Davis and Januszkiewicz [11], which thereby coincides with our moment-angle complex.

3.2 Homotopy fibre construction

The classifying space for the circle S^1 can be identified with the infinite-dimensional projective space $\mathbb{C}P^\infty$. The classifying space BT^m of the m-torus is a product of m copies of $\mathbb{C}P^\infty$. The cohomology of BT^m is the polynomial ring $\mathbb{Z}[v_1, \ldots, v_m]$, $\deg v_i = 2$ (the cohomology is taken with integer coefficients, unless another coefficient ring is explicitly specified). The total space ET^m of the universal principal T^m-bundle over BT^m can be identified with the product of m infinite-dimensional spheres.

In [11] Davis and Januszkiewicz considered the *homotopy quotient* of $\mathcal{Z}_K$ by the T^m-action (also known as the *Borel construction*). We refer to it as the *Davis-Januszkiewicz space*:

$$DJ(K) := ET^m \times_{T^m} \mathcal{Z}_K = ET^m \times \mathcal{Z}_K / \sim,$$

where $(e, z) \sim (et^{-1}, tz)$. There is a a a fibration $p\colon DJ(K) \to BT^m$ with fibre $\mathcal{Z}_K$. The cohomology of the Borel construction of a T^m-space X is called the *equivariant cohomology* and denoted by $H^*_{T^m}(X)$.

A theorem of [11] states that the cohomology ring of $DJ(K)$ (or the equivariant cohomology of $\mathcal{Z}_K$) is isomorphic to $\mathbb{Z}[K]$. This result can be clarified by an alternative construction of $DJ(K)$ [8], which we review below.

The space BT^m has the canonical cell decomposition in which each factor $\mathbb{C}P^\infty$ has one cell in every even dimension. Given $\omega \subseteq [m]$, define the subproduct

$$BT^\omega := \{(x_1, \ldots, x_m) \in BT^m \colon x_i = * \text{ if } i \notin \omega\},$$

where $*$ is the basepoint (zero-cell) of $\mathbb{C}P^\infty$. Now for a simplicial complex K on $[m]$ define the following cellular subcomplex:

$$BT^K := \bigcup_{\sigma \in K} BT^\sigma \subseteq BT^m. \tag{3.3}$$

Proposition 3.5. *The cohomology of BT^K is isomorphic to the Stanley-Reisner ring $\mathbb{Z}[K]$. Moreover, the inclusion $i\colon BT^K \hookrightarrow BT^m$ of cellular complexes induces the following quotient epimorphism on cohomology:*

$$i^*\colon \mathbb{Z}[v_1, \ldots, v_m] \to \mathbb{Z}[K] = \mathbb{Z}[v_1, \ldots, v_m]/\mathcal{I}_K .$$

Proof. Let B_i^{2k} denote the $2k$-dimensional cell in the ith factor of BT^m, and $C^*(BT^m)$ the cellular cochain module. A monomial $v_{i_1}^{k_1} \ldots v_{i_p}^{k_p}$ represents the cellular cochain $(B_{i_1}^{2k_1} \ldots B_{i_p}^{2k_p})^*$ in $C^*(BT^m)$. Under the cochain homomorphism induced by the inclusion $BT^K \subset BT^m$ the cochain $(B_{i_1}^{2k_1} \ldots B_{i_p}^{2k_p})^*$ maps identically if $\{i_1, \ldots, i_p\} \in K$ and to zero otherwise, whence the statement follows. $\square$

Theorem 3.6. *There is a deformation retraction $DJ(K) \to BT^K$ such that the diagram*

$$
\begin{array}{ccc}
DJ(K) & \overset{p}{\longrightarrow} & BT^m \\
\downarrow & & \| \\
BT^K & \overset{i}{\longrightarrow} & BT^m
\end{array}
$$

is commutative.

Proof. We have $\mathcal{Z}_K = \bigcup_{\sigma \in K} B_\sigma$, and each B_σ is T^m-invariant. Hence, there is the corresponding decomposition of the Borel construction:

$$
DJ(K) = ET^m \times_{T^m} \mathcal{Z}_K = \bigcup_{\sigma \in K} ET^m \times_{T^m} B_\sigma.
$$

Suppose $|\sigma| = s$. Then $B_\sigma \cong (D^2)^s \times T^{m-s}$, so we have

$$
ET^m \times_{T^m} B_\sigma \cong (ET^s \times_{T^s} (D^2)^s) \times ET^{m-s}.
$$

The space $ET^s \times_{T^s} (D^2)^s$ is the total space of a $(D^2)^s$-bundle over BT^s, and ET^{m-s} is contractible. It follows that there is a deformation retraction $ET^m \times_{T^m} B_\sigma \to BT^\sigma$. These homotopy equivalences corresponding to different simplices fit together to yield the required homotopy equivalence between $p \colon DJ(K) \to BT^m$ and $i \colon BT^K \hookrightarrow BT^m$. $\square$

Corollary 3.7. *The space $\mathcal{Z}_K$ is the homotopy fibre of the cellular inclusion $i \colon BT^K \hookrightarrow BT^m$. Hence [11] there are ring isomorphisms*

$$
H^*(DJ(K)) = H_{T^m}^*(\mathcal{Z}_K) \cong \mathbb{Z}[K].
$$

In view of the last two statements we shall also use the notation $DJ(K)$ for BT^K, and refer to the whole class of spaces homotopy equivalent to $DJ(K)$ as the *Davis-Januszkiewicz homotopy type*.

An important question arises: to what extent does the isomorphism of the cohomology ring of a space X with the face ring $\mathbb{Z}[K]$ determine the homotopy type of X? In other words, for given K, does there exist a 'fake' Davis-Januszkiewicz space, whose cohomology is isomorphic to $\mathbb{Z}[K]$, but which is not homotopy equivalent to $DJ(K)$? This question is addressed in [21]. It is shown there [21, Prop. 5.11] that if $\mathbb{Q}[K]$ is a *complete intersection ring* and X is a nilpotent cell complex of finite type whose rational cohomology is isomorphic to $\mathbb{Q}[K]$, then X is rationally homotopy equivalent to $DJ(K)$. Using the formality of $DJ(K)$, this can be rephrased by saying that the complete intersection face rings are *intrinsically formal* in the sense of Sullivan.

Note that the class of simplicial complexes K for which the face ring $\mathbb{Q}[K]$ is a complete intersection has a transparent geometrical interpretation: such a K is a join of simplices and boundaries of simplices.

3.3 Coordinate subspace arrangements

Yet another interpretation of the moment-angle complex $\mathcal{Z}_K$ comes from its identification up to homotopy with the complement of the complex coordinate subspace arrangement corresponding to K. This leads to an application of toric topology in the theory of arrangements, and allows us to describe and effectively calculate the cohomology rings of coordinate subspace arrangement complements and in certain cases identify their homotopy types.

A *coordinate subspace* in $\mathbb{C}^m$ can be written as

$$L_\omega = \{(z_1, \ldots, z_m) \in \mathbb{C}^m : z_{i_1} = \cdots = z_{i_k} = 0\} \qquad (3.4)$$

for some subset $\omega = \{i_1, \ldots, i_k\} \subseteq [m]$. Given a simplicial complex K, we may define the corresponding *coordinate subspace arrangement* $\{L_\omega : \omega \notin K\}$ and its *complement*

$$U(K) = \mathbb{C}^m \setminus \bigcup_{\omega \notin K} L_\omega.$$

Note that if $K' \subset K$ is a subcomplex, then $U(K') \subset U(K)$. It is easy to see [8, Prop. 8.6] that the assignment $K \mapsto U(K)$ defines a one-to-one order preserving correspondence between the set of simplicial complexes on $[m]$ and the set of coordinate subspace arrangement complements in $\mathbb{C}^m$.

The subset $U(K) \subset \mathbb{C}^m$ is invariant with respect to the coordinatewise T^m-action. It follows from (3.1) that $\mathcal{Z}_K \subset U(K)$.

Proposition 3.8. *There is a T^m-equivariant deformation retraction*

$$U(K) \xrightarrow{\simeq} \mathcal{Z}_K.$$

Proof. In analogy with (3.3), we may write

$$U(K) = \bigcup_{\sigma \in K} U_\sigma, \tag{3.5}$$

where

$$U_\sigma := \{(z_1, \dots, z_m) \in \mathbb{C}^m : z_i \neq 0 \text{ for } i \notin \sigma\}.$$

Then there are obvious homotopy equivalences (deformation retractions)

$$\mathbb{C}^\sigma \times (\mathbb{C} \setminus 0)^{[m] \setminus \sigma} \cong U_\sigma \xrightarrow{\simeq} B_\sigma \cong (D^2)^\sigma \times (S^1)^{[m] \setminus \sigma}.$$

These patch together to give the required map $U(K) \to \mathcal{Z}_K$. □

Example 3.9.

(i) Let $K = \partial \Delta^{m-1}$. Then $U(K) = \mathbb{C}^m \setminus 0$ (recall that $\mathcal{Z}_K \cong S^{2m-1}$ in this case).

(ii) Let $K = \{v_1, \dots, v_m\}$ (m points). Then

$$U(K) = \mathbb{C}^m \setminus \bigcup_{1 \leq i < j \leq m} \{z_i = z_j = 0\},$$

the complement to the set of all codimension 2 coordinate planes.

(iii) More generally, if K is the i-skeleton of Δ^{m-1}, then $U(K)$ is the complement to the set of all coordinate planes of codimension $(i+2)$.

The reader may have noticed a similar pattern in several constructions of toric spaces which appeared above; compare (3.1), (3.3) and (3.5). The following general framework was suggested to the author by Neil Strickland in a private communication.

Construction 3.10 (*K-power*). Let X be a space and $W \subset X$ a subspace. For a simplicial complex K on $[m]$ and $\sigma \in K$, we set

$$(X, W)^\sigma := \{(x_1, \dots, x_m) \in X^m : x_j \in W \text{ for } j \notin \sigma\}$$

and

$$(X, W)^K := \bigcup_{\sigma \in K} (X, W)^\sigma = \bigcup_{\sigma \in K} \left(\prod_{i \in \sigma} X \times \prod_{i \notin \sigma} W \right).$$

We refer to the space $(X, W)^K \subseteq X^m$ as the *K-power* of (X, W). If X is a pointed space and $W = pt$ is the basepoint, then we shall use the abbreviated notation $X^K := (X, pt)^K$. Examples considered above include $\mathcal{Z}_K = (D^2, S^1)^K$, $cc(K) = (I^1, S^0)^K$, $DJ(K) = (\mathbb{C}P^\infty)^K$ and $U(K) = (\mathbb{C}, \mathbb{C}^*)^K$.

Homotopy theorists would recognize the K-power as an example of the *colimit* of a diagram of topological spaces over the *face category* of K (objects are simplices and morphisms are inclusions). The diagram assigns the space $(X, W)^\sigma$ to a simplex σ; its colimit is $(X, W)^K$. These observations are further developed and used to construct models of loop spaces of toric spaces as well as for homotopy and homology calculations in [23] and [22].

3.4 Toric varieties, quasitoric manifolds, and torus manifolds

Several important classes of manifolds with torus action emerge as quotients of moment-angle complexes by appropriate freely acting subtori.

First we give the following characterization of lsops in the face ring. Let K^{n-1} be a simplicial complex and $t_1, \ldots, t_n$ a sequence of degree-two elements in $\mathbf{k}[K]$. We may write

$$t_i = \lambda_{i1} v_1 + \cdots + \lambda_{im} v_m, \quad i = 1, \ldots, n. \tag{3.6}$$

For an arbitrary simplex $\sigma \in K$, we have $K_\sigma = \Delta^{|\sigma|-1}$ and $\mathbf{k}[K_\sigma]$ is the polynomial ring $\mathbf{k}[v_i : i \in \sigma]$ on $|\sigma|$ generators. The inclusion $K_\sigma \subset K$ induces the restriction homomorphism r_σ from $\mathbf{k}[K]$ to the polynomial ring, mapping v_i identically if $i \in \sigma$ and to zero otherwise.

Lemma 3.11. *A degree-two sequence $t_1, \ldots, t_n$ is an lsop in $\mathbf{k}[K^{n-1}]$ if and only if for every $\sigma \in K$ the elements $r_\sigma(t_1), \ldots, r_\sigma(t_n)$ generate the positive ideal $\mathbf{k}[v_i : i \in \sigma]_+$.*

Proof. Suppose (3.6) is an lsop. For simplicity we denote its image under any restriction homomorphism by the same letters. Then the restriction induces an epimorphism of the quotient rings:

$$\mathbf{k}[K]/(t_1, \ldots, t_n) \to \mathbf{k}[v_i : i \in \sigma]/(t_1, \ldots, t_n).$$

Since (3.6) is an lsop, $\mathbf{k}[K]/(t_1, \ldots, t_n)$ is a finitely generated $\mathbf{k}$-module. Hence, so is $\mathbf{k}[v_i : i \in \sigma]/(t_1, \ldots, t_n)$. But the latter can be finitely generated only if $t_1, \ldots, t_n$ generates $\mathbf{k}[v_i : i \in \sigma]_+$.

The 'if' part may be proved by considering the sum of restrictions

$$\mathbf{k}[K] \to \bigoplus_{\sigma \in K} \mathbf{k}[v_i : i \in \sigma]$$

which turns out to be a monomorphism; see [6, Th. 5.1.16] for details. $\square$

Obviously, it is enough to consider only restrictions to the maximal simplices in the previous lemma.

Suppose now that K is Cohen-Macaulay (e.g. K is a sphere triangulation). Then every lsop is a regular sequence (however, for $\mathbf{k} = \mathbb{Z}$ or a field of finite characteristic an lsop may fail to exist).

Now we restrict to the case $\mathbf{k} = \mathbb{Z}$ and organize the coefficients in (3.6) into an $n \times m$-matrix $\Lambda = (\lambda_{ij})$. For an arbitrary maximal simplex $\sigma \in K$ denote by Λ_σ the square submatrix formed by the elements λ_{ij} with $j \in \sigma$. The matrix Λ defines a linear map $\mathbb{Z}^m \to \mathbb{Z}^n$ and a homomorphism $T^m \to T^n$. We denote both by λ and denote the kernel of the latter map by T_Λ.

Theorem 3.12. *The following conditions are equivalent:*

(a) *the sequence (3.6) is an lsop in $\mathbb{Z}[K^{n-1}]$;*
(b) $\det \Lambda_\sigma = \pm 1$ *for every maximal simplex $\sigma \in K$;*
(c) $T_\Lambda \cong T^{m-n}$ *and T_Λ acts freely on $\mathcal{Z}_K$.*

Proof. The equivalence of (a) and (b) is a reformulation of Lemma 3.11. Let us prove the equivalence of (b) and (c). Every isotropy subgroup of the T^m-action on $\mathcal{Z}_K$ has the form

$$T^\sigma = \{(z_1, \ldots, z_m) \in T^m : z_i = 1 \text{ if } i \notin \sigma\}$$

for some simplex $\sigma \in K$. Now, (b) is equivalent to the condition that $T_\Lambda \cap T^\sigma = \{e\}$ for arbitrary maximal σ, whence the statement follows. $\square$

We denote the quotient $\mathcal{Z}_K/T_\Lambda$ by $M_K^{2n}(\Lambda)$, and abbreviate it to M_K^{2n} or to M^{2n} when the context allows. If K is a triangulated sphere, then $\mathcal{Z}_K$ is a manifold, hence, so is M_K^{2n}. The n-torus $T^n = T^m/T_\Lambda$ acts on M_K^{2n}. This construction produces the following two important classes of T^n-manifolds as particular examples.

Let $K = K_P$ be a polytopal triangulation, dual to the boundary complex of a simple polytope P. Then the map λ determined by the matrix Λ may be regarded as an assignment of an integer vector to every facet of P. The map λ coming from a matrix satisfying the condition of Theorem 3.12(b) was called a *characteristic map* by Davis and Januszkiewicz [11]. We refer to the corresponding quotient $M_P^{2n}(\Lambda) = \mathcal{Z}_{K_P}/T_\Lambda$ as a *quasitoric manifold* (a toric manifold in the terminology of Davis-Januszkiewicz).

Let us assume further that P is realized in $\mathbb{R}^n$ with integer coordinates of vertices, so we can write

$$P^n = \{ \boldsymbol{x} \in \mathbb{R}^n : \langle \boldsymbol{l}_i, \boldsymbol{x} \rangle \geq -a_i, \ i = 1, \ldots, m \}, \tag{3.7}$$

where $\boldsymbol{l}_i$ are inward pointing normals to the facets of P^n (we may further assume these vectors to be primitive), and $a_i \in \mathbb{Q}$. Let Λ be the matrix formed by the column vectors $\boldsymbol{l}_i$, $i = 1, \ldots, m$. Then $\mathcal{Z}_{K_P}/T_\Lambda$ can be identified with the *projective toric variety* [10, 13] determined by the polytope P. The condition of Theorem 3.12(b) is equivalent to the requirement that the toric variety is nonsingular. Thereby a nonsingular projective toric variety is a quasitoric manifold (but there are many quasitoric manifolds which are not toric varieties).

We also note that smooth projective toric varieties provide examples of *symplectic* $2n$-dimensional manifolds with *Hamiltonian* T^n-action. These symplectic manifolds can be obtained via the process of *symplectic reduction* from the standard Hamiltonian T^m-action on $\mathbb{C}^m$. A choice of an $(m - n)$-dimensional toric subgroup provides a *moment map*

$$\mu \colon \mathbb{C}^m \to \mathbb{R}^{m-n}$$

and the corresponding moment-angle complex $\mathcal{Z}_{K_P}$ can be identified with the level surface $\mu^{-1}(a)$ of the moment map for any of its regular values a. The details of this construction can be found in [8, p. 130].

Finally, we mention that if K is an arbitrary (not necessarily polytopal) triangulation of sphere, then the manifold $M_K^{2n}(\Lambda)$ is a *torus manifold* in the sense of Hattori-Masuda [19]. The corresponding multi-fan has K as the underlying simplicial complex. This particular class of torus manifolds has many interesting properties.

4 Cohomology of moment-angle complexes

The main result of this section (Theorem 4.7) identifies the *integral* cohomology algebra of the moment-angle complex $\mathcal{Z}_K$ with the Tor-algebra of the face ring of the simplicial complex K. Over the rationals this result was proved in [7] by studying the Eilenberg-Moore spectral sequence of the fibration $\mathcal{Z}_K \to DJ(K) \to BT^m$; a more detailed account of applications of the Eilenberg-Moore spectral sequence to toric topology can be found in [8]. The new proof, which works with integer coefficients as well, relies upon a construction of a special cellular decomposition of $\mathcal{Z}_K$ and subsequent analysis of the corresponding cellular cochains.

One of the key ingredients here is a specific cellular approximation of the diagonal map $\Delta\colon \mathcal{Z}_K \to \mathcal{Z}_K \times \mathcal{Z}_K$. Cellular cochains do not admit a functorial associative multiplication because a proper cellular diagonal approximation does not exist in general. The construction of moment-angle complexes is given by a functor from the category of simplicial complexes to the category of spaces with a torus action. We show that in this special case the cellular approximation of the diagonal is functorial with respect to those maps of moment-angle complexes which are induced by simplicial maps. The corresponding cellular cochain algebra is isomorphic to a quotient of the Koszul complex for $\mathbf{k}[K]$ by an acyclic ideal, and its cohomology is isomorphic to the Tor-algebra. The proofs have been sketched in [5]; here we follow the more detailed exposition of [9]. Another proof of Theorem 4.7 follows from recent independent work of M. Franz [12, Th. 1.2].

4.1 Cell decomposition

The polydisc $(D^2)^m$ has a cell decomposition in which each D^2 is subdivided into cells 1, T and D of dimensions 0, 1 and 2 respectively, see Figure 3. Each cell of this complex is a product of cells of 3 different

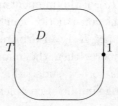

Fig. 3.

types and we encode it by a word $T \in \{D, T, 1\}^m$ in a three-letter alphabet. Assign to each pair of subsets $\sigma, \omega \subseteq [m]$, $\sigma \cap \omega = \varnothing$, the word $T(\sigma, \omega)$ which has the letter D on the positions indexed by σ and letter T on the positions with indices from ω.

Lemma 4.1. $\mathcal{Z}_K$ *is a cellular subcomplex of* $(D^2)^m$. *A cell* $T(\sigma, \omega) \subset (D^2)^m$ *belongs to* $\mathcal{Z}_K$ *if and only if* $\sigma \in K$.

Proof. We have $\mathcal{Z}_K = \cup_{\sigma \in K} B_\sigma$ and each B_σ is the closure of the cell $T(\sigma, [m] \setminus \sigma)$. $\qquad\square$

Therefore we can consider the cellular cochain complex $C^*(\mathcal{Z}_K)$, which has an additive basis consisting of the cochains $T(\sigma, \omega)^*$. It has a natural bigrading defined by

$$\operatorname{bideg} T(\sigma, \omega)^* = (-|\omega|, 2|\sigma| + 2|\omega|),$$

so $\operatorname{bideg} D = (0, 2)$, $\operatorname{bideg} T = (-1, 2)$ and $\operatorname{bideg} 1 = (0, 0)$. Moreover, since the cellular differential does not change the second grading, $C^*(\mathcal{Z}_K)$ splits into the sum of its components having fixed second degree:

$$C^*(\mathcal{Z}_K) = \bigoplus_{j=1}^{m} C^{*, 2j}(\mathcal{Z}_K).$$

The cohomology of $\mathcal{Z}_K$ thereby acquires an additional grading, and we may define the *bigraded Betti numbers* $b^{-i, 2j}(\mathcal{Z}_K)$ by

$$b^{-i, 2j}(\mathcal{Z}_K) := \operatorname{rank} H^{-i, 2j}(\mathcal{Z}_K), \quad i, j = 1, \ldots, m.$$

For the ordinary Betti numbers we have $b^k(\mathcal{Z}_K) = \sum_{2j-i=k} b^{-i, 2j}(\mathcal{Z}_K)$.

Lemma 4.2. *Let* $\varphi \colon K_1 \to K_2$ *be a simplicial map between simplicial complexes on the sets* $[m_1]$ *and* $[m_2]$ *respectively. Then there is an equivariant cellular map* $\varphi_{\mathcal{Z}} \colon \mathcal{Z}_{K_1} \to \mathcal{Z}_{K_2}$ *covering the induced map* $|\operatorname{cone} K_1'| \to |\operatorname{cone} K_2'|$.

Proof. Define a map of polydiscs

$$\varphi_D \colon (D^2)^{m_1} \to (D^2)^{m_2}, \quad (z_1, \ldots, z_{m_1}) \mapsto (w_1, \ldots, w_{m_2}),$$

where

$$w_j := \prod_{i \in \varphi^{-1}(j)} z_i, \quad j = 1, \ldots, m_2$$

(we set $w_j = 1$ if $\varphi^{-1}(j) = \varnothing$). Assume $\tau \in K_1$. In the notation of (3.1), we have $\varphi_D(B_\tau) \subseteq B_{\varphi(\tau)}$. Since φ is a simplicial map, $\varphi(\tau) \in K_2$ and $B_{\varphi(\tau)} \subset \mathcal{Z}_{K_2}$, so the restriction of φ_D to $\mathcal{Z}_{K_1}$ is the required map. $\square$

Corollary 4.3. *The correspondence* $K \mapsto \mathcal{Z}_K$ *gives rise to a functor from the category of simplicial complexes and simplicial maps to the category of spaces with torus actions and equivariant maps. It induces a natural transformation between the simplicial cochain functor of* K *and the cellular cochain functor of* $\mathcal{Z}_K$.

We also note that the maps respect the bigrading, so the bigraded Betti numbers are also functorial.

4.2 Koszul algebras

Our algebraic model for the cellular cochains of $\mathcal{Z}_K$ is obtained by taking the quotient of the Koszul algebra $[\Lambda[u_1, \ldots, u_m] \otimes \mathbf{k}[K], d]$ from Lemma 2.11 by a certain acyclic ideal. Namely, we introduce a factor algebra

$$R^*(K) := \Lambda[u_1, \ldots, u_m] \otimes \mathbb{Z}[K]/(v_i^2 = u_i v_i = 0, \ i = 1, \ldots, m),$$

where the differential and bigrading are as in (2.4). Let

$$\varrho \colon \Lambda[u_1, \ldots, u_m] \otimes \mathbb{Z}[K] \to R^*(K)$$

be the quotient projection. The algebra $R^*(K)$ has a finite additive basis consisting of the monomials of the form $u_\omega v_\sigma$ where $\omega \subseteq [m]$, $\sigma \in K$ and $\omega \cap \sigma = \varnothing$ (remember that we are using the notation $u_\omega = u_{i_1} \ldots u_{i_k}$ for $\omega = \{i_1, \ldots, i_k\}$). Therefore, we have an additive inclusion (a monomorphism of bigraded differential modules)

$$\iota \colon R^*(K) \to \Lambda[u_1, \ldots, u_m] \otimes \mathbb{Z}[K]$$

which satisfies $\varrho \cdot \iota = \mathrm{id}$.

The following statement shows that the finite-dimensional quotient $R^*(K)$ has the same cohomology as the Koszul algebra.

Lemma 4.4. *The quotient map* $\varrho \colon \Lambda[u_1, \ldots, u_m] \otimes \mathbb{Z}[K] \to R^*(K)$ *induces an isomorphism in cohomology.*

Proof. The argument is similar to that used in the proof of the acyclicity of the Koszul resolution. We construct a cochain homotopy between the maps id and $\iota \cdot \varrho$ from $\Lambda[u_1, \ldots, u_m] \otimes \mathbb{Z}[K]$ to itself, that is, a map s satisfying

$$ds + sd = \mathrm{id} - \iota \cdot \varrho. \tag{4.1}$$

First assume that $K = \Delta^{m-1}$. We denote the corresponding bigraded algebra $\Lambda[u_1, \ldots, u_m] \otimes \mathbb{Z}[K]$ by

$$E = E_m := \Lambda[u_1, \ldots, u_m] \otimes \mathbb{Z}[v_1, \ldots, v_m], \tag{4.2}$$

while $R^*(K)$ is isomorphic to

$$\left(\Lambda[u] \otimes \mathbb{Z}[v]/(v^2 = uv = 0)\right)^{\otimes m} = R^*(\Delta^0)^{\otimes m}. \tag{4.3}$$

For $m = 1$, the map $s_1 \colon E^{0,*} = \mathbf{k}[v] \to E^{-1,*}$ given by

$$s_1(a_0 + a_1 v + \ldots + a_j v^j) = (a_2 v + a_3 v^2 + \ldots + a_j v^{j-1})u$$

is a cochain homotopy. Indeed, we can write an element of E as either

x or xu with $x = a_0 + a_1v + \ldots + a_jv^j \in E^{0,2j}$. In the former case, $ds_1x = x - a_0 - a_1v = x - \iota\varrho x$ and $s_1dx = 0$. In the latter case, $xu \in E^{-1,2j}$, then $ds_1(xu) = 0$ and $s_1d(xu) = xu - a_0u = xu - \iota\varrho(xu)$. In both cases (4.1) holds. Now we may assume by induction that for $m = k - 1$ there is a cochain homotopy operator $s_{k-1}: E_{k-1} \to E_{k-1}$. Since $E_k = E_{k-1} \otimes E_1$, $\varrho_k = \varrho_{k-1} \otimes \varrho_1$ and $\iota_k = \iota_{k-1} \otimes \iota_1$, a direct check shows that the map

$$s_k = s_{k-1} \otimes \mathrm{id} + \iota_{k-1}\varrho_{k-1} \otimes s_1$$

is a cochain homotopy between id and $\iota_k\varrho_k$, which finishes the proof for $K = \Delta^{m-1}$.

In the case of arbitrary K the algebras $\Lambda[u_1, \ldots, u_m] \otimes \mathbb{Z}[K]$ are $R^*(K)$ are obtained from (4.2) and (4.3) respectively by factoring out the Stanley-Reisner ideal $\mathcal{I}_K$. This factorization does not affect the properties of the constructed map s, which finishes the proof. □

Now comparing the additive structure of $R^*(K)$ with that of the cellular cochains $C^*(K)$, we see that the two coincide:

Lemma 4.5. *The map*

$$g: R^*(K) \to C^*(\mathcal{Z}_K),$$
$$u_\omega v_\sigma \mapsto T(\sigma, \omega)^*$$

is an isomorphism of bigraded differential modules. In particular, we have an additive isomorphism

$$H[R^*(K)] \cong H^*(\mathcal{Z}_K).$$

Having identified the algebra R^* with the cellular cochains of $\mathcal{Z}_K$, we can also interpret the cohomology isomorphism from Lemma 4.4 topologically. To do this we shall identify the Koszul algebra $\Lambda[u_1, \ldots, u_m] \otimes \mathbb{Z}[K]$ with the cellular cochains of a space homotopy equivalent to $\mathcal{Z}_K$.

Let S^∞ be an infinite-dimensional sphere obtained as a direct limit (union) of standardly embedded odd-dimensional spheres. The space S^∞ is contractible and has a cell decomposition with one cell in every dimension. The boundary of an even-dimensional cell is the closure of the appropriate odd-dimensional cell, while the boundary of an odd cell is zero. The 2-skeleton of this cell decomposition is a 2-disc decomposed as shown on Figure 3, while the 1-skeleton is the circle $S^1 \subset S^\infty$. The cellular cochain complex of S^∞ can be identified with the algebra

$$\Lambda[u] \otimes \mathbb{Z}[v], \quad \deg u = 1, \deg v = 2, \quad du = v, dv = 0.$$

From the obvious functorial properties of Construction 3.10 we obtain a deformation retraction

$$\mathcal{Z}_K = (D^2, S^1)^K \hookrightarrow (S^\infty, S^1)^K \longrightarrow (D^2, S^1)^K$$

onto a cellular subcomplex.

The cellular cochains of the K-power $(S^\infty, S^1)^K$ can be identified with the Koszul algebra $\Lambda[u_1, \ldots, u_m] \otimes \mathbb{Z}[K]$. Since $\mathcal{Z}_K \subset (S^\infty, S^1)^K$ is a deformation retract, the cellular cochain map

$$\Lambda[u_1, \ldots, u_m] \otimes \mathbb{Z}[K] = C^*\big((S^\infty, S^1)^K\big) \to C^*(\mathcal{Z}_K) = R^*(K),$$

induces an isomorphism in cohomology. In fact, the algebraic homotopy map s constructed in the proof of Lemma 4.4 is the map induced on the cochains by the topological homotopy.

4.3 Cellular cochain algebras

Here we introduce a multiplication for cellular cochains of $\mathcal{Z}_K$ and establish a ring isomorphism in Lemma 4.5. This task runs into a complication because cellular cochains in general do not carry a functorial associative multiplication; the classical definition of the cohomology multiplication involves a diagonal map, which is not cellular. However, in our case there is a way to construct a canonical cellular approximation of the diagonal map $\Delta \colon \mathcal{Z}_K \to \mathcal{Z}_K \times \mathcal{Z}_K$ in such a way that the resulting multiplication in cellular cochains coincides with that in $R^*(K)$.

The standard definition of the multiplication in cohomology of a cell complex X via cellular cochains is as follows. Consider a composite map of cellular cochain complexes:

$$C^*(X) \otimes C^*(X) \xrightarrow{\ \times\ } C^*(X \times X) \xrightarrow{\ \widetilde{\Delta}^*\ } C^*(X). \qquad (4.4)$$

Here the map $\times$ assigns to a cellular cochain $c_1 \otimes c_2 \in C^{q_1}(X) \otimes C^{q_2}(X)$ the cochain $c_1 \times c_2 \in C^{q_1+q_2}(X \times X)$ whose value on a cell $e_1 \times e_2 \in X \times X$ is $(-1)^{q_1 q_2} c_1(e_1) c_2(e_2)$. The map $\widetilde{\Delta}^*$ is induced by a cellular approximation $\widetilde{\Delta}$ of the diagonal map $\Delta \colon X \to X \times X$. In cohomology, the map (4.4) induces a multiplication $H^*(X) \otimes H^*(X) \to H^*(X)$ which does not depend on a choice of cellular approximation and is functorial. However, the map (4.4) is not itself functorial because of the arbitrariness in the choice of a cellular approximation.

In the special case $X = \mathcal{Z}_K$ we may apply the following construction. Consider a map $\widetilde{\Delta} \colon D^2 \to D^2 \times D^2$, defined using polar coordinates

$z = \rho e^{i\varphi} \in D^2$, $0 \le \rho \le 1$, $0 \le \varphi < 2\pi$ as follows:

$$\rho e^{i\varphi} \mapsto \begin{cases} (1 + \rho(e^{2i\varphi} - 1), 1) & \text{for } 0 \le \varphi \le \pi, \\ (1, 1 + \rho(e^{2i\varphi} - 1)) & \text{for } \pi \le \varphi < 2\pi. \end{cases}$$

This is a cellular map taking ∂D^2 to $\partial D^2 \times \partial D^2$ and homotopic to the diagonal $\Delta\colon D^2 \to D^2 \times D^2$ in the class of such maps. Taking an m-fold product, we obtain a cellular approximation

$$\widetilde{\Delta}\colon (D^2)^m \to (D^2)^m \times (D^2)^m$$

which restricts to a cellular approximation for the diagonal map of $\mathcal{Z}_K$ for arbitrary K, as described by the following commutative diagram:

$$
\begin{array}{ccc}
\mathcal{Z}_K & \longrightarrow & (D^2)^m \\
\widetilde{\Delta} \downarrow & & \downarrow \widetilde{\Delta} \\
\mathcal{Z}_K \times \mathcal{Z}_K & \longrightarrow & (D^2)^m \times (D^2)^m
\end{array}
$$

Note that this diagonal approximation is functorial with respect to those maps $\mathcal{Z}_{K_1} \to \mathcal{Z}_{K_2}$ of moment-angle complexes that are induced by simplicial maps $K_1 \to K_2$ (see Lemma 4.2).

Lemma 4.6. *The cellular cochain algebra $C^*(\mathcal{Z}_K)$ defined by the diagonal approximation $\widetilde{\Delta}\colon \mathcal{Z}_K \to \mathcal{Z}_K \times \mathcal{Z}_K$ and (4.4) is isomorphic to $R^*(K)$. Therefore, we get an isomorphism of cohomology algebras:*

$$H[R^*(K)] \cong H^*(\mathcal{Z}_K; \mathbb{Z}).$$

Proof. We first consider the case $K = \Delta^0$, that is, $\mathcal{Z}_K = D^2$. The cellular cochain complex of D^2 is additively generated by the cochains $1 \in C^0(D^2)$, $T^* \in C^1(D^2)$ and $D^* \in C^2(D^2)$ dual to the corresponding cells, see Figure 3. The multiplication defined in $C^*(D^2)$ by (4.4) is trivial, so we get a multiplicative isomorphism

$$R^*(\Delta^0) = \Lambda[u] \otimes \mathbb{Z}[v]/(v^2 = uv = 0) \to C^*(D^2).$$

Now, for $K = \Delta^{m-1}$ we obtain a multiplicative isomorphism

$$f\colon R^*(\Delta^{m-1}) = \frac{\Lambda[u_1, \ldots, u_m] \otimes \mathbb{Z}[v_1, \ldots, v_m]}{(v_i^2 = u_i v_i = 0)} \to C^*((D^2)^m)$$

by taking the tensor product. Since $\mathcal{Z}_K \subseteq (D^2)^m$ is a cell subcomplex for arbitrary K we obtain a multiplicative map $q\colon C^*((D^2)^m) \to C^*(\mathcal{Z}_K)$.

Now consider the commutative diagram

$$
\begin{array}{ccc}
R^*(\Delta^{m-1}) & \xrightarrow{\ f\ } & C^*((D^2)^m) \\
{\scriptstyle p}\downarrow & & \downarrow{\scriptstyle q} \\
R^*(K) & \xrightarrow{\ g\ } & C^*(\mathcal{Z}_K).
\end{array}
$$

Here the maps p, f and q are multiplicative, while g is an additive isomorphism by Lemma 4.5. Take $\alpha, \beta \in R^*(K)$. Since p is onto, we have $\alpha = p(\alpha')$ and $\beta = p(\beta')$. Then

$$
g(\alpha\beta) = gp(\alpha'\beta') = qf(\alpha'\beta') = gp(\alpha')gp(\beta') = g(\alpha)g(\beta),
$$

and g is also a multiplicative isomorphism, which finishes the proof. $\square$

Combining the results of Lemmas 2.11, 2.12, 4.4 and 4.6, we come to the main result of this section.

Theorem 4.7. *There is an isomorphism, functorial in K, of bigraded algebras*

$$
H^{*,*}(\mathcal{Z}_K; \mathbb{Z}) \cong \mathrm{Tor}_{\mathbb{Z}[v_1,\ldots,v_m]}(\mathbb{Z}[K], \mathbb{Z}) \cong H\big[\Lambda[u_1, \ldots, u_m] \otimes \mathbb{Z}[K], d\big],
$$

where the bigrading and the differential in the last algebra are defined by (2.4).

As an illustration we give two examples of particular cohomology calculations, which have a transparent geometrical interpretation. More examples of calculations may be found in [8].

Example 4.8.

- Let $K = \partial\Delta^{m-1}$. Then

$$
\mathbb{Z}[K] = \mathbb{Z}[v_1, \ldots, v_m]/(v_1 \cdots v_m).
$$

The fundamental class of $\mathcal{Z}_K \cong S^{2m-1}$ is represented by the bideg $(-1, 2m)$ cocycle $u_1 v_2 v_3 \cdots v_m \in \Lambda[u_1, \ldots, u_m] \otimes \mathbb{Z}[K]$.

- Let $K = \{v_1, \ldots, v_m\}$ (m points). Then $\mathcal{Z}_K$ is homotopy equivalent to the complement in $\mathbb{C}^m$ of the set of all codimension-two coordinate planes (see Example 3.9). Then

$$
\mathbb{Z}[K] = \mathbb{Z}[v_1, \ldots, v_m]/(v_i v_j, \ i \neq j).
$$

The subspace of cocycles in $R^*(K)$ is generated by

$$
v_{i_1} u_{i_2} u_{i_3} \cdots u_{i_k}, \quad k \geq 2 \text{ and } i_p \neq i_q \text{ for } p \neq q,
$$

and has dimension $m\binom{m-1}{k-1}$. The subspace of coboundaries is generated by the elements of the form

$$d(u_{i_1} \cdots u_{i_k})$$

and is $\binom{m}{k}$-dimensional. Therefore

$$\dim H^0(\mathcal{Z}_K) = 1,$$

$$\dim H^1(\mathcal{Z}_K) = H^2(U(K)) = 0,$$

$$\dim H^{k+1}(\mathcal{Z}_K) = m\binom{m-1}{k-1} - \binom{m}{k} = (k-1)\binom{m}{k}, \quad 2 \le k \le m,$$

and multiplication in the cohomology of $\mathcal{Z}_K$ is trivial. Note that in general multiplication in the cohomology of $\mathcal{Z}_K$ is far from being trivial; for example, if K is a sphere triangulation then $\mathcal{Z}_K$ is a manifold by Lemma 3.3.

The above cohomology calculation suggests that the complement of the subspace arrangement from the previous example is homotopy equivalent to a wedge of spheres. This is indeed the case, as the following theorem shows.

Theorem 4.9 (Grbić-Theriault [16]). *The complement of the set of all codimension-two coordinate subspaces in $\mathbb{C}^m$ has the homotopy type of the wedge of spheres*

$$\bigvee_{k=2}^{m} (k-1)\binom{m}{k} S^{k+1}.$$

The proof is based on an analysis of the homotopy fibre of the inclusion $DJ(K) \hookrightarrow BT^m$, which is homotopy equivalent to $\mathcal{Z}_K$ (or $U(K)$) by the first part of Corollary 3.7. We shall return to coordinate subspace arrangements once again in the next section.

5 Applications to combinatorial commutative algebra

5.1 A multiplicative version of Hochster's theorem

As a first application we give a proof of a generalization of Hochster's theorem (Theorem 2.9) obtained by Baskakov in [3].

The bigraded structure in the cellular cochains of $\mathcal{Z}_K$ can be further refined as

$$C^*(\mathcal{Z}_K) = \bigoplus_{\omega \subseteq [m]} C^{*, 2\omega}(\mathcal{Z}_K)$$

where $C^{*,\,2\omega}(\mathcal{Z}_K)$ denotes the subcomplex generated by the cochains $\mathcal{T}(\sigma, \omega \setminus \sigma)^*$ with $\sigma \subseteq \omega$ and $\sigma \in K$. Thus, $C^*(\mathcal{Z}_K)$ now becomes a $\mathbb{Z} \oplus \mathbb{Z}^m$-graded module, and the bigraded cohomology groups decompose accordingly as

$$H^{-i,\,2j}(\mathcal{Z}_K) = \bigoplus_{\omega \subseteq [m]\,:\,|\omega|=j} H^{-i,\,2\omega}(\mathcal{Z}_K) \tag{5.1}$$

where $H^{-i,\,2\omega}(\mathcal{Z}_K) := H^{-i}[C^{*,\,2\omega}(\mathcal{Z}_K)]$.

Given two simplicial complexes K_1 and K_2 with vertex sets V_1 and V_2 respectively, their *join* is the following complex on $V_1 \sqcup V_2$:

$$K_1 * K_2 := \{\sigma \subseteq V_1 \sqcup V_2 : \sigma = \sigma_1 \cup \sigma_2,\ \sigma_1 \in K_1,\ \sigma_2 \in K_2\}.$$

Now we introduce a multiplication in the sum

$$\bigoplus_{\substack{p \geq -1,\\ \omega \subseteq [m]}} \widetilde{H}^p(K_\omega)$$

where K_ω is the full subcomplex and $\widetilde{H}^{-1}(\varnothing) = \mathbb{Z}$, as follows. Take two elements $\alpha \in \widetilde{H}^p(K_{\omega_1})$ and $\beta \in \widetilde{H}^q(K_{\omega_2})$. Assume that $\omega_1 \cap \omega_2 = \varnothing$. Then we have an inclusion of subcomplexes

$$i \colon K_{\omega_1 \cup \omega_2} = K_{\omega_1} \sqcup K_{\omega_2} \hookrightarrow K_{\omega_1} * K_{\omega_2}$$

and an isomorphism of reduced simplicial cochains

$$f \colon \widetilde{C}^p(K_{\omega_1}) \otimes \widetilde{C}^q(K_{\omega_2}) \xrightarrow{\ \cong\ } \widetilde{C}^{p+q+1}(K_{\omega_1} * K_{\omega_2}).$$

Now set

$$\alpha \cdot \beta := \begin{cases} 0, & \omega_1 \cap \omega_2 \neq \varnothing, \\ i^* f(a \otimes b) \in \widetilde{H}^{p+q+1}(K_{\omega_1 \sqcup \omega_2}), & \omega_1 \cap \omega_2 = \varnothing. \end{cases}$$

Theorem 5.1 (Baskakov [3, Th. 1]). *There are isomorphisms*

$$\widetilde{H}^p(K_\omega) \xrightarrow{\ \cong\ } H^{p+1-|\omega|,\,2\omega}(\mathcal{Z}_K)$$

which are functorial with respect to simplicial maps and induce a ring isomorphism

$$\gamma \colon \bigoplus_{\substack{p \geq -1,\\ \omega \subseteq [m]}} \widetilde{H}^p(K_\omega) \xrightarrow{\ \cong\ } H^*(\mathcal{Z}_K).$$

Proof. Define a map of cochain complexes

$$\widetilde{C}^*(K_\omega) \to C^{*+1-|\omega|,\,2\omega}(\mathcal{Z}_K), \quad \sigma^* \mapsto \mathcal{T}(\sigma, \omega \setminus \sigma)^*.$$

It is a functorial isomorphism by observation, whence the isomorphism of the cohomology groups follows.

The statement about the ring isomorphism follows from the isomorphism $H^*(\mathcal{Z}_K) \cong H[R^*(K)]$ established in Lemma 4.5 and analysing the ring structure in $R^*(K)$. □

Corollary 5.2. *There is an isomorphism*

$$H^{-i,2j}(\mathcal{Z}_K) \cong \bigoplus_{\omega \subseteq [m]:\ |\omega|=j} \widetilde{H}^{j-i-1}(K_\omega).$$

As a further corollary we obtain Hochster's theorem (Theorem 2.9):

$$\mathrm{Tor}^{-i,*}_{\mathbb{Z}[v_1,\ldots,v_m]}(\mathbb{Z}[K],\mathbb{Z}) \cong \bigoplus_{\omega \subseteq [m]} \widetilde{H}^{|\omega|-i-1}(K_\omega).$$

5.2 Alexander duality and coordinate subspace arrangements revisited

The multiplicative version of Hochster's theorem can also be applied to cohomology calculations of subspace arrangement complements.

A coordinate subspace can be defined either by setting some coordinates to zero as in (3.4), or as the linear span of a subset of the standard basis in $\mathbb{C}^m$. This gives an alternative way to parametrize coordinate subspace arrangements by simplicial complexes. Namely, we can write

$$\{L_\omega \colon \omega \notin K\} = \{\mathrm{span}\langle e_{i_1},\ldots,e_{i_k}\rangle \colon \{i_1,\ldots,i_k\} \in \widehat{K}\}$$

where $\widehat{K}$ is the simplicial complex given by

$$\widehat{K} := \{\omega \subseteq [m] \colon [m] \setminus \omega \notin K\}.$$

$\widehat{K}$ is called the *dual complex* of K. The cohomology of full subcomplexes in K is related to the homology of links in $\widehat{K}$ by means of the following combinatorial version of the Alexander duality theorem.

Theorem 5.3 (Alexander duality). *Let $K \neq \Delta^{m-1}$ be a simplicial complex on the set $[m]$ and $\sigma \notin K$, that is, $\widehat{\sigma} = [m] \setminus \sigma \in \widehat{K}$. Then there are isomorphisms*

$$\widetilde{H}_j(K_\sigma) \cong \widetilde{H}^{|\sigma|-3-j}(\mathrm{link}_{\widehat{K}}\, \widehat{\sigma}).$$

In particular, for $\sigma = [m]$ we get

$$\widetilde{H}_j(K) \cong \widetilde{H}^{m-3-j}(\widehat{K}), \quad -1 \le j \le m-2.$$

A proof can be found in [9, §2.2]. Using the duality between the full subcomplexes of K and links of $\widehat{K}$ we can reformulate the cohomology calculation of $U(K)$ as follows.

Proposition 5.4. *We have*

$$\widetilde{H}_i(U(K)) \cong \bigoplus_{\sigma \in \widehat{K}} \widetilde{H}^{2m-2|\sigma|-i-2}(\text{link}_{\widehat{K}} \sigma).$$

Proof. From Proposition 3.8 and Corollary 5.2 we obtain

$$H_p(U(K)) = \bigoplus_{\tau \subseteq [m]} \widetilde{H}_{p-|\tau|-1}(K_\tau).$$

Nonempty simplices $\tau \in K$ do not contribute to the above sum, since the corresponding subcomplexes K_τ are contractible. Since $\widetilde{H}^{-1}(\varnothing) = \mathbf{k}$ the empty subset of $[m]$ only contributes $\mathbf{k}$ to $H^0(U(K))$. Hence we may rewrite the above formula as

$$\widetilde{H}_p(U(K)) = \bigoplus_{\tau \notin K} \widetilde{H}_{p-|\tau|-1}(K_\tau).$$

Using Theorem 5.3, we calculate

$$\widetilde{H}_{p-|\tau|-1}(K_\tau) = \widetilde{H}^{|\tau|-3-p+|\tau|+1}(\text{link}_{\widehat{K}} \widehat{\tau}) = \widetilde{H}^{2m-2|\widehat{\tau}|-p-2}(\text{link}_{\widehat{K}} \widehat{\tau}),$$

where $\widehat{\tau} = [m] \setminus \tau$ is a simplex in $\widehat{K}$, as required.   □

Remark. Proposition 5.4 is a particular case of the well-known *Goresky-Macpherson formula* [15, Part III], which calculates the dimensions of the (co)homology groups of an arbitrary subspace arrangement in terms of its *intersection poset* (which coincides with the poset of faces of $\widehat{K}$ in the case of coordinate arrangements). Thus the study of moment-angle complexes not only allows us to retrieve the multiplicative structure of the cohomology of complex coordinate subspace arrangement complements, but also connects two seemingly unrelated results, namely the Goresky-Macpherson formula from the theory of arrangements and Hochster's formula from combinatorial commutative algebra.

5.3 Massey products in the cohomology of $\mathcal{Z}_K$

Here we address the question of existence of nontrivial Massey products in the Koszul complex

$$[\Lambda[u_1, \ldots, u_m] \otimes \mathbb{Z}[K], d]$$

of the face ring. Massey products constitute a series of higher-order operations (or *brackets*) in the cohomology of a differential graded algebra, with the second-order operation coinciding with the cohomology multiplication, while the higher-order brackets are only defined for certain tuples of cohomology classes. A geometrical approach to constructing nontrivial triple Massey products in the Koszul complex of the face ring has been developed by Baskakov in [4] as an extension of the cohomology calculation in Theorem 5.1. It is well-known that nontrivial higher Massey products obstruct the *formality* of a differential graded algebra, which in our case leads to a family of nonformal moment-angle manifolds $\mathcal{Z}_K$.

Massey products in the cohomology of the Koszul complex of a local ring R were studied by Golod [14] in connection with the calculation of the Poincaré series of $\mathrm{Tor}_R(\mathbf{k}, \mathbf{k})$. The main result of Golod is a calculation of the Poincaré series for the class of rings with vanishing Massey products in the Koszul complex (including the cohomology multiplication). Such rings were called *Golod* in [17], where the reader can find a detailed exposition of Golod's theorem together with several further applications. The Golod property of face rings was studied in [20], where several combinatorial criteria for Goldness were given.

The difference between our situation and that of Golod is that we are mainly interested in the cohomology of the Koszul complex for the face ring of a sphere triangulation K. The corresponding face ring $\mathbf{k}[K]$ does not qualify for Goldness, as the corresponding moment-angle complex $\mathcal{Z}_K$ is a manifold, and therefore, the cohomology of the Koszul complex of $\mathbf{k}[K]$ must possess many nontrivial products. Our approach aims at identifying a class of simplicial complexes with nontrivial cohomology product but vanishing higher-order Massey operations in the cohomology of the Koszul complex.

Let K_i be a triangulation of a sphere S^{n_i-1} with $|V_i| = m_i$ vertices, $i = 1, 2, 3$. Set $m := m_1 + m_2 + m_3$, $n := n_1 + n_2 + n_3$, and

$$K := K_1 * K_2 * K_3, \quad \mathcal{Z}_K = \mathcal{Z}_{K_1} \times \mathcal{Z}_{K_2} \times \mathcal{Z}_{K_3}.$$

Note that K is a triangulation of S^{n-1} and $\mathcal{Z}_K$ is an $(m+n)$-manifold.

Given $\sigma \in K$, the *stellar subdivision* of K at σ is obtained by replacing the star of σ by the cone over its boundary:

$$\zeta_\sigma(K) = (K \setminus \mathrm{star}_K \sigma) \cup (\mathrm{cone}\, \partial \, \mathrm{star}_K \sigma).$$

Now choose maximal simplices $\sigma_1 \in K_1$ and $\sigma_2', \sigma_2'' \in K_2$ such that $\sigma_2' \cap \sigma_2'' = \varnothing$, and $\sigma_3 \in K_3$. Set

$$\widetilde{K} := \zeta_{\sigma_1 \cup \sigma_2'}(\zeta_{\sigma_2'' \cup \sigma_3}(K)).$$

Then $\widetilde{K}$ is a triangulation of S^{n-1} with $m+2$ vertices. Take generators

$$\beta_i \in \widetilde{H}^{n_i-1}(\widetilde{K}_{V_i}) \cong \widetilde{H}^{n_i-1}(S^{n_i-1}), \quad i = 1, 2, 3,$$

where $\widetilde{K}_{V_i}$ is the restriction of $\widetilde{K}$ to the vertex set of K_i, and set

$$\alpha_i := \gamma(\beta_i) \in H^{n_i-m_i,2m_i}(\mathcal{Z}_{\widetilde{K}}) \subseteq H^{m_i+n_i}(\mathcal{Z}_{\widetilde{K}}),$$

where γ is the isomorphism from Theorem 5.1. Then

$$\beta_1\beta_2 \in \widetilde{H}^{n_1+n_2-1}(\widetilde{K}_{V_1 \cup V_2}) \cong \widetilde{H}^{n_1+n_2-1}(S^{n_1+n_2-1} \setminus \mathrm{pt}) = 0,$$

and therefore, $\alpha_1\alpha_2 = \gamma(\beta_1\beta_2) = 0$, and similarly $\alpha_2\alpha_3 = 0$. In these circumstances the triple Massey product $\langle \alpha_1, \alpha_2, \alpha_3 \rangle \subset H^{m+n-1}(\mathcal{Z}_{\widetilde{K}})$ is defined. Recall that $\langle \alpha_1, \alpha_2, \alpha_3 \rangle$ is the set of cohomology classes represented by the cocycles $(-1)^{\deg a_1+1} a_1 f + e a_3$ where a_i is a cocycle representing α_i, $i = 1, 2, 3$, while e and f are cochains satisfying $de = a_1 a_2$, $df = a_2 a_3$. A Massey product is called *trivial* if it contains zero.

Theorem 5.5. *The triple Massey product*

$$\langle \alpha_1, \alpha_2, \alpha_3 \rangle \subset H^{m+n-1}(\mathcal{Z}_{\widetilde{K}})$$

in the cohomology of the $(m+n+2)$-manifold $\mathcal{Z}_{\widetilde{K}}$ is nontrivial.

Proof. Consider the subcomplex of $\widetilde{K}$ consisting of those two new vertices added to K in the process of stellar subdivision. By Lemma 4.2, the inclusion of this subcomplex induces an embedding of a 3-dimensional sphere $S^3 \subset \mathcal{Z}_{\widetilde{K}}$. Since the two new vertices are not joined by an edge in $\mathcal{Z}_{\widetilde{K}}$, the embedded 3-sphere defines a nontrivial class in $H^3(\mathcal{Z}_{\widetilde{K}})$. By construction the dual cohomology class is contained in the Massey product $\langle \alpha_1, \alpha_2, \alpha_3 \rangle$. On the other hand, this Massey product is defined up to elements from the subspace

$$\alpha_1 \cdot H^{m_2+m_3+n_2+n_3-1}(\mathcal{Z}_{\widetilde{K}}) + \alpha_3 \cdot H^{m_1+m_2+n_1+n_2-1}(\mathcal{Z}_{\widetilde{K}}).$$

The multigraded components of the group $H^{m_2+m_3+n_2+n_3-1}(\mathcal{Z}_{\widetilde{K}})$ different from that determined by the full subcomplex $\widetilde{K}_{V_2 \cup V_3}$ do not affect the nontriviality of the Massey product, while the multigraded component corresponding to $\widetilde{K}_{V_2 \cup V_3}$ is zero since this subcomplex is contractible. The group $H^{m_1+m_2+n_1+n_2-1}(\mathcal{Z}_{\widetilde{K}})$ is treated similarly. It follows that the Massey product contains a unique nonzero element in its multigraded component and so is nontrivial. $\qquad\square$

As is well known, the nontriviality of Massey products obstructs formality of manifolds, see e.g. [2].

Corollary 5.6. *For every sphere triangulation $\widetilde{K}$ obtained from another triangulation by applying two stellar subdivisions as described above, the 2-connected moment-angle manifold $\mathcal{Z}_{\widetilde{K}}$ is nonformal.*

In the proof of Theorem 5.5 the nontriviality of the Massey product is established geometrically. A parallel argument may be carried out algebraically in terms of the algebra $R^*(K)$, as illustrated in the following example.

Example 5.7. Consider the simple polytope P^3 shown on Figure 4.

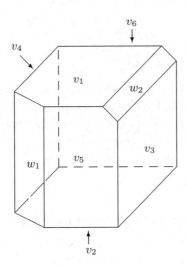

Fig. 4.

This polytope is obtained by cutting two non-adjacent edges off a cube and has 8 facets. The dual triangulation K_P is obtained from an octahedron by applying stellar subdivisions at two non-adjacent edges. The face ring is

$$\mathbb{Z}[K_P] = \mathbb{Z}[v_1, \ldots, v_6, w_1, w_2]/\mathcal{I}_{K_P},$$

where v_i, $i = 1, \ldots, 6$, are the generators coming from the facets of the cube and w_1, w_2 are the generators corresponding to the two new facets, see Figure 4, and

$$\mathcal{I}_P = (v_1 v_2, v_3 v_4, v_5 v_6, w_1 w_2, v_1 v_3, v_4 v_5, w_1 v_3, w_1 v_6, w_2 v_2, w_2 v_4).$$

The corresponding algebra $R^*(K_P)$ has additional generators $u_1, \ldots, u_6$, t_1 and t_2 of total degree 1 satisfying $du_i = v_i$ and $dt_i = w_i$. Consider the cocycles

$$a = v_1 u_2, \quad b = v_3 u_4, \quad c = v_5 u_6$$

and the corresponding cohomology classes $\alpha, \beta, \gamma \in H^{-1,4}[R^*(K)]$. The equations

$$ab = de, \quad bc = df$$

have a solution $e = 0$, $f = v_5 u_3 u_4 u_6$, so the triple Massey product $\langle \alpha, \beta, \gamma \rangle \in H^{-4,12}[R^*(K)]$ is defined. This Massey product is nontrivial by Theorem 5.5. The cocycle

$$af + ec = v_1 v_5 u_2 u_3 u_4 u_6$$

represents a nontrivial cohomology class $[v_1 v_5 u_2 u_3 u_4 u_6] \in \langle \alpha, \beta, \gamma \rangle$ and so the algebra $R^*(K_P)$ and the manifold $\mathcal{Z}_{K_P}$ are not formal.

In view of Theorem 5.5, the question arises of describing the class of simplicial complexes K for which the algebra $R^*(K)$ (equivalently, the Koszul algebra $[\Lambda[u_1, \ldots, u_m] \otimes \mathbb{Z}[K], d]$ or the space $\mathcal{Z}_K$) is formal (in particular, does not contain nontrivial Massey products). For example, a direct calculation shows that this is the case if K is the boundary of a polygon.

5.4 Toral rank conjecture

Here we relate our cohomological calculations with moment-angle complexes to an interesting conjecture in the theory of transformation groups. This 'toral rank conjecture' has strong links with rational homotopy theory, as described in [1]. Therefore this last subsection, although not containing new results, aims at encouraging rational homotopy theorists to turn their attention to combinatorial commutative algebra of simplicial complexes.

A torus action on a space X is called *almost free* if all isotropy subgroups are finite. The *toral rank* of X, denoted $\mathrm{trk}(X)$, is the largest k for which there exists an almost free T^k-action on X.

The *toral rank conjecture* of Halperin [18] suggests that

$$\dim H^*(X; \mathbb{Q}) \geq 2^{\mathrm{trk}(X)}$$

for any finite dimensional space X. Equality is achieved, for example, if $X = T^k$.

Moment angle complexes provide a wide class of almost free torus actions:

Theorem 5.8 (Davis-Januszkiewicz [11, 7.1]). *Let K be an $(n-1)$-dimensional simplicial complex with m vertices. Then* $\operatorname{trk} \mathcal{Z}_K \geq m - n$.

Proof. Choose an lsop in $t_1, \ldots, t_n$ in $\mathbb{Q}[K]$ according to Lemma 2.3 and write

$$t_i = \lambda_{i1} v_1 + \ldots + \lambda_{im} v_m, \quad i = 1, \ldots, n.$$

Then the matrix $\Lambda = (\lambda_{ij})$ defines a linear map $\lambda \colon \mathbb{Q}^m \to \mathbb{Q}^n$. Changing λ to $k\lambda$ for a sufficiently large k if necessary, we may assume that λ is induced by a map $\mathbb{Z}^m \to \mathbb{Z}^n$, which we also denote by λ. It follows from Lemma 3.11 that for every simplex $\sigma \in K$ the restriction $\lambda|_{\mathbb{Z}^\sigma} \colon \mathbb{Z}^\sigma \to \mathbb{Z}^n$ of the map λ to the coordinate subspace $\mathbb{Z}^\sigma \subseteq \mathbb{Z}^m$ is injective.

Denote by T_Λ the subgroup in T^m corresponding to the kernel of the map $\lambda \colon \mathbb{Z}^m \to \mathbb{Z}^n$. Then T_Λ is a product of an $(m - n)$-dimensional torus N and a finite group. The intersection of the torus N with the coordinate subgroup $T^\sigma \subseteq T^m$ is a finite subgroup. Since the isotropy subgroups of the T^m-action on $\mathcal{Z}_K$ are of the form T^σ (see the proof of Theorem 3.12), the torus N acts on $\mathcal{Z}_K$ almost freely. $\square$

Note that by construction the space $\mathcal{Z}_K$ is 2-connected.

In view of Theorem 5.1, we get the following reformulation of the toral rank conjecture for $\mathcal{Z}_K$:

$$\dim \bigoplus_{\omega \subseteq [m]} \widetilde{H}^*(K_\omega; \mathbb{Q}) \geq 2^{m-n}$$

for any simplicial complex K^{n-1} on m vertices.

Example 5.9. Let K be the boundary of an m-gon. Then the calculation of [8, Exam. 7.22] shows that

$$\dim H^*(\mathcal{Z}_K) = (m - 4)2^{m-2} + 4 \geq 2^{m-2}.$$

Bibliography

[1] C. Allday and V. Puppe. *Cohomological methods in transformation groups*, volume 32 of *Cambridge Studies in Adv. Math.* Cambridge Univ. Press, Cambridge, 1993.

[2] I. K. Babenko and I. A. Taimanov. Massey products in symplectic manifolds. *Mat. Sbornik*, 191(8):3–44, 2000. Translation in *Sb. Math.*, 191(8):1107–1146, 2000.

[3] I. V. Baskakov. Cohomology of K-powers of spaces and the combinatorics of simplicial subdivisions. *Uspekhi Mat. Nauk,* 57(5):147–148, 2002. Translation in *Russian Math. Surveys,* 57(5):989–990, 2002.

[4] I. V. Baskakov. Massey triple products in the cohomology of moment-angle complexes. *Uspekhi Mat. Nauk,* 58(5):199–200, 2003. Translation in *Russian Math. Surveys,* 58(5):1039–1041, 2003.

[5] I. V. Baskakov, V. M. Buchstaber, and T. E. Panov. Cellular cochain algebras and torus actions. *Uspekhi Mat. Nauk,* 59(3):159–160, 2004. Translation in *Russian Math. Surveys,* 59(3):562–563, 2004; cf. arXiv: math.AT/0407189.

[6] W. Bruns and J. Herzog. *Cohen-Macaulay rings (revised edition),* volume 39 of *Cambridge Studies in Adv. Math.* Cambridge Univ. Press, Cambridge, 1998.

[7] V. M. Buchstaber and T. E. Panov. Torus actions and combinatorics of polytopes. *Trudy Mat. Inst. Steklova,* 225:96–131, 1999. Translation in *Proc. Steklov Inst. Math.,* 225:87–120, 1999; cf. arXiv: math.AT/9909166.

[8] V. M. Buchstaber and T. E. Panov. *Torus actions and their applications in topology and combinatorics,* volume 24 of *University Lecture Series.* Amer. Math. Soc., Providence, R.I., 2002.

[9] V. M. Bukhshtaber and T. E. Panov. *Toricheskie deistviya v topologii i kombinatorike [Torus actions in topology and combinatorics].* Moskovskiĭ Tsentr Nepreryvnogo Matematicheskogo Obrazovaniya, Moscow, 2004.

[10] V. I. Danilov. The geometry of toric varieties. *Uspekhi Mat. Nauk,* 33(2):85–134, 1978. Translation in *Russian Math. Surveys,* 33(2):97–154, 1978.

[11] M. W. Davis and T. Januszkiewicz. Convex polytopes, Coxeter orbifolds and torus actions. *Duke Math. J.,* 62(2):417–451, 1991.

[12] M. Franz. On the integral cohomology of smooth toric varieties. *Trudy Mat. Inst. Steklova,* 252:61–70, 2006. Translation in *Proc. Steklov Inst. Math.,* 252:53–62, 2006; cf. arXiv: math.AT/0308253.

[13] W. Fulton. *Introduction to toric varieties,* volume 131 of *Annals of Math. Stud.* Princeton Univ. Press, Princeton, 1993.

[14] E. Golod. On the cohomology of some local rings. *Soviet Math. Dokl.,* 3:745–749, 1962.

[15] M. Goresky and R. MacPherson. *Stratified Morse theory,* volume 14 of *Ergebnisse der Mathematik und ihrer Grenzgebiete (3).* Springer-Verlag, Berlin, 1988.

[16] J. Grbić and S. Theriault. The homotopy type of the complement of the codimension-two coordinate subspace arrangement. *Uspekhi Mat. Nauk,* 59(6):203–204, 2004. Translation in *Russian Math. Surveys,* 59(3):1207–1209, 2004.

[17] T. Gulliksen and G. Levin. *Homology of local rings.* Queen's Paper in Pure and Applied Mathematics, No. 20. Queen's University, Kingston, Ont., 1969.

[18] S. Halperin. Rational homotopy and torus actions. In *Aspects of topology,* volume 93 of *London Math. Soc. Lecture Note Ser.,* pages 293–306. Cambridge Univ. Press, Cambridge, 1985.

[19] A. Hattori and M. Masuda. Theory of multi-fans. *Osaka J. Math.,* 40:1–68, 2003.

[20] J. Herzog, V. Reiner, and V. Welker. Componentwise linear ideals and Golod rings. *Michigan Math. J.,* 46(2):211–223, 1999.

[21] D. Notbohm and N. Ray. On Davis-Januszkiewicz homotopy types, I. Formality and rationalisation. *Algebraic & Geometric Topology*, 5:31–51, 2005.

[22] T. E. Panov and N. Ray. Categorical aspects of toric topology. University of Manchester, *in preparation*.

[23] T. E. Panov, N. Ray, and R. Vogt. Colimits, Stanley-Reisner algebras, and loop spaces. In *Categorical decomposition techniques in algebraic topology (Isle of Skye, 2001)*, volume 215 of *Progr. Math.*, pages 261–291. Birkhäuser, Basel, 2004.

[24] R. P. Stanley. *Combinatorics and commutative algebra*, volume 41 of *Progress in Mathematics*. Birkhäuser Boston Inc., Boston, MA, second edition, 1996.

Department of Geometry and Topology

Faculty of Mathematics and Mechanics

Moscow State University

Leninskiye Gory

Moscow 119992

Russia

and

Institute for Theoretical and Experimental Physics

Moscow 117259

Russia

tpanov@mech.math.msu.su

Three lectures on the Borsuk partition problem

Andreii M. Raigorodskii

To A. I. Averbuch, my grandfather

1 Borsuk's conjecture: a historical overview and high-dimensional counterexamples

In this lecture we discuss one of the most famous conjectures of combinatorial geometry – Borsuk's conjecture:

every bounded set of points $\Omega \subset \mathbb{R}^d$ having nonzero diameter can be partitioned into $d + 1$ subsets of smaller diameter.

First of all we present a detailed history of the problem. Then we proceed to counterexamples to Borsuk's conjecture in high dimensions and to asymptotic lower bounds for the value $f(d)$, where $f(d)$ is the minimal number such that every bounded set of points can be partitioned into $f(d)$ parts of smaller diameter. In particular, we describe Nilli's method and its far-reaching modifications and extensions obtained by the author. Finally, we propose new approaches to the multi-dimensional Borsuk problem and formulate some corresponding conjectures.

1.1 The history of the Borsuk partition problem.

In 1933 K. Borsuk conjectured (see [8]) that every bounded set of points $\Omega \subset \mathbb{R}^d$ with nonzero diameter can be partitioned into $d + 1$ subsets of smaller diameter. (In this text $|\mathbf{x} - \mathbf{y}|$ denotes the standard Euclidean distance between two vectors $\mathbf{x}, \mathbf{y} \in \mathbb{R}^d$ and diam Ω denotes the diameter of a point set $\Omega \subset \mathbb{R}^d$ with respect to this distance.) Thus Borsuk's conjecture states that every Ω such that $0 < \operatorname{diam}\Omega < \infty$ can be represented as a disjoint union

$$\Omega = \Omega_1 \sqcup \cdots \sqcup \Omega_{d+1}$$

where $\operatorname{diam}\Omega_i < \operatorname{diam}\Omega$ for all $i = 1,\ldots,d+1$. In other words, if $f(d)$ is the minimal number such that every d-dimensional set of points can be partitioned into $f(d)$ subsets of smaller diameter, then Borsuk's conjecture states that $f(d) = d+1$.

In fact, the lower bound $f(d) \geq d+1$ is almost evident: one can consider the set of all vertices of a regular d-simplex, which of course (by the pigeonhole principle) cannot be partitioned into d parts of smaller diameter. There exist more general and sophisticated examples of sets providing the estimate $f(d) \geq d+1$, such as a sphere (see [7]) or an arbitrary set of constant width (see [37]).

At the same time, Borsuk's conjecture appears quite justified. Indeed, Borsuk himself proved it for the case $d = 2$ and for the (almost trivial) case of a Euclidean sphere (see [8] and Lecture 3). Then H. Hadwiger succeeded in generalizing the last result to the case of an arbitrary body with a smooth boundary (see [24, 5] and Lecture 2). Because of this, Borsuk's conjecture was widely believed to be true and various approaches to proving it were proposed by numerous authors. The conjecture was proved in dimension 3 (see [45, 12, 22, 27, 33, 41, 40, 53] and Lecture 3) and for any d-dimensional set with a sufficiently rich group of symmetries (see [57] and Lecture 2). Moreover, different upper bounds for $f(d)$ were obtained, the best ones due to O. Schramm [59] and to J. Bourgain and J. Lindenstrauss [9]:

$$f(d) \leq \left(\sqrt{\frac{3}{2}} + o(1)\right)^d = (1.224\ldots + o(1))^d$$

(see [53, 6] for details about the preceding results).

Nevertheless, the general case remained open until 1993, when the conjecture was dramatically disproved in dimensions $d \geq 2015$ by J. Kahn and G. Kalai [31]. Moreover, Kahn and Kalai showed that

$$f(d) \geq (1.203\ldots + o(1))^{\sqrt{d}}.$$

This result has since been improved more than once: first A. Nilli [42] found counterexamples in all dimensions $d \geq 946$; then J. Grey and B. Weissbach [21] showed that Borsuk's conjecture was false for all $d \geq 903$, and the present author [48, 49] succeeded in turn in disproving the conjecture for $d \geq 561$. Finally, A. Hinrichs has very recently discovered counterexamples for $d \geq 323$ [29]. As for lower bounds for $f(d)$, the only improvement of Kahn and Kalai's estimate was provided by this author in [49]:

$$f(d) \geq (2/\sqrt{3} + o(1))^{\sqrt{2}\sqrt{d}} = (1.2255 + o(1))^{\sqrt{d}}.$$

Note that the possibility of constructing a counterexample to Borsuk's conjecture was suggested even before 1993: see [57, 14, 35].

Now we have several possible directions for investigation and consequently for discussion. The first concerns different counterexamples to Borsuk's conjecture and lower bounds for the value $f(d)$. We discuss this in the rest of this lecture; a more detailed history of the corresponding results concerning Borsuk's problem will be also given. The second possible direction is to look for solutions for special cases of Borsuk's problem. We return to this subject in the second lecture, where we discuss the approaches of Hadwiger and Rogers and also some results of the author on $(0, 1)$-*polytopes* and *cross-polytopes*. The third lecture is devoted to the 'low-dimensional' case, i.e. dimensions 2, 3 and 4. We present new approaches to bridging the gap between the dimension 3 and dimension 323, where nothing is known for sure.

Further details concerning the history of the Borsuk partition problem can be found in [5, 53, 6, 23, 25, 11, 1].

1.2 *Counterexamples to Borsuk's conjecture, and lower bounds for $f(d)$*

Kahn and Kalai's approach

As we mentioned above, the first counterexample to Borsuk's conjecture was proposed in [31] by J. Kahn and G. Kalai. The original construction was based on a famous result by P. Frankl and R. Wilson [18] concerning some extremal properties of hypergraphs (families of finite sets) whose edges (finite sets) satisfy special intersection conditions. We start by formulating Frankl and Wilson's results.

Theorem 1.1 (P. Frankl, R. Wilson). *Put $q = p^\alpha$, where p is a prime number and $\alpha \geq 1$. Consider an n-element set $\Re_n$ and a family of its k-element subsets $\mathcal{M} = \{M_1, \ldots, M_s\} \subset 2^{\Re_n}$. Assume that for every two different i and j the cardinality of the intersection of the sets M_i and M_j is not congruent to k modulo q:*

$$\mathrm{card}\,(M_i \cap M_j) \not\equiv k \pmod{q} \quad \text{for all } i \neq j, i, j = 1, \ldots, s.$$

Then the cardinality of the whole family of sets $\mathcal{M}$ is bounded from above by the value $\binom{n}{q-1}$, i.e. $s \leq \binom{n}{q-1}$.

For our purposes we need only a special case of this result, formulated in the following immediate corollary.

Corollary 1.2. *Let* $q = p^\alpha$, *where* p *and* α *are as in Theorem 1.1. Let* $n = 4q$. *Consider an* n-*element set* $\Re_n$ *and a family of its* $\frac{n}{2} = 2q$-*element subsets* $\mathcal{M} = \{M_1, \ldots, M_s\} \subset 2^{\Re_n}$. *Assume that for every two different* i *and* j *the cardinality of the intersection of the sets* M_i *and* M_j *is not equal to* $\frac{n}{4} = q$:

$$\operatorname{card}(M_i \cap M_j) \neq \frac{n}{4} \quad \text{for all } i \neq j, i, j = 1, \ldots, s.$$

Then the cardinality of the whole family of sets $\mathcal{M}$ *does not exceed* $2\binom{n}{q-1}$, *i.e.* $s \leq 2\binom{n}{q-1}$.

To deduce this corollary from Frankl and Wilson's theorem, it suffices to divide an arbitrary family of sets $\mathcal{M} \subset 2^{\Re_n}$ into two disjoint parts

$$\mathcal{M} = (\mathcal{M} \cap \mathcal{K}_1) \sqcup (\mathcal{M} \cap \mathcal{K}_2),$$

where $\mathcal{K}_1$ consists of all possible $\frac{n}{2}$-element subsets of $\Re_n = \{1, \ldots, n\}$ containing 1 and $\mathcal{K}_2$ consists of all $\frac{n}{2}$-element subsets of the set $\{2, \ldots, n\}$. Theorem 1.1 is obviously applicable to each of the two parts, and this proves the corollary.

As for the theorem itself, we shall prove it even in a more general form while presenting Nilli's method (see the next part of the lecture and specifically Remark 1.6).

Now we proceed to the construction of Kahn and Kalai's counterexample. First of all, we show how to disprove Borsuk's conjecture in dimension $d = 2015$ by using Kahn and Kalai's approach. To this end, we fix $n = 64 = 4 \times 2^4$, so that in the notation of Frankl and Wilson's theorem we have $q = 16$, $p = 2$ and $\alpha = 4$. Then we consider the set $\Re_n = \{1, \ldots, n\}$ and take W to be the collection of all possible pairs of different elements in $\Re_n$, i.e.

$$W = \{(1,2), (1,3), \ldots, (1,n), (2,3), (2,4), \ldots, (2,n), \ldots, (n-1,n)\},$$

$$\operatorname{card} W = \binom{n}{2} = \frac{n(n-1)}{2} = 2016.$$

For each disjoint union $\Re_n = A \sqcup B$ let

$$S(A, B) = \{w \in W : \operatorname{card}(w \cap A) = 1; \operatorname{card}(w \cap B) = 1\};$$

the cardinality of $S(A, B)$ is obviously equal to $(\mathrm{card}\,A) \times (\mathrm{card}\,B)$. Finally, let $\mathcal{M} = \{S(A, B) : \mathrm{card}\,A = 32 = 2q\}$. In this case

$$\mathrm{card}\,S(A, B) = 32^2 = 1024 \quad, \quad \mathrm{card}\,\mathcal{M} = \frac{1}{2}\binom{n}{n/2} = \frac{1}{2}\binom{64}{32}$$

and $\mathcal{M}$ may be naturally interpreted as a family of some 1024-element subsets of the set $\Re_{2016} = \{1, \dots, 2016\}$.

Note that the cardinality of the intersection of the sets $S(A, B) \in \mathcal{M}$ and $S(C, D) \in \mathcal{M}$ is minimal if and only if $\mathrm{card}(A \cap C) = q = 16$ or, equivalently, $\mathrm{card}(B \cap D) = q = 16$. Thus by Corollary 1.2, in any subfamily of sets $\mathcal{L} \subset \mathcal{M}$ satisfying the condition $\mathrm{card}\,\mathcal{L} > 2\binom{n}{q-1}$, one can find a pair of sets realizing the minimal cardinality of the intersection.

To every $S(A, B) \in \mathcal{M}$ assign the $(0, 1)$-vector

$$\mathbf{x}(A, B) = (x_1, \dots, x_{2016}),$$

where $x_i = 1$ if $i \in S(A, B)$ and $x_i = 0$ otherwise. Clearly, for every such vector the linear relation $x_1 + \cdots + x_{2016} = \mathrm{card}\,S(A, B) = 1024$ holds and therefore the set X of all vectors $\mathbf{x}(A, B)$ lies in an affine subspace of $\mathbb{R}^{2016}$ whose dimension is equal to 2015. On the other hand, it is easy to see that the diameter of X is attained exactly on those pairs of vectors $\mathbf{x}(A, B)$ and $\mathbf{x}(C, D)$ whose preimages $S(A, B)$ and $S(C, D)$ have intersection of minimal cardinality.

Suppose that X can be partitioned into f parts of smaller diameter, say $X = X_1 \sqcup \cdots \sqcup X_f$, where

$$f < \frac{\frac{1}{2}\binom{n}{n/2}}{2\binom{n}{q-1}}.$$

Then by the pigeonhole principle there is a part X_i with cardinality exceeding $2\binom{n}{q-1}$. Consequently the corresponding family of preimages $\mathcal{L}$ also has cardinality exceeding $2\binom{n}{q-1}$, which immediately leads to a contradiction. Thus we have found a set $X \subset \mathbb{R}^{2015}$ that cannot be partitioned into f parts of smaller diameter, so that

$$f(2015) \geq \frac{\frac{1}{2}\binom{n}{n/2}}{2\binom{n}{q-1}} = \frac{\frac{1}{2}\binom{64}{32}}{2\binom{64}{15}} = 2872.109\dots > 2016$$

and the construction of Kahn and Kalai's counterexample in dimension $d = 2015$ is complete.

To obtain the general result, one should repeat the above construction step-by-step for every $n = 4q = 4p^\alpha$, so that the corresponding dimension of the counterexample will be equal to $d = \binom{n}{2} - 1$ and $f(d)$ will be bounded from below by

$$\frac{\frac{1}{2}\binom{n}{n/2}}{2\binom{n}{q-1}} = (1.203\ldots + o(1))^{\sqrt{d}}.$$

Finally, one should apply the Prime Number Theorem (see, for example, [47]) and approximate an arbitrary d by some $d' = \binom{n}{2} - 1$ with $n = 4p^\alpha$: one can see that the form of the estimate will not change and $f(d)$ will remain greater than $(1.203\ldots + o(1))^{\sqrt{d}}$.

Remark 1.3. The best dimension that can be achieved within the framework of Kahn and Kalai's approach is $d = 2015$. Indeed, in this approach we take an arbitrary n of the form $n = 4p^\alpha$ and then apply Frankl and Wilson's theorem. In principle, we could consider, say, $n = 52 = 4 \times 13$, which is the nearest to 64 of all the numbers of the appropriate form. However already in this case $d = 1325$ whereas $f(d)$ can be bounded from below only by 601.

Nilli's method

The main idea of the method that we present in this part of our lecture goes back to a paper by N. Alon, L. Babai and H. Suzuki (see [2] and also [4]). The method consists of a straightforward generalization of Kahn and Kalai's approach, obtained by slightly extending the Frankl and Wilson theorem. A. Nilli was the first who applied this method to disproving Borsuk's conjecture [42].

Fix $n = 44 = 4 \times 11$ and consider the following family of vectors:

$$\Sigma = \{\mathbf{x} = (x_1, \ldots, x_n) \in \{-1, 1\}^n : x_1 = 1, x_1 \ldots x_n = 1\},$$

i.e. Σ consists of all possible vertices of the cube $[-1, 1]^n$ with an even number of negative coordinates and the first coordinate equal to one. Clearly $\operatorname{card} \Sigma = 2^{n-2} = 2^{42}$. At the same time, it is easy to see that for every two vectors $\mathbf{a}, \mathbf{b} \in \Sigma$ the scalar product $(\mathbf{a}, \mathbf{b})$ is congruent to 0 modulo 4. Moreover, $(\mathbf{a}, \mathbf{b}) \equiv 0 \pmod{11}$ if and only if either $(\mathbf{a}, \mathbf{b}) = 0$ or $(\mathbf{a}, \mathbf{b}) = n$, i.e. $\mathbf{a} = \mathbf{b}$. Now the following lemma can be formulated.

Lemma 1.4. *Let $Q = \{\mathbf{a}_1, \ldots, \mathbf{a}_s\} \subset \Sigma$ be any subfamily of vectors with the property that*

$$(\mathbf{a}_i, \mathbf{a}_j) \not\equiv 0 \pmod{11} \quad \text{for all } i \neq j, \ i, j = 1, \ldots, s. \tag{1.1}$$

Then the cardinality of Q does not exceed

$$\sum_{k=0}^{10} \binom{n-1}{k}. \tag{1.2}$$

Proof. To each $\mathbf{a} \in \Sigma$ assign a polynomial $\tilde{P}_{\mathbf{a}} \in (\mathbb{Z}/11\mathbb{Z})[x_2, \ldots, x_n]$ defined as follows. First take the polynomial

$$P_{\mathbf{a}}(\mathbf{x}) = \prod_{k=1}^{10} (k - (\mathbf{a}, \mathbf{x})) \in (\mathbb{Z}/11\mathbb{Z})[x_2, \ldots, x_n]$$

(we suppose that $P_{\mathbf{a}}$ exists only for vectors $\mathbf{x} = (x_1, \ldots, x_n) \in \mathbb{R}^n$ with the first coordinate x_1 equal to one; in particular, $P_{\mathbf{a}}$ is defined on Σ). Then we represent this polynomial as a linear combination of monomials $x_{i_1}^{\delta_{i_1}} \cdots x_{i_k}^{\delta_{i_k}}$. Finally, we substitute 1 for each $\delta_{i_\nu} = 2m + 1$ and 0 for each $\delta_{i_\nu} = 2m$. We denote the resulting linear combination by $\tilde{P}_{\mathbf{a}}$. Note that for every $\mathbf{a}, \mathbf{b} \in \Sigma$, the equality $\tilde{P}_{\mathbf{a}}(\mathbf{b}) = P_{\mathbf{a}}(\mathbf{b})$ holds (since $x_i = \pm 1$ and consequently $x_i^2 = 1$). Moreover, it is easy to see that for $\mathbf{a}, \mathbf{b} \in \Sigma$,

$$\tilde{P}_{\mathbf{a}}(\mathbf{b}) \equiv 0 \pmod{11} \quad \text{if and only if} \quad (\mathbf{a}, \mathbf{b}) \not\equiv 0 \pmod{11}. \tag{1.3}$$

Now we use (1.3) to show that the polynomials $\tilde{P}_{\mathbf{a}_1}, \ldots, \tilde{P}_{\mathbf{a}_s}$ corresponding to the vectors $\mathbf{a}_i \in Q$ are linearly independent over the field $\mathbb{Z}/11\mathbb{Z}$. Assume the contrary. Let $c_1, \ldots, c_s \in \mathbb{Z}/11\mathbb{Z}$ be constants such that

$$(c_1, \ldots, c_s) \not\equiv (0, \ldots, 0) \pmod{11} \tag{1.4}$$

and

$$c_1 \tilde{P}_{\mathbf{a}_1}(\mathbf{x}) + \cdots + c_s \tilde{P}_{\mathbf{a}_s}(\mathbf{x}) \equiv 0 \pmod{11} \text{for all } \mathbf{x}. \tag{1.5}$$

We can substitute an arbitrary $\mathbf{a}_i \in Q$ for $\mathbf{x}$ in (1.5). Then (1.3) and property (1.1) imply that $\tilde{P}_{\mathbf{a}_i}(\mathbf{a}_i) \not\equiv 0 \pmod{11}$. Hence for every i the constant c_i must be congruent to zero modulo 11, which contradicts (1.4).

It suffices to notice that the estimate

$$\operatorname{card} Q = s \leq \dim\{\tilde{P}_{\mathbf{a}}\} = \sum_{k=0}^{10} \binom{n-1}{k}$$

follows immediately from the above argument concerning linear independence and from the definition of the polynomials $\tilde{P}_{\mathbf{a}}$. This completes the proof of the lemma. $\square$

To each $\mathbf{x} = (x_1, \ldots, x_n) \in \Sigma$ assign the vector $\mathbf{x} * \mathbf{x} = (y_{ij}) = (x_i x_j)$, $i, j = 1, \ldots, n$, i.e.

$$\mathbf{x} * \mathbf{x} = (x_1^2, x_1 x_2, \ldots, x_1 x_n, \ldots, x_{n-1} x_n, x_n^2).$$

This correspondence is obviously a bijection and it provides us with the family of vectors $\Sigma^* = \{\mathbf{x} * \mathbf{x}\} \subset \mathbb{R}^{n^2}$, where $\operatorname{card} \Sigma^* = \operatorname{card} \Sigma = 2^{42}$. Furthermore, Σ^* lies in an affine subspace of $\mathbb{R}^{n^2}$ with dimension $d = \binom{n}{2} = 946$. Indeed, the linear relations $y_{ij} = y_{ji}$ and $y_{ii} = x_i^2 = 1$ hold.

It is easy to check that the scalar product of any two vectors $\mathbf{x} * \mathbf{x}$, $\mathbf{z} * \mathbf{z}$ in Σ^* is equal to the square of the scalar product of the preimages $\mathbf{x}, \mathbf{z} \in \Sigma$. Therefore the diameter of the family of vectors Σ^* is attained exactly on those pairs of vectors $\mathbf{x} * \mathbf{x}, \mathbf{z} * \mathbf{z}$ whose preimages have zero scalar product.

Suppose that Σ^* can be partitioned into f parts of smaller diameter, $\Sigma^* = \Omega_1^* \sqcup \cdots \sqcup \Omega_f^*$ say, where

$$f < \frac{2^{42}}{\sum_{k=0}^{10} \binom{n-1}{k}}.$$

Then by the pigeonhole principle there is a part Ω_i^* with $\operatorname{card} \Omega_i^* > \sum_{k=0}^{10} \binom{n-1}{k}$. Thus the corresponding family of preimages $\Omega_i \subset \Sigma$ also has cardinality greater than $\sum_{k=0}^{10} \binom{n-1}{k}$. Lemma 1.4 and the simple remarks made before it imply that Ω_i contains two vectors with zero scalar product and consequently the diameter of Ω_i^* equals the diameter of the whole family Σ^*. This contradicts our supposition and thus

$$f(d) = f(946) \geq \frac{2^{42}}{\sum_{k=0}^{10} \binom{n-1}{k}} = 1649.87\ldots > 947.$$

This completes Nilli's construction in dimension 946.

Remark 1.5. The above construction can be performed for any dimension $d = \binom{n}{2}$ with $n = 4p$. One can easily check that this disproves Borsuk's conjecture in all dimensions $d \geq 946$. However this construction does not lead to any substantial improvement of the bound $f(d) \geq (1.203\ldots + o(1))^{\sqrt{d}}$: one can only refine the explicit expression for $o(1)$. Later we shall discuss various modifications and extensions of Nilli's method that lead to improvements of the constant $1.203\ldots$ and allow us to deal with the dimensions $d = \binom{n}{2}$, where $n = 4p^\alpha$.

Remark 1.6. Note that the above construction is actually a very straightforward and natural generalization of that by Kahn and Kalai. Indeed,

Lemma 1.4 (more precisely, its extensions; see the previous remark) corresponds to Frankl and Wilson's theorem, and if for a moment we fix
the number of ones in every $\mathbf{x} \in \Sigma$, then we obtain exactly the assertion
of Frankl and Wilson: (1.1) is just a restriction on the cardinalities of
intersections and the sum in (1.2) can be replaced by the maximal bi-
nomial coefficient, since the number of ones is fixed and therefore in the
polynomials $\tilde{P}_{\mathbf{x}}$ monomials of lower degree can be represented as linear
combinations of monomials of maximal degree. Moreover, the construc-
tion of Σ^* is completely parallel to that of $\mathcal{M}$ in Kahn and Kalai's
approach, e.g. pairs (x_i, x_j) correspond to pairs in W, and so on.

At the same time, the bounds in Lemma 1.4 and in its direct extensions
are tight. Indeed, let us consider the family of vectors Q, consisting of
all $(-1, 1)$-vectors whose first coordinate is positive and such that the
number of -1s among the last $n - 1$ coordinates equals successively

$$0, (n - 1) - 1, 2, (n - 1) - 3, \ldots, p - 3, (n - 1) - (p - 2), p - 1,$$

where as before $n = 4p$. Then $\operatorname{card} Q = \sum_{k=0}^{p-1} \binom{n-1}{k}$ and the vectors
from Q are pairwise non-orthogonal modulo p.

Remark 1.7. The value $n = 44$ in Nilli's construction is almost the
best possible (cf. Remark 1.3). One can also easily construct a perfectly
analogous counterexample with $n = 43$, $d = 903$ (in fact this was done
by J. Grey and B. Weissbach in [21]) and even with $n = 42$, $d = 861$.
However, further improvements need the introduction of new ideas; we
shall speak about them below.

A modification of Nilli's method

In this part of our lecture we present an important modification of Nilli's
method, proposed by this author in 1997 (see [48]). This modification
allows us to disprove Borsuk's conjecture in dimension $d = 561 (= \binom{34}{2})$
and even in all dimensions $d \geq 561$.

As in the previous part of the lecture, we begin with constructing a
family of vectors Σ. However in this case the construction will be slightly
different. For $n = 36 = 4 \times 9 = 4 \times 3^2$, consider

$$\Sigma = \{\mathbf{x} = (x_1, \ldots, x_n) \in \{-1, 1\}^n : x_1 = x_2 = x_3 = 1, x_1 \ldots x_n = 1\}.$$

Thus Σ is still a family of vertices of the cube $[-1, 1]^n$, but in this case the
first three coordinates are fixed and, more importantly, the dimension n
does not have the form $n = 4p$, where p is a prime†. We shall show that

---

† Of course, in Frankl and Wilson's theorem the corresponding n could be equal to
$4p^\alpha$; nevertheless, the theorem is not applicable in this case.

it is in fact still possible to work with such a family (cf. Remarks 1.5 and 1.7 above).

First of all, we note that as before the cardinality of Σ can be easily computed and it is equal to $2^{n-4} = 2^{32}$. Moreover, the scalar product of any two vectors $\mathbf{a}, \mathbf{b} \in \Sigma$ is still congruent to zero modulo four. Consequently we have the following arithmetic property, similar to (but not coinciding with) the analogous property mentioned before Lemma 1.4.

$$(\mathbf{a}, \mathbf{b}) \equiv 0 \pmod{9} \quad \text{if and only if} \quad (\mathbf{a}, \mathbf{b}) = 0 \ \text{for all} \ \mathbf{a} \neq \mathbf{b} \in \Sigma; \tag{1.6}$$

$$(\mathbf{a}, \mathbf{b}) \equiv 4 \pmod{9} \quad \text{if and only if} \quad (\mathbf{a}, \mathbf{b}) = 4 \ \text{for all} \ \mathbf{a}, \mathbf{b} \in \Sigma. \tag{1.7}$$

(Here the equalities $x_1 = x_2 = x_3 = 1$ are essential.)

Lemma 1.8. *The cardinality of Let $Q = \{\mathbf{a}_1, \ldots, \mathbf{a}_s\} \subset \Sigma$ be any subfamily of vectors with the property that*

$$(\mathbf{a}_i, \mathbf{a}_j) \not\equiv 0 \pmod{9} \ \text{and} \ (\mathbf{a}_i, \mathbf{a}_j) \not\equiv 4 \pmod{9} \tag{1.8}$$

for all $i \neq j$, $i, j = 1, \ldots, s$. Then the cardinality of Q does not exceed

$$\sum_{k=0}^{7} \binom{n-3}{k}. \tag{1.9}$$

Note that Property (1.8) is completely parallel to Property (1.1) from the previous part of the lecture and the difference between them is caused by the form of the corresponding arithmetic properties. Moreover, the bound in Lemma 1.8 is still tight.

The proof of this lemma is very close to that of Lemma 1.4. It suffices to outline the modifications. The polynomials $P_{\mathbf{a}}$ now have the form

$$P_{\mathbf{a}}(\mathbf{x}) = \frac{1}{9} \prod_{k=1}^{3} (k - (\mathbf{a}, \mathbf{x})) \times \prod_{k=5}^{8} (k - (\mathbf{a}, \mathbf{x}))$$

and we suppose that they are defined only for $\mathbf{x} = (1, 1, 1, x_4, \ldots, x_n)$. We can consider them as elements of $\mathbb{Q}[x_4, \ldots, x_n]$. The polynomials $\tilde{P}_{\mathbf{a}}$ are obtained from the polynomials $P_{\mathbf{a}}$ just as before.

We consider linear independence over the field of rational numbers $\mathbb{Q}$ rather than over a finite field. We can do this because the polynomials $\tilde{P}_{\mathbf{a}}$ can be interpreted as integer functions on Q whose values, due to property (1.8), are congruent to zero modulo three except in the case of $\tilde{P}_{\mathbf{a}_i}(\mathbf{a}_i)$.

The inequality

$$\operatorname{card} Q = s \le \sum_{k=0}^{7} \binom{n-3}{k}$$

completes the proof of the lemma.   □

Now we proceed to find the family Σ^*. The previous approach also needs some essential refinements. To each $\mathbf{x} = (x_1, \ldots, x_n) \in \Sigma$ assign the vector $\mathbf{x} * \mathbf{x} = (y_{ij}) = (x_i x_j)$, $i = 2, \ldots, n$, $j = 4, \ldots, n$. This correspondence is obviously a bijection and its form implies the equality $(\mathbf{x} * \mathbf{x}, \mathbf{z} * \mathbf{z}) = ((\mathbf{x}, \mathbf{z}) - 1)((\mathbf{x}, \mathbf{z}) - 3)$. Hence the diameter of Σ^* is attained on those and only those pairs of vectors whose preimages have scalar product equal to 0 or 4. The actual dimension of Σ^* is equal to $d = \binom{n-3}{2} + n - 3 = 561$. Further steps – namely, applying the pigeon-hole principle, Lemma 1.8 and the arithmetic properties (1.6)–(1.7) – coincide precisely with Nilli's and we do not repeat them. Finally, we get

$$f(561) \ge \frac{2^{32}}{\sum_{k=0}^{7} \binom{n-3}{k}} > 758.$$

This completes the construction of a counterexample in dimension 561.

Remark 1.9. The above argument needs only a slight modification in order to obtain a counterexample even for $d = 560$ (see the paper [62] by B. Weissbach). One should introduce slight changes into the construction of an initial family of vectors; in place of Σ, one should consider

$$\Sigma_1 = \left\{ \begin{array}{l} \mathbf{x} = (x_1, \ldots, x_n) \in \{-1, 1\}^n : x_1 = x_2 = x_3 = 1, \\ \qquad\qquad\qquad x_4 x_5 + x_4 x_6 + x_5 x_6 = -1, \\ \qquad\qquad\qquad x_1 \cdots x_n = 1 \end{array} \right\}.$$

Then $\operatorname{card} \Sigma_1 = 6 \times 2^{29}$, and the dimension of the corresponding family Σ_1^* equals 560 since there is an additional linear relation between the coordinates y_{ij}. All the other steps of the construction are precisely as before. Thus

$$f(560) \ge \frac{6 \times 2^{29}}{\sum_{k=0}^{7} \binom{n-3}{k}} \ge 0.75 \times 758 = 568.5 > 561$$

as required.

In order to complete this part of the lecture it remains to show how Borsuk's conjecture can be disproved for all $d \geq 561$. We present the argument proposed by this author in [49].

Put $n = 36$ and $d_0 = 561$. Let Σ^* be the family of vectors described while discussing a counterexample in dimension d_0. Clearly, the elements of Σ^* can be rewritten as vectors of the form $\mathbf{x} = (x_1, \ldots, x_{(n-1)(n-3)})$. It follows from the construction of Σ^* that there are indices $i_1, \ldots, i_{n-3}$ such that $x_{i_j} = 1$ for all $j = 1, \ldots, n-3$ and all $\mathbf{x} \in \Sigma^*$.

Fix an arbitrary dimension $d \geq 564$ and put

$$t = d - d_0, \, m = \sqrt{(n-1)(n-3) - 3} + 1.$$

Consider the family of vectors $\Sigma_d = \{(x_1, \ldots, x_{(n-1)(n-3)+t})\}$ such that $x_{i_1} = x_{i_2} = x_{i_3} = 1$; there is i such that $2 \leq i \leq t-1$, $x_{(n-1)(n-3)+i} = m$ and

$$x_{(n-1)(n-3)+1} = \cdots = x_{(n-1)(n-3)+i-1} = 1 \, ,$$
$$x_{(n-1)(n-3)+i+1} = \cdots = x_{(n-1)(n-3)+t} = 1 \, ;$$

and all other components are equal to zero. We have $\operatorname{card} \Sigma_d = t - 2$. Let

$$\tilde{\Sigma}_d = \{(x_1, \ldots, x_{(n-1)(n-3)}, 1, \ldots, 1)\} \subset \mathbb{R}^{(n-1)(n-3)+t}$$

be the family of vectors obtained from those belonging to Σ^* by adding t components equal to 1. Finally, let $\Sigma_d^* = \Sigma_d \sqcup \tilde{\Sigma}_d$.

One can easily see that the dimension of an affine subspace containing Σ_d^* equals $d_0 + (t - 2) = d - 2$. Moreover,

$$\operatorname{diam} \tilde{\Sigma}_d = \operatorname{diam} \Sigma^* = \sqrt{2(n-1)(n-3) - 6} \, .$$

At the same time, the distance between any two vectors from Σ_d equals $\sqrt{2(n-1)(n-3) - 6}$, as does the distance between any $\mathbf{x} \in \Sigma_d$ and $\mathbf{y} \in \tilde{\Sigma}_d$. Thus $f(d-2) \geq f(d_0) + t - 2 \geq 756 + t > d$. This completes the proof. □

Remark 1.10. The above method of 'lifting' the dimension of a counterexample can be easily applied in a more general context. The following assertion was proved by Gayfullin and Gurevich in [20].

Let M be a subset of $\mathbb{R}^k$ of diameter D which lies on a $(k-1)$-dimensional sphere of radius r such that $r^2 \leq 3D^2/4$. If M is a counterexample to Borsuk's conjecture, then there is a counterexample to the conjecture for any $d > k$.

It is clear that in our case $M = \Sigma^*$ satisfies the condition of this assertion: here $D = \sqrt{2(n-1)(n-3) - 6}$ and $r = \sqrt{(n-1)(n-3)}$.

Remark 1.11. Finally, note that we essentially worked with $n = 4 \times 3^2$. As we have already mentioned, it is possible to repeat all the arguments for the case of an arbitrary $n = 4p^\alpha$. This is described in detail in [62] (see also [53]) and will not be discussed further.

At the same time, even by applying the general construction to obtaining a lower bound for $f(d)$, we get almost the same result as before: $f(d) \geq (1.203\ldots + o(1))^{\sqrt{d}}$. All the new refinements are hidden in the function $o(1)$ and they do not provide us with any improvement of the main exponent. The goal of the rest of this lecture is to improve the constant $1.203\ldots$

Another modification of Nilli's method

We complete this lecture by presenting another important extension of Nilli's method, which leads to an improvement of the lower bound $f(d) \geq (1.203\ldots + o(1))^{\sqrt{d}}$ and to some far-reaching conjectures. First of all, we prove the author's result

$$f(d) \geq (2/\sqrt{3} + o(1))^{\sqrt{2}\sqrt{d}} = (1.2255\ldots + o(1))^{\sqrt{d}}$$

(see also [53] and [49]).

Let $n \in \mathbb{N}$ be sufficiently large. Fix $\delta = 1 - 1/\sqrt{3}$ and let p be the odd prime number nearest to the value δn. Note that for any sufficiently large real x and for $\alpha < 1$ (e.g. for $\alpha = 38/61$), there is a prime number between x and $x + x^\alpha$ (see [47] for example). Hence we may assume that $\delta n - (\delta n)^\alpha < p < \delta n + (\delta n)^\alpha$. Note that the choice of α will influence only the form of the function $o(1)$ from our bound.

We continue the proof in the usual way by considering the family of vectors

$$\Sigma = \left\{ \begin{array}{l} \mathbf{x} = (x_1, \ldots, x_n) \in \{0, 1, -1\}^n : \mathrm{card}\{i : x_i = \pm 1\} = p; \\ \qquad\qquad\qquad\qquad \mathrm{card}\{i : x_i = 0\} = n - p \end{array} \right\}$$

In addition, we suppose that the first nonzero coordinate of any $\mathbf{x} \in \Sigma$ is always positive. Thus we generalize Nilli's method, considering not only the vertices of the unit cube in $\mathbb{R}^n$ but also the centres of some faces of the cube. Let us show that Σ provides us with the desired construction.

Indeed, it is clear that $\mathrm{card}\,\Sigma = 2^{p-1} \binom{n}{p}$. Moreover, the usual *arithmetic property* holds for every two different vectors $\mathbf{a}, \mathbf{b} \in \Sigma$:

$$(\mathbf{a}, \mathbf{b}) \equiv 0 \pmod{p} \quad \text{if and only if} \quad (\mathbf{a}, \mathbf{b}) = 0. \tag{1.10}$$

We have the following lemma.

Lemma 1.12. *The cardinality of any subfamily* $Q = \{\mathbf{a}_1, \ldots, \mathbf{a}_s\} \subset \Sigma$ *with the property*

$$(\mathbf{a}_i, \mathbf{a}_j) \not\equiv 0 \pmod{p} \quad \textit{for all } i \neq j, \ i, j = 1, \ldots, s, \tag{1.11}$$

does not exceed

$$\sum_{l=0}^{p-1} \sum_{k=0}^{\left[\frac{p-1-l}{2}\right]} \binom{n}{k} \binom{n-k}{p-1-l-2k}. \tag{1.12}$$

Once again, the main idea of the proof of Lemma 1.12 consists of using linear independence of some polynomials; these polynomials are defined over the finite field $\mathbb{Z}/p\mathbb{Z}$ as in Nilli's original approach. The transition from a polynomial $P_\mathbf{a}(\mathbf{x}) = \prod_{k=1}^{p-1}(k - (\mathbf{a}, \mathbf{x}))$ to the corresponding polynomial $\tilde{P}_\mathbf{a}$ is by applying successively the equalities $x_i^3 = x_i$ to the standard representation of $P_\mathbf{a}$. These equalities have been chosen because they hold for 0, 1 and -1. It is easy to check that (1.11) implies the linear independence, and to see that the dimension of the space of the polynomials $\tilde{P}_\mathbf{a}$ is equal to (1.12).

The final part of the construction is just as in Nilli's method: take

$$\Sigma^* = \{\mathbf{x} * \mathbf{x} = (x_i x_j) : \mathbf{x} = (x_1, \ldots, x_n) \in \Sigma, i, j = 1, \ldots, n\}$$

and $d = \frac{n(n+1)}{2} - 1$. The pigeonhole principle, Lemma 1.12 and (1.10) imply, after a detailed calculation, that

$$f(d) \geq 2^{p-1} \binom{n}{p} \left(\sum_{l=0}^{p-1} \sum_{k=0}^{\left[\frac{p-1-l}{2}\right]} \binom{n}{k} \binom{n-k}{p-1-l-2k} \right)^{-1}$$

$$\geq \left(\frac{2}{\sqrt{3}} + o(1) \right)^{\sqrt{2}\sqrt{d}}$$

$$= (1.2255 + o(1))^{\sqrt{d}}.$$

This completes the proof of the bound. $\qquad\qquad\square$

Remark 1.13. The above considerations naturally imply the following conclusion. The Frankl and Wilson type or, say, the Alon, Babai and Suzuki (Nilli) type theorems allow us to deal with some finite d-dimensional sets of points and to obtain good counterexamples to Borsuk's conjecture as well as nontrivial lower bounds for the value $f(d)$. To be more precise, all the results from this lecture involve constructing a set of vertices of the unit cube in $\mathbb{R}^d$, or a set of vertices and centres of faces of this cube. In other words, we could consider the convex hulls of those sets as well and work ultimately with either so-called $(0,1)$-*polytopes* or with so-called *cross-polytopes* (also called $(0,1,-1)$-polytopes).

One can ask whether it is possible to extend this approach further by using general *integer polytopes*, i.e. convex hulls of arbitrary sets of points in a lattice in $\mathbb{R}^n$ (say, in $\mathbb{Z}^n$). Unfortunately the author does not know of any really useful extension of this type; however, the corresponding approach is discussed in detail in the paper [53].

On the other hand, it is perfectly natural to ask how 'strong' the counterexamples constructed using $(0,1)$-and cross-polytopes could be. In the next lecture we follow this line of investigation and prove some partial results concerning *upper bounds* for the minimal number of parts of smaller diameter necessary to partition a $(0,1)$-polytope or cross-polytope from a sufficiently large class.

Let us discuss the bound (1.12) from the last lemma in more detail. First of all, note that by varying the definition of the corresponding polynomials one can work not only with prime moduli, but also with moduli equal to some p^α as well. On the other hand, as we mentioned above, the bounds in Nilli's method as well and in its first modification by the author are obviously tight. Now it is easy to understand that the bound (1.12) is definitely not the best possible. Indeed, consider $n = 8$ and the family of vectors

$$\Sigma = \left\{ \begin{array}{l} \mathbf{x} = (x_1, \ldots, x_n) \in \{0, 1, -1\}^n : \mathrm{card}\{i : x_i = \pm 1\} = 4; \\ \qquad\qquad \mathrm{card}\{i : x_i = 0\} = 4 \end{array} \right\} \quad (1.13)$$

(Here, as above, the first nonzero coordinate is supposed to be positive.) The standard linear algebra method of obtaining an upper bound for the maximal cardinality of a family of pairwise non-orthogonal vectors $Q \subset \Sigma$ corresponds in fact to the use of the polynomials

$$P_{\mathbf{a}}(\mathbf{x}) = \frac{1}{2} \prod_{k=1}^{3} (k - (\mathbf{a}, \mathbf{x}))$$

and the respective polynomials $\tilde{P}_{\mathbf{a}}$ (see also [53, 62]). Therefore $\mathrm{card}\, Q$ is bounded above by

(the number of monomials of the form $x_i x_j x_k$)

+(the number of monomials of the form $x_i^2 x_j$)

+(the number of monomials of the form x_i^2)

$$\vdots$$

+(the number of monomials of the form x_i) + 1

$$= 157.$$

However, it is easy to obtain the estimate card $Q \leq 70$ by splitting Σ into 70 subfamilies consisting each of eight pairwise orthogonal vectors. For instance, the first such subfamily can be chosen as follows:

$$(1,1,1,1,0,0,0,0), \qquad (1,1,-1,-1,0,0,0,0),$$
$$(1,-1,-1,1,0,0,0,0), \qquad (1,-1,1,-1,0,0,0,0),$$
$$(0,0,0,0,1,1,1,1), \qquad (0,0,0,0,1,1,-1,-1),$$
$$(0,0,0,0,1,-1,-1,1), \qquad (0,0,0,0,1,-1,1,-1).$$

More careful analysis leads to the following far-reaching conjecture.

Conjecture. *If in the above definition we replace the number* 8 *by an arbitrary* $n = 2^k$ *and the number* 4 *(i.e. the number of nonzero coordinates) by* 2^{k-1}*, then we obtain the bound* card $Q \leq R(n)$*, where* $R(n) \leq 2^n$.

If this conjecture (in a slightly stronger form) is true, then it is possible to disprove the Borsuk conjecture for all $d \geq 135$ and to show that $f(d) \geq (\sqrt{2} + o(1))^{\sqrt{2}\sqrt{d}} = (1.6325\ldots + o(1))^{\sqrt{d}}$ (see also [53, 50] for some details).

Let us look once again at the case $n = 8$. As noted above, in this case the standard (and usually powerful!) linear algebra method is substantially worse than a simple 'subdivision' argument. It is still possible to make the following modification, which is probably far-reaching too, of the basic polynomial approach. Consider (1.13): instead of polynomials, to each vector $\mathbf{a} = (a_1, \ldots, a_n) \in \Sigma$ assign the *function*

$$F_{\mathbf{a}}(\mathbf{x}) = \frac{1}{4}((\mathbf{a}, \mathbf{x}) - 2)^2 - \frac{1}{2}\left\{\frac{1}{2}\sum_{i=1}^{8} a_i^2 x_i^2\right\}.$$

Here the curly brackets $\{y\}$ denote the fractional part of a real number y. The last term differs from zero (and is equal to $\frac{1}{4}$) if and only if the nonzero parts of vectors $\mathbf{a}$ and $\mathbf{x}$, when interpreted as four-subsets of $\Re_n = \{1, \ldots, n\}$, have intersection consisting of an odd number of elements (1 or 3).

Now we can translate the standard method of finding a linearly independent system from the language of polynomials into that of our functions, by proving that for any fixed subfamily $Q = \{\mathbf{a}_1, \ldots, \mathbf{a}_s\} \subset \Sigma$ satisfying the ordinary condition of non-orthogonality, the corresponding functions $F_{\mathbf{a}_1}, \ldots, F_{\mathbf{a}_s}$ are linearly independent over the field of rationals. Indeed, $F_{\mathbf{a}}(\mathbf{x})$ is integer over Q and is congruent to 1 modulo 2 only when $\mathbf{a} = \mathbf{x}$; independence follows. (Here we use the following fact: the scalar product of any two vectors from Q is even if and only if the cardinality of the intersection of the nonzero parts of these vectors is even.)

It remains to compute the dimension of the space of our functions. This dimension is equal to the following sum:

$$
\begin{aligned}
&(\text{number of monomials of the form } x_i^2) \\
+&(\text{number of monomials of the form } x_i x_j) \\
+&(\text{number of monomials of the form } x_i) + 1 \\
+&\left(\begin{array}{l} \text{number of independent functions of the form} \\ (a_1^2 x_1^2 + \cdots + a_n^2 x_n^2)/2 \text{ with } a_i, x_i \in \{0, -1, 1\} \end{array}\right).
\end{aligned}
$$

In our case, the last term in this sum equals $\frac{1}{2}\binom{8}{4} = 35$ and therefore the whole sum equals 80. This gives the bound[†] $s = \operatorname{card} Q \leq 80$. Unfortunately, this bound is also worse than the trivial one, but it has important advantages: on the one hand, it is substantially better than the bound $s \leq 157$ obtained with the help of the standard linear algebra method; and on the other hand, the trivial approach does not lead to any good results in the dimensions $n = 2^k$, whereas we believe strongly that one can 'lift' the new argument so that it will still work at least for $n = 2^k$.

This suggestion completes the first lecture. We would like to account for not discussing the result of [29]: of course, it does better in dimension $d = 323$, and it even works with an appropriate *integer polytope* (cf. Remark 1.13); but it cannot be generalized, since it is based exclusively on the properties of the Leech lattices (see [10]) and those properties cannot be 'lifted' to higher dimensions.

† Moreover, $s \leq 79$, since we have, in addition, the linear relation $x_1^2 + \cdots + x_8^2 = 4$.

2 Borsuk's conjecture: partial solutions and other partial results

In this lecture we discuss some special cases where one can find a solution to Borsuk's problem. Furthermore, we present various new asymptotic upper bounds (discovered by the author) for the value $p(d)$, where $p(d)$ is the minimal number such that every d-dimensional $(0,1)$-polytope or cross-polytope from a sufficiently large class can be partitioned into $p(d)$ parts of smaller diameter.

2.1 Hadwiger's solution in the 'smooth' case

As we mentioned in the first lecture while presenting the history of Borsuk's problem, there are some special cases where the corresponding conjecture can be proved in any dimension d, i.e. there are particular classes of sets such that every bounded point set $\Omega \subset \mathbb{R}^d$ from such a class can actually be partitioned into $d + 1$ subsets of smaller diameter. Of course, in a single lecture it is impossible to give a detailed survey of all such classes that have ever been discovered. We shall only discuss those which we regard as the most important in two respects: on the one hand they illustrate some essential ideas and methods; and on the other hand they provide us with additional motivation for working at partial results in Borsuk's problem, e.g. those concerning $(0,1)$-polytopes and cross-polytopes (cf. Lecture 1 and see §2.3 below; see also [52]). More detailed surveys (though emphasizing some other aspects of the problem) can be found in [53] and [6].

In this section we restrict attention to the class of convex bodies $\Omega \subset \mathbb{R}^d$ with a smooth boundary $\partial\Omega$, so that there exists a unique support *(tangent)* plane at every point $P \in \partial\Omega$. (Here we take the word 'body' to mean a compact set with nonempty interior.) Note that convexity is not a necessary condition, since a set can be partitioned into some number of parts of smaller diameter whenever its convex hull can (see [13]). However we assume this condition for convenience.

Now we are ready to formulate the following result due to H. Hadwiger (see also [24, 5, 6]).

Theorem 2.1. *Every convex d-dimensional body Ω with a smooth boundary can be partitioned into $d + 1$ parts of smaller diameter.*

We start the proof by considering a Euclidean sphere; as far as the Borsuk problem is concerned, it is not different from a Euclidean ball (cf. the first part of Lecture 1). A ball in $\mathbb{R}^d$ is an example of a convex body with a smooth boundary. It can be partitioned into the required number of parts as follows.

We can without loss of generality restrict attention to the ball B of diameter 1. Let R be a regular d-simplex inscribed into B and let $R_1, \ldots, R_{d+1}$ be its faces of maximal dimension $d-1$. For a face R_i let $\mathbf{e}_1, \ldots, \mathbf{e}_d$ be the set of vertices corresponding to this face. Let O be the centre of the ball. The face R_i specifies the part $B_i = B \cap A_i$, where A_i is the angle generated by the vectors $O\mathbf{e}_1, \ldots, O\mathbf{e}_d$. Thus we partition B into $B = B_1 \sqcup \cdots \sqcup B_{d+1}$. It remains to calculate the diameters of the parts B_i and to check that all these diameters are smaller than 1. For instance, if $d = 2$, then each diam $B_i = \frac{\sqrt{3}}{2} = 0.866 \ldots < 1$, and if $d = 3$, then diam $B_i = \sqrt{\frac{3+\sqrt{3}}{6}} = 0.888 \ldots < 1$.

This simple result is due to Borsuk (see [8]). We shall return to this example in the next lecture, where we discuss the 'small-dimensional' case of Borsuk's conjecture.

Now we proceed to the proof of Theorem 2.1. In fact, almost everything has already been done. Indeed, the partition $B = B_1 \sqcup \ldots \sqcup B_{d+1}$ can be reconstructed for an arbitrary convex body Ω of diameter 1 with a smooth interior by using the well-known *Gauss transformation*. To each point $P \in \partial\Omega$ assign the point $g(P) \in \partial B$ such that the tangent plane to B at $g(P)$ is parallel to the tangent plane to Ω at P and the corresponding normal vectors have the same direction. To each set $B'_i = B_i \cap \partial B$ assign, in turn, the set $\Omega'_i = \{P \in \partial\Omega : g(P) \in B'_i\}$. Clearly diam $\Omega'_i < 1$, since diam $B'_i < 1$. Finally, we fix an arbitrary point Q from the (nonempty) interior of the body Ω and let Ω_i be the convex hull of Ω'_i and Q. It is clear that diam $\Omega_i < 1$. The theorem follows.   □

Theorem 2.1 was refined by Hadwiger himself (see [24]). The following result holds:

Theorem 2.2. *If a convex body $\Omega \subset \mathbb{R}^d$ of diameter 1 is such that on the interior side of its boundary a d-dimensional ball of radius r can 'roll over' without obstacles, then*

$$\Omega = \Omega_1 \sqcup \cdots \sqcup \Omega_{d+1} \quad , \quad \text{diam}\,\Omega_i \leq 1 - 2r\left(1 - \sqrt{1 - \frac{1}{d^2}}\right).$$

For instance, consider the unit ball B. Certainly, Theorem 2.2 is applicable to $\Omega = B$ with $r = \frac{1}{2}$ (in a way, B is a limiting case of the set Ω in the theorem). Hence by the theorem, $B = B_1 \sqcup \cdots \sqcup B_{d+1}$ where $\operatorname{diam} B_i \leq \sqrt{1 - 1/d^2}$. Consequently, if $d = 2$ then $\operatorname{diam} B_i \leq \sqrt{3/4}$. This bound coincides with the bound obtained in the beginning of the proof of Theorem 2.1, *especially for the ball*. However for $d = 3$, we only get $\operatorname{diam} B_i \leq \sqrt{8/9} = 0.9428\ldots$ and this is already worse than the corresponding estimate $\operatorname{diam} B_i \leq 0.888\ldots$ We do not prove Theorem 2.2 here.

2.2 The 'centrally symmetric' case and C. A. Rogers' partial solution

In this part of the lecture, we would like to say a few words about the cases when a body of diameter 1 is symmetric in some sense. First of all, a simple result due to A. S. Riesling [56] should be mentioned: *every centrally symmetric compact convex d-dimensional body Ω can be partitioned into $d + 1$ parts of smaller diameter.* The partition can be constructed in the same manner as the one for the ball in the previous part of the lecture: one should take a regular d-simplex whose centroid coincides with the symmetry centre of Ω, etc. Moreover, the number $d+1$ of parts of smaller diameter is attained on a ball. In fact, the main results of Riesling deal with Borsuk's problem in some spaces of constant curvature, but we do not dwell on this here (see e.g. [6] for details). We only note that this result of Riesling is already very important. Since it does not require any smoothness at all, it remains true for polytopes, for example.

Now it is natural to proceed to another 'symmetric' partial solution of Borsuk's problem due to C. A. Rogers. The following theorem was discovered "*in an unsuccessful attempt to disprove Borsuk's conjecture*" [57].

Theorem 2.3. *If a set Ω of diameter 1 in $\mathbb{R}^d$ is invariant under the group of congruences that leave a regular d-simplex invariant, then Ω can be covered by a system of $d + 1$ closed convex sets each of diameter not exceeding*

$$\sqrt{\left(1 - \frac{6 - 4\sqrt{2}}{t(t+1)d}\right)}$$

where $t = [(d+1)/2]$, provided $d \geq 3$.

We do not give a complete proof of this theorem; however we present a concise sketch of the proof.

Step 1. It is possible (see [57] and cf. [13]) to prove the following lemma.

Lemma 2.4. *Let G be a group of congruences that leave the origin of $\mathbb{R}^d$ invariant, and let Ω be a set of diameter 1 that is invariant under G. Then there is a closed convex set K of constant width 1 that contains Ω and is invariant under G.*

Step 2. Suppose that Ω is a set of constant width 1 and thus of diameter 1. We may assume that Ω lies in the d-dimensional plane

$$x_0 + x_1 + \cdots + x_d = 0$$

and is invariant under the group of permutations of the $d+1$ coordinates (it is more convenient to work in $\mathbb{R}^{d+1}$). Finally, note that Ω is contained in a ball of radius

$$\sqrt{\left(\frac{d}{2(d+1)}\right)}$$

(see [30]) and contains the ball of points $\mathbf{x} = (x_0, \ldots, x_d)$ with

$$|\mathbf{x}| \leq 1 - \sqrt{\left(\frac{d}{2(d+1)}\right)}, x_0 + \cdots + x_d = 0$$

(see [13]).

Step 3. Let us describe a system of sets $F_0, \ldots, F_d$ covering Ω. Let F_i be the collection of all points $\mathbf{x} \in \Omega$ such that

$$x_{i+1} - x_i \geq x_{i+j+1} - x_{i+j}, j = 1, \ldots, n,$$

where the indices of the coordinates are reduced modulo $d+1$ if necessary.

Step 4. It remains to calculate the diameters of the parts $F_0, \ldots, F_d$ and to check that the bound from the statement of Theorem holds for each diam F_i. The theorem follows.   □

Remark. It is natural to ask whether the covering obtained in this theorem is invariant under the group of congruences of the simplex. Rogers has shown [57] that such coverings do not necessarily exist by constructing a counterexample in $\mathbb{R}^8$.

2.3 Results for $(0,1)$-polytopes and cross-polytopes

In this section we discuss in detail some results of the author, which concern *upper* bounds for the minimal number of parts of smaller diameter required to partition an arbitrary $(0,1)$-polytope or a cross-polytope from a sufficiently large class. The motivation for this was given in the first lecture (see Remark 1.13 and also [53, 52]). In order to formulate the corresponding theorems and to prove them we need some definitions and notation.

Additional definitions and notation

Consider Euclidean space $\mathbb{R}^d$. Let $\mathcal{F}_{k,l}^d = \{\mathbf{v}_1, \ldots, \mathbf{v}_s\} \subset \mathbb{R}^d$ be an arbitrary family of $(0,1)$-vectors such that

$$(\operatorname{diam} \mathcal{F}_{k,l}^d)^2 = k \text{ and } v_i^1 + \cdots + v_i^d = l \text{ for all } i$$

(here $\mathbf{v}_i = (v_i^1, \ldots, v_i^d)$). In other words, $\mathcal{F}_{k,l}^d$ is a subset of the set of vertices of the cube $[0,1]^d$ such that the maximal distance between its elements is equal to $\sqrt{k}$ and the number of nonzero coordinates of every vector $\mathbf{v} \in \mathcal{F}_{k,l}^d$ is equal to l. Hence it is clear that the value k is even, the inequalities $0 \leq k \leq d$ and $0 \leq l \leq d$ hold. Moreover, $k \leq 2l$ provided $2l \leq d$ and $k \leq 2d - 2l$ provided $2l > d$. Note that the Euclidean distance between vectors from such a family coincides with the square root of the Hamming distance between them and therefore k is the Hamming diameter of this family.

Now let $\mathcal{F}_{k,l_1,l_2}^d = \{\mathbf{w}_1, \ldots, \mathbf{w}_t\} \subset \mathbb{R}^d$ be an arbitrary family of $(0, 1, -1)$-vectors such that

$$(\operatorname{diam} \mathcal{F}_{k,l_1,l_2}^d)^2 = k$$

and

$$(w_i^1)^2 + \cdots + (w_i^d)^2 = l_1 + l_2 \quad , \quad w_i^1 + \cdots + w_i^d = l_1 - l_2 \text{ for all } i$$

(here $\mathbf{w}_i = (w_i^1, \ldots, w_i^d)$ as before). In other words, $\mathcal{F}_{k,l_1,l_2}^d$ is a subset of the set of vertices and centres of faces of all the intermediate dimensions of the cube $[-1,1]^d$ such that the maximal distance between its elements is equal to $\sqrt{k}$, the number of unit coordinates in each vector $\mathbf{w} \in \mathcal{F}_{k,l_1,l_2}^d$ is l_1, and, moreover, the number of negative components of any vector is l_2. Certainly, this definition also imposes some natural restrictions on the parameters k, l_1, and l_2.

Finally, to each family of vectors $\mathcal{F}_{k,l}^d = \{\mathbf{v}_1, \ldots, \mathbf{v}_s\}$ assign the $(0,1)$-polytope $\Omega_{k,l}^d = \operatorname{conv}\{\mathbf{v}_1, \ldots, \mathbf{v}_s\}$, and to each family of vectors $\mathcal{F}_{k,l_1,l_2}^d = \{\mathbf{w}_1, \ldots, \mathbf{w}_t\}$ assign the *cross-polytope* $\Omega_{k,l_1,l_2}^d = \operatorname{conv}\{\mathbf{w}_1, \ldots, \mathbf{w}_t\}$.

Let $f_{k,l}(d)$ be the minimal number such that every family of vectors $\mathcal{F}_{k,l}^d$ (every $(0,1)$-polytope $\Omega_{k,l}^d$) can be partitioned into $f_{k,l}(d)$ parts of smaller diameter (i.e. of diameter less than $\sqrt{k}$). Similarly, let $f_{k,l_1,l_2}(d)$ be the minimal number such that every family of vectors $\mathcal{F}_{k,l_1,l_2}^d$ (every cross-polytope Ω_{k,l_1,l_2}^d) can be partitioned into $f_{k,l_1,l_2}(d)$ parts of smaller diameter.

Let $\mathcal{H}_{k,l}^d$ be the graph $\mathcal{H}_{k,l}^d = (\mathcal{V}_{k,l}^d, \mathcal{E}_{k,l}^d)$ whose vertex set

$$\mathcal{V}_{k,l}^d = \left\{ \mathbf{v}_1, \ldots, \mathbf{v}_{\binom{d}{l}} \right\}$$

consists of all possible d-dimensional $(0,1)$-vectors with exactly l nonzero coordinates, and whose edge set $\mathcal{E}_{k,l}^d$ consists of all possible pairs $(\mathbf{v}_i, \mathbf{v}_j) \in \mathcal{V}_{k,l}^d \times \mathcal{V}_{k,l}^d$ satisfying the condition $|\mathbf{v}_1 - \mathbf{v}_2|^2 = k$, where the parameters k and l are, as above, subject to natural restrictions.

Similarly, let $\mathcal{H}_{k,l_1,l_2}^d = (\mathcal{V}_{k,l_1,l_2}^d, \mathcal{E}_{k,l_1,l_2}^d)$ be the graph whose vertex set

$$\mathcal{V}_{k,l_1,l_2}^d = \left\{ \mathbf{w}_1, \ldots, \mathbf{w}_{\binom{d}{l_1,l_2}} \right\}$$

consists of all possible d-dimensional $(0,1,-1)$-vectors with exactly l_1 positive and l_2 negative coordinates, and whose edge set $\mathcal{E}_{k,l_1,l_2}^d$ is constructed like the edge set of $\mathcal{H}_{k,l}^d$.

Finally, let $\chi(\mathcal{H}_{k,l}^d)$ and $\chi(\mathcal{H}_{k,l_1,l_2}^d)$ be the chromatic numbers of the graphs $\mathcal{H}_{k,l}^d$ and $\mathcal{H}_{k,l_1,l_2}^d$ (see, e.g. [1, 17, 3, 26] for a detailed discussion of related matters; see also [38] for some results concerning the value of $\chi(\mathcal{H}_{k,l}^d)$).

In this part of the lecture, we use the combinatorial methods elaborated for the set covering problem to obtain upper bounds for $f_{k,l}(d)$ and $f_{k,l_1,l_2}(d)$. These bounds are substantially better for a sufficiently large class of cases than the general estimates of $f(d)$ given in [59] (or, which is almost the same, in [9]). We also find upper bounds for the chromatic numbers $\chi(\mathcal{H}_{k,l}^d)$ and $\chi(\mathcal{H}_{k,l_1,l_2}^d)$.

Note that in small dimensions the same problems have been thoroughly investigated by several authors, including F. Schiller [58], A. Schrijver and C. Payan [44], J. Petersen [46] and G. M. Ziegler [63, 64]. They have succeeded in proving that $f_{k,l}(d) \leq d + 1$ for all $d \leq 9$ and all admissible values of k and l. Moreover, they proved Borsuk's conjecture for $(0,1)$-polytopes in all dimensions $d \leq 9$. However the case of cross-polytopes in small dimensions has never been investigated and it would be very interesting to discuss it somewhere as well (see also [53]).

Now we can proceed to formulating the results.

Statements of results

In this lecture we prove the following four theorems, formulated in terms of the notation introduced above.

It is convenient to introduce the function

$$F(s, x) = \max\left(x, \log \frac{s}{x}\right) + x + 1 \qquad (0 < s, 1 \le x) \qquad (*)$$

to simplify notation in the following estimates.

Theorem 2.5.

$$f_{k,l}(d) \le d \min\left\{ \binom{l}{k/2}, \ F\left(\binom{d}{l}, \binom{d}{l-k/2}\binom{l}{k/2}^{-1}\right)\right\} \qquad (2.1)$$

Theorem 2.6.

$$\chi(\mathcal{H}_{k,l}^d) \le F\left(\binom{d}{l}, \binom{d}{l+1-k/2}\binom{l}{l+1-k/2}^{-1}\right) \qquad (2.2)$$

Theorem 2.7.

$$\begin{aligned}
f_{k,l_1,l_2}(d) &\le \chi(\mathcal{H}_{k,l_1,l_2}^d) \\
&\le \min_{m_1,m_2} F\left(\binom{d}{l_1,l_2}, \binom{d}{m_1,m_2}\binom{l_1}{m_1}^{-1}\binom{l_2}{m_2}^{-1}\right).
\end{aligned} \qquad (2.3)$$

Here the minimum is taken over all natural numbers $m_1 \le l_1$ and $m_2 \le l_2$ such that $|\mathbf{v}_1 - \mathbf{v}_2|^2 < k$, where $\mathbf{v}_1, \mathbf{v}_2 \in \mathbb{R}^{d-m_1-m_2}$ are two arbitrary $(0, 1, -1)$-vectors satisfying the conditions

$$(v_i^1)^2 + \cdots + (v_i^{d-m_1-m_2})^2 = l_1 + l_2 - m_1 - m_2,$$
$$v_i^1 + \cdots + v_i^{d-m_1-m_2} = l_1 - l_2 - m_1 + m_2$$

for $\mathbf{v}_i = (v_i^1, \ldots, v_i^{d-m_1-m_2})$, $i = 1, 2$.

Theorem 2.8. *Let $\mathcal{X} = \{(x_1, x_2)\}$ be the set of all pairs of natural numbers (x_1, x_2) $(x_1 \le l_1, x_2 \le l_2)$ such that a pair of $(0, 1, -1)$-vectors $\mathbf{v}_1, \mathbf{v}_2 \in \mathbb{R}^{d-x_1-x_2}$ satisfying the following restrictions can be found:*

- *if $\mathbf{v}_i = (v_i^1, \ldots, v_i^{d-x_1-x_2})$, $i = 1, 2$, then*

$$(v_i^1)^2 + \cdots + (v_i^{d-x_1-x_2})^2 = l_1 + l_2 - x_1 - x_2 \text{ and}$$
$$v_i^1 + \cdots + v_i^{d-x_1-x_2} = l_1 - l_2 - x_1 + x_2 \ ;$$

- $\{\nu : v_1^\nu = 1\} \cap \{\nu : v_2^\nu = 1\} = \emptyset;$
- $\{\nu : v_1^\nu = -1\} \cap \{\nu : v_2^\nu = -1\} = \emptyset;$
- $|\mathbf{v}_1 - \mathbf{v}_2|^2 \le k.$

Put $m_i = \min_{(x_1,x_2) \in X} x_i$, $i = 1, 2$. Then $f_{k,l_1,l_2}(d)$ is bounded above by

$$\min_{h_1,h_2} F\left(\binom{d-m_1-m_2}{l_1-m_1,\, l_2-m_2}, \binom{d-m_1-m_2}{h_1,\, h_2}\binom{l_1-m_1}{h_1}^{-1}\binom{l_2-m_2}{h_2}^{-1}\right) \qquad (2.4)$$

Here the minimum is taken over all natural numbers $h_1 \le l_1 - m_1$ and $h_2 \le l_2 - m_2$ such that $|\mathbf{w}_1 - \mathbf{w}_2|^2 < k$, where the $(0, 1, -1)$-vectors $\mathbf{w}_1, \mathbf{w}_2 \in \mathbb{R}^{d-m_1-h_1-m_2-h_2}$ must satisfy

$$(w_i^1)^2 + \cdots + (w_i^{d-m_1-h_1-m_2-h_2})^2 = l_1 + l_2 - m_1 - h_1 - m_2 - h_2,$$

$$w_i^1 + \cdots + w_i^{d-m_1-h_1-m_2-h_2} = l_1 - l_2 - m_1 - h_1 + m_2 + h_2.$$

with $\mathbf{w}_i = (w_i^1, \ldots, w_i^{d-m_1-h_1-m_2-h_2})$, $i = 1, 2$.

Remark. Clearly, the estimates (2.1), (2.3) and (2.4) are most interesting when $k \asymp d$, $l \asymp d$ and $l_1 + l_2 \asymp d$. Indeed, if we do not assume any of these conditions, then the cardinality of an arbitrary family of vectors $\mathcal{F}_{k,l}^d$ or $\mathcal{F}_{k,l_1,l_2}^d$ is bounded from above as follows: $s = \operatorname{card} \mathcal{F}_{k,l}^d < e^{cd}$ and $t = \operatorname{card} \mathcal{F}_{k,l_1,l_2}^d < e^{cd}$ for all $c > 0$; and therefore it suffices to consider the trivial bounds $f_{k,l}(d) \le s$ and $f_{k,l_1,l_2}(d) \le t$ in order to obtain an estimate better than that from [59].

Before we proceed to the proof of Theorems 2.5–2.8, we present some examples illustrating the relation between the bounds (2.1)–(2.4) and the estimates from [59].

Examples.
1) Let $d = 4d'$ and $k = l = \frac{d}{2}$. Then the bound (2.1) can be readily transformed into the inequality

$$f_{k,l}(d) \le P_1(d) \min\{(1.414\ldots)^d, (1.240\ldots)^d\}$$

where $P_1(d)$ is a polynomial. In this case (2.1) is worse than the bound $f_{k,l}(d) \le P(d)(1.5)^{d/2} = P(d)(1.224\ldots)^d$ (again $P(d)$ is a polynomial) from the paper [59].

2) Let $d = 12d'$, $k = \frac{d}{6}$ and $l = \frac{d}{3}$. Then (2.1) implies the inequality

$$f_{k,l}(d) \le P_2(d) \min\{(1.206\ldots)^d, (1.454\ldots)^d\}.$$

Consequently in this case the estimate (2.1) already becomes better than the upper bound discovered in [59]. The minimum in (2.1) is attained on the first of the two arguments.

3) Let $d = 12d'$, $k = \frac{d}{2}$ and $l = \frac{d}{3}$. Then the bound (2.1) has the form

$$f_{k,l}(d) \le P_3(d) \min\{(1.206\ldots)^d, (1.104\ldots)^d\}.$$

which is also better than the corresponding bound from [59]. However the minimum in (2.1) is now attained on the second of the two arguments.

4) Let $d = 32d'$, $k = \frac{d}{4}$ and $l_1 = l_2 = \frac{d}{16}$. Put $m_1 = m_2 = \frac{d}{32} + 1$. Then the values m_1 and m_2 satisfy all the conditions of Theorem 2.7, which yields

$$f_{k,l_1,l_2}(d) \leq P_4(d)(1.209\ldots)^d.$$

Thus the bound (2.3) leads to a better result than the inequality from [59]. On the other hand, it is clear that the values m_1 and m_2 in Theorem 2.8 are equal to zero, and therefore in the case under consideration the bound (2.4) coincides with that in (2.3).

5) Finally, let $d = 24d'$ and $k = l_1 = l_2 = \frac{d}{24}$. Then by applying Theorem 2.8 we get

$$f_{k,l_1,l_2}(d) \leq P_5(d)(1.171\ldots)^d,$$

which again is better then the bound from [59]. At the same time, Theorem 2.7 only allows us to show that $f_{k,l_1,l_2}(d) \leq P_5'(d)(1.277\ldots)^d$, which is already worse than the analogous inequality in [59].

The proofs of Theorems 2.5–2.8

In order to prove Theorems 2.5–2.8, we need some additional definitions relating to the set covering problem.

Consider an arbitrary n-element set; for sake of definiteness, put $\Re_n = \{1,\ldots,n\}$. Consider a collection $\mathcal{M} = \{M_1,\ldots,M_s\} \subset 2^{\Re_n}$ of subsets of $\Re_n$ such that card $M_j \geq m$ for all j. A set $S \subset \Re_n$ such that $S \cap M_j \neq \emptyset$ for all j is called a *system of common representatives* (*s.c.r.*) for the collection $\mathcal{M}$. (It is clear that in general such a set $S = S(\mathcal{M})$ is not unique for the collection $\mathcal{M}$.) Put $\tau(\mathcal{M}) = \min \operatorname{card} S(\mathcal{M})$, where the minimum is taken over all s.c.r.'s $S = S(\mathcal{M})$ for the collection of sets $\mathcal{M}$. The value $\tau(\mathcal{M})$ is uniquely defined. For every set of parameters n, m and s, and for each collection $\mathcal{M}$ corresponding to this set, the following upper bound is well known [17, 34, 51]:

$$\tau(\mathcal{M}) \leq \max\left\{\frac{n}{m}\log\frac{sm}{n}, \frac{n}{m}\right\} + \frac{n}{m} + 1. \tag{2.5}$$

Thus in the notation defined above in ($*$), $\tau(\mathcal{M}) \leq F\left(s, \frac{n}{m}\right)$. We are going to use this bound very often in what follows. On the other hand, various lower bounds for the quantity $\tau(\mathcal{M})$ have been established by different authors (see [34, 51], for instance). However we do not present these lower bounds, since we shall not use them in the proofs that follow.

Now we proceed to the proof of Theorem 2.8. The proof is in four steps.

Step 1. First of all, we fix some values of the parameters k, l_1, and l_2 and consider an arbitrary family of vectors $\mathcal{F}^d_{k,l_1,l_2} = \{\mathbf{w}_1, \ldots, \mathbf{w}_t\} \subset \mathbb{R}^d$. To each $\mathbf{w}_i = (w_i^1, \ldots, w_i^d) \in \mathcal{F}^d_{k,l_1,l_2}$ we assign the ordered pair of sets (M_i^1, M_i^2), where $M_i^j = \{\nu \in \Re_d : w_i^\nu = (-1)^{j+1}\}$. We have card $M_i^j = l_j$ and $M_i^1 \cap M_i^2 = \emptyset$. Further, let $\mathcal{M}^d_{k,l_1,l_2} = \{(M_i^1, M_i^2)\}_{i=1}^t$ be the collection of all pairs (M_i^1, M_i^2). Let m_1 and m_2 be the values defined in the statement of the theorem. In the rest of the proof we assume that $m_1 > 0$ and $m_2 > 0$; the cases when at least one of the numbers m_1, m_2 is equal to zero can be considered similarly.

Step 2. Let $\mathcal{K}$ be the collection of all possible ordered pairs of subsets (K_i^1, K_i^2) in $2^{\Re_d}$ such that

$$\operatorname{card} K_i^1 = m_1 \quad , \quad \operatorname{card} K_i^2 = m_2 \quad \text{and} \quad \operatorname{card}(K_i^1 \cap K_i^2) = 0.$$

There is a one-to-one correspondence between $\mathcal{K}$ and the set $\Re_{\binom{d}{m_1,m_2}}$. At the same time, to each pair $(M_i^1, M_i^2) \in \mathcal{M}^d_{k,l_1,l_2}$ assign (in one-to-one fashion) the set

$$\mathcal{L}_i = \left\{\nu \in \Re_{\binom{d}{m_1,m_2}} : K_\nu^1 \subset M_i^1; K_\nu^2 \subset M_i^2\right\}.$$

Hence card $\mathcal{L}_i = \binom{l_1}{m_1}\binom{l_2}{m_2}$. Finally, let

$$\mathfrak{L} = \{\mathcal{L}_i\}_{i=1}^t \subset 2^{\Re_{\binom{d}{m_1,m_2}}}.$$

Note that the definition of the values m_1 and m_2 implies $\mathcal{L}_i \cap \mathcal{L}_j \neq \emptyset$ for all i, j. Indeed, if we fix any two pairs $(M_i^1, M_i^2), (M_j^1, M_j^2) \in \mathcal{M}^d_{k,l_1,l_2}$ and put $x_1 = \operatorname{card}(M_i^1 \cap M_j^1)$ and $x_2 = \operatorname{card}(M_i^2 \cap M_j^2)$, then we get $(x_1, x_2) \in \mathcal{X}$, where $\mathcal{X}$ is the set from the statement of the theorem. This is equivalent to the existence of at least one pair of sets $(K_\nu^1, K_\nu^2) \in \mathcal{K}$ such that $K_\nu^1 \subset (M_i^1 \cap M_j^1)$ and $K_\nu^2 \subset (M_i^2 \cap M_j^2)$. It follows from the definition of the collection of sets $\mathfrak{L}$ that $\mathcal{L}_i \cap \mathcal{L}_j \neq \emptyset$ as claimed.

Therefore each set $\mathcal{L}_i \in \mathfrak{L}$ is an s.c.r. for $\mathfrak{L}$ and consequently†

$$\tau = \tau(\mathfrak{L}) \leq \binom{l_1}{m_1}\binom{l_2}{m_2}.$$

Let $\{\sigma_1, \ldots, \sigma_\tau\}$ be an arbitrary s.c.r. of the minimal cardinality for $\mathfrak{L}$ and let $\{(K^1_{\sigma_1}, K^2_{\sigma_1}), \ldots, (K^1_{\sigma_\tau}, K^2_{\sigma_\tau})\} \subset \mathcal{K}$ be the collection of pairs corresponding to $\sigma_1, \ldots, \sigma_\tau$. Finally, let

$$\mathcal{M}_i = \{(M^1, M^2) \in \mathcal{M}^d_{k, l_1, l_2} : K^1_{\sigma_i} \subset M^1; K^2_{\sigma_i} \subset M^2\};$$

then $\mathcal{M}^d_{k, l_1, l_2} = \mathcal{M}_1 \cup \cdots \cup \mathcal{M}_\tau$ is a covering, since $\{\sigma_1, \ldots, \sigma_\tau\}$ is an s.c.r. for $\mathfrak{L}$.

Step 3. Consider an arbitrary collection of pairs $\mathcal{M}_i$ as defined in the previous part of the proof. For each i let

$$M^{1,*} = M^1 \setminus K^1_{\sigma_i} \quad, \quad M^{2,*} = M^2 \setminus K^2_{\sigma_i} \quad \text{for every } (M^1, M^2) \in \mathcal{M}_i.$$

Put $\Re_{d-m_1-m_2} = \Re_d \setminus (K^1_{\sigma_i} \cup K^2_{\sigma_i})$ and let

$$\mathcal{M}^*_i = \left\{(M^{1,*}, M^{2,*}) \subset 2^{\Re_{d-m_1-m_2}} : (M^1, M^2) \in \mathcal{M}_i\right\}.$$

For each pair $(M^{1,*}, M^{2,*}) \in \mathcal{M}^*_i$, the equalities $\operatorname{card} M^{j,*} = l_j - m_j$, and $\operatorname{card}(M^{1,*} \cap M^{2,*}) = 0$ hold. Fix two arbitrary quantities h_1, h_2 satisfying the restrictions from the statement of the theorem and assume that $h_1 > 0$ and $h_2 > 0$ (the cases when $h_1 h_2 = 0$ can be considered analogously).

Let $\tilde{\mathcal{K}}$ be the collection of all ordered pairs of sets $(\tilde{K}^1_\nu, \tilde{K}^2_\nu)$ in $2^{\Re_{d-m_1-m_2}}$ such that

$$\operatorname{card} \tilde{K}^1_\nu = h_1 \quad, \quad \operatorname{card} \tilde{K}^2_\nu = h_2 \quad \text{and} \quad \operatorname{card}(\tilde{K}^1_\nu \cap \tilde{K}^2_\nu) = 0.$$

Arguing as in the second part of the proof, we can find a one-to-one correspondence between the collection of pairs $\tilde{\mathcal{K}}$ and the set $\Re_{\binom{d-m_1-m_2}{h_1, h_2}}$; at the same time, each pair $(M^{1,*}_\nu, M^{2,*}_\nu) \in \mathcal{M}^*_i$ corresponds (in one-to-one fashion) to the set

$$\tilde{\mathcal{L}}_\nu = \left\{\lambda \in \Re_{\binom{d-m_1-m_2}{h_1, h_2}} : \tilde{K}^1_\lambda \subset M^{1,*}_\nu; \tilde{K}^2_\lambda \subset M^{2,*}_\nu\right\}.$$

We get $\operatorname{card} \tilde{\mathcal{L}}_\nu = \binom{l_1-m_1}{h_1}\binom{l_2-m_2}{h_2}$. Finally, put

$$\tilde{\mathfrak{L}} = \{\tilde{\mathcal{L}}_\nu\}^{\operatorname{card} \mathcal{M}^*_i}_{\nu=1} \subset 2^{\Re\binom{d-m_1-m_2}{h_1, h_2}}.$$

and set

† Note that the bound for $\tau(\mathfrak{L})$ was obtained with the help of a kind of Erdős-Ko-Rado-type argument for systems of common representatives, rather than by the use of the inequality (2.5). See [16].

- $n = \binom{d-m_1-m_2}{h_1,\,h_2}$,
- $m = \binom{l_1-m_1}{h_1}\binom{l_2-m_2}{h_2}$, and
- $s = \operatorname{card}\mathcal{M}_i^* = \operatorname{card}\mathcal{M}_i \le \binom{d-m_1-m_2}{l_1-m_1,\,l_2-m_2}$.

Then (2.5) implies the inequality $\tilde{\tau} = \tau(\tilde{\mathfrak{L}}) \le F\left(s, \frac{n}{m}\right)$, which does not depend on i.

Let $\{\lambda_1, \ldots, \lambda_{\tilde{\tau}}\}$ be an arbitrary s.c.r. of minimal cardinality for the collection of sets $\tilde{\mathfrak{L}}$. Let $\{(\tilde{K}^1_{\lambda_1}, \tilde{K}^2_{\lambda_1}), \ldots, (\tilde{K}^1_{\lambda_{\tilde{\tau}}}, \tilde{K}^2_{\lambda_{\tilde{\tau}}})\} \subset \tilde{\mathcal{K}}$ be the collection of pairs corresponding to the elements $\lambda_1, \ldots, \lambda_{\tilde{\tau}}$. Again, if we let

$$\mathcal{M}_{i\nu}^* = \left\{(M^{1,*}, M^{2,*}) \in \mathcal{M}_i^* : \tilde{K}^1_{\lambda_\nu} \subset M^{1,*};\ \tilde{K}^2_{\lambda_\nu} \subset M^{2,*}\right\},$$

then $\mathcal{M}_i^* = \mathcal{M}_{i1}^* \cup \cdots \cup \mathcal{M}_{i\tilde{\tau}}^*$ is a covering, since $\{\lambda_1, \ldots, \lambda_{\tilde{\tau}}\}$ is an s.c.r. for $\tilde{\mathfrak{L}}$. Finally, consider one more covering $\mathcal{M}_i = \mathcal{M}_{i1} \cup \cdots \cup \mathcal{M}_{i\tilde{\tau}}$, where

$$\mathcal{M}_{i\nu} = \left\{(M^{1,*} \cup K^1_{\sigma_i}, M^{2,*} \cup K^2_{\sigma_i}) \ : \ (M^{1,*}, M^{2,*}) \in \mathcal{M}_{i\nu}^*\right\}.$$

Step 4. Let us summarize the results of Steps 1–3 of the proof. We have constructed the covering $\mathcal{M}_{k,l_1,l_2}^d = \bigcup_i \bigcup_j \mathcal{M}_{ij}$. Fix an arbitrary element $\mathcal{M}_{ij}$ of this covering and an arbitrary pair of sets (M^1, M^2) in $\mathcal{M}_{ij}$. We can reconstruct the vector $\mathbf{w} \in \mathcal{F}_{k,l_1,l_2}^d$ corresponding to the pair of sets (M^1, M^2); let $\mathcal{F}_{ij}$ be the family of all vectors reconstructed in this way from the pairs of sets $(M^1, M^2) \in \mathcal{M}_{ij}$. We get a covering $\mathcal{F}_{k,l_1,l_2}^d = \bigcup_i \bigcup_j \mathcal{F}_{ij}$. The restrictions on h_1 and h_2 from the statement of the theorem imply that if $\mathbf{w}_1, \mathbf{w}_2 \in \mathcal{F}_{ij}$, then $|\mathbf{w}_1 - \mathbf{w}_2|^2 < k$, and therefore $\operatorname{diam}\mathcal{F}_{ij} < \operatorname{diam}\mathcal{F}_{k,l_1,l_2}^d$. In other words, we have constructed, for every $h_1 > 0$ and $h_2 > 0$, a partition† of the family of vectors $\mathcal{F}_{k,l_1,l_2}^d$ into parts of smaller diameter, such that the number of parts is bounded from above by

$$\binom{l_1}{m_1}\binom{l_2}{m_2}F\left(s, \frac{n}{m}\right).$$

Note that in the case when $h_1 h_2 = 0$ a partition of the family $\mathcal{F}_{k,l_1,l_2}^d$ into parts of smaller diameter, with a similar bound on the number of these parts, can also be proposed.

It remains to replace the values n and m in the inequality above by their explicit expressions in terms of $k, l_1, l_2, m_1, m_2, h_1$, and h_2, to bound the quantity s as in the third part of the proof and to take

---

† By having constructed a covering we have also obtained a partition.

the minimum over all possible h_1 and h_2. This implies (2.4) and thus completes the proof. $\qquad\square$

To prove Theorem 2.7, it suffices to repeat all the arguments from parts 1, 2 and 4 of the proof of Theorem 2.8. The difference is that the values m_1 and m_2 must be interpreted as in the statement of Theorem 2.7 and the bound on $\tau(\mathfrak{L})$ is obtained by means of the estimate (2.5).

It remains to prove Theorems 2.5 and 2.6. The methods we have just developed can be used to do this. The necessary arguments are closely connected with Turán's problem concerning set covering (see [17, 32, 34, 60]). Nevertheless we present these arguments in detail, at least for Theorem 2.5. The proof of this theorem is in several places similar to the proof of Theorem 2.8.

Step 1. First fix some values of the parameters k and l and consider an arbitrary family of vectors $\mathcal{F}^d_{k,l} = \{\mathbf{v}_1, \ldots, \mathbf{v}_r\} \subset \mathbb{R}^d$. To each vector $\mathbf{v}_i = (v_i^1, \ldots, v_i^d) \in \mathcal{F}^d_{k,l}$ assign the set $M_i = \{\nu \in \Re_d : v_i^\nu = 1\} \subset \Re_d$. This yields $\operatorname{card} M_i = l$. Finally, let $\mathcal{M}^d_{k,l} = \{M_1, \ldots, M_r\}$ be the collection of all sets M_i.

Step 2. Let $\mathcal{K}$ be the collection of all sets $K_i \subset \Re_d$ such that $\operatorname{card} K_i = l - \frac{k}{2}$. There is a one-to-one correspondence between the collection $\mathcal{K}$ and the set $\Re_{\left(l - k/2\right)}^d$. To each set $M_i \in \mathcal{M}^d_{k,l}$ assign (in one-to-one fashion) the set

$$\mathcal{L}_i = \left\{ \nu \in \Re_{\left(l-k/2\right)}^d : K_\nu \subset M_i \right\}.$$

Thus we get $\operatorname{card} \mathcal{L}_i = \binom{l}{l-k/2}$. Finally, let $\mathfrak{L}$ be the collection of sets

$$\mathfrak{L} = \{\mathcal{L}_i\}_{i=1}^r \subset 2^{\Re_{\left(l-k/2\right)}^d}.$$

The relation $\mathcal{L}_i \cap \mathcal{L}_j \neq \emptyset$ holds for all admissible values of i and j. Therefore each set $\mathcal{L}_i \in \mathfrak{L}$ is an s.c.r. for $\mathfrak{L}$ and consequently

$$\tau = \tau(\mathfrak{L}) \leq \binom{l}{l-k/2}.$$

On the other hand, by putting $n = \binom{d}{l-k/2}$, $m = \binom{l}{l-k/2}$ and $s = r \leq \binom{d}{l}$, we obtain the bound (see (2.5)) $\tau \leq F\left(s, \frac{n}{m}\right)$.

Let $\{\sigma_1, \ldots, \sigma_\tau\}$ be an arbitrary s.c.r. of minimal cardinality for $\mathfrak{L}$. Let $\{K_{\sigma_1}, \ldots, K_{\sigma_\tau}\} \subset \mathcal{K}$ be the collection of sets corresponding to the elements $\sigma_1, \ldots, \sigma_\tau$. Finally, if we let

$$\mathcal{M}_i = \{M \in \mathcal{M}^d_{k,l} : K_{\sigma_i} \subset M\}$$

then $\mathcal{M}_{k,l}^d = \mathcal{M}_1 \cup \cdots \cup \mathcal{M}_\tau$ is a covering, since $\{\sigma_1, \ldots, \sigma_\tau\}$ is an s.c.r. for $\mathfrak{L}$.

Step 3. For an arbitrary collection of sets $\mathcal{M}_i$ defined in the previous part of the proof consider a covering $\mathcal{M}_i = \cup_\nu \mathcal{M}_{i\nu}$, where

$$\mathcal{M}_{i\nu} = \{M \in \mathcal{M}_i : (K_{\sigma_i} \cup \{\nu\}) \subset M\}$$

and $\{\nu\} \cap K_{\sigma_i} = \emptyset$. The number of elements in this covering does not depend of i and is bounded from above by d.

Step 4. To complete the proof of the theorem it remains to repeat almost literally the arguments from the fourth part of the proof of Theorem 2.8. The theorem follows.   □

The proof of Theorem 2.6 coincides almost completely with the proof of Theorem 2.5. It is only necessary to consider the sets $K_\nu \subset \Re_d$ such that $\operatorname{card} K_\nu = l - \frac{k}{2} + 1$, to apply (2.5) in order to bound the value $\tau(\mathfrak{L})$, and to construct the first covering.

Remark. Once can obtain slight improvements of the estimates presented in Theorems 2.5–2.8, by calculating the value $\tau(\mathcal{M})$ more carefully (see [34]). However, all such improvements influence only the relations inside the logarithmic term, so we do not discuss them here.† Note also that the techniques we have developed can be used to deal with particular values of the parameters k, l, l_1 and l_2 (see e.g. [17]).

3 Borsuk's conjecture: the low-dimensional case

In this lecture we mainly consider Borsuk's problem in dimensions 3 and 4. First we describe some old approaches to the proof of the conjecture in the 2- and 3-dimensional cases, namely Eggleston's non-elementary approach, Grünbaum's and Heppes' idea of constructing a *universal cover*, and so on. We then discuss a new method (developed recently by the author and, in its computational part, by Yuri Kalnishkan), which allows us to obtain deep results in dimensions 3 and 4.

---

† The relations inside the logarithmic term can also be diminished by replacing factors of the form $\binom{d}{l}$ with something based on Frankl and Wilson's theorem (see Lecture 1 and [18]), e.g. by $\binom{d-1}{l-k-1}$.

3.1 The history of Borsuk's conjecture in the low-dimensional case

In Lecture 1 we presented the complete history of the Borsuk partition problem, except for the (very important) cases $d = 2$, 3 and 4, since these cases deserve to be considered in a separate lecture. Now we close this gap and discuss low-dimensional aspects of the problem.

Before proceeding directly to various results concerning Borsuk's conjecture for $2 \leq d \leq 4$, note that in the case $d = 1$ everything is evident. Indeed, as in the previous lectures we may restrict ourselves to the case of unit diameter, and any 1-dimensional point set of diameter 1 can be trivially *covered* by a segment of the same diameter (length). Then this segment can be divided into two parts of length $\frac{1}{2}$, which is half as long as the initial diameter. Later we shall see that this idea of covering every set by a *universal* one (e.g. by a segment when $d = 1$) is very useful.

Let us consider the planar case, i.e. the case $d = 2$, which is the next simplest. The conjecture was proved for this case by Borsuk himself [8]. Borsuk's proof uses a simple lemma first discovered by J. Pál [43]:

Lemma 3.1. *Every planar point set of diameter 1 can be inscribed into a regular hexagon such that the distance between its parallel edges (sides) is equal to 1.*

The proof of Lemma 3.1 can be also found in [5] and we shall not give it here. Note that we use an approach similar to that described for $d = 1$: a hexagon plays the part of a '2-dimensional segment', i.e. it is a *universal cover* in the following sense.

Definition 3.2. By a *universal cover in $\mathbb{R}^d$* we mean a set Ω such that any set A of fixed diameter lies in some rigid copy of Ω.

To prove the conjecture, it remains to partition a regular hexagon of the dimensions described in Lemma 3.1 into three parts of diameter less than 1 (and this was actually done by Borsuk). Although the diameter of the hexagon exceeds 1, there exists a natural partition of the hexagon into three parts which each have diameter $\frac{\sqrt{3}}{2} = 0.866\ldots < 1$.

This bound is the best possible: one can easily construct an example of a set $\Omega \subset \mathbb{R}^2$, $\operatorname{diam}\Omega = 1$, that cannot be partitioned into three parts of diameter less than $\frac{\sqrt{3}}{2}$. (For instance one can take a unit circle, whose best partition has already been described in our previous lecture.) In other words, if (see [53]) we let

$$\alpha_2 = \sup_{\Omega \subset \mathbb{R}^2} \inf_{\Omega_1, \ldots, \Omega_3} \max_{i=1,2,3} \operatorname{diam}\Omega_i$$

where the supremum is taken over the sets of diameter 1 and the infimum over all possible partitions of a fixed set Ω into parts of smaller diameter, then $\alpha_2 = \frac{\sqrt{3}}{2}$. Note that the corresponding value α_1 defined for $d = 1$ equals $\frac{1}{2}$ and is attained on a 1-dimensional 'circle', i.e. a segment.

We have seen – at least in the planar case, which is the first non-trivial one – that the idea of constructing an appropriate universal cover is quite fruitful. Apart from proving Borsuk's conjecture, it provides us with some more information, such as the exact value of α_2. However, there exists another, *partial* approach to the case of the Euclidean plane (cf. Lecture 2). This approach deals only with polygons (or with finite point sets), but it is worth mentioning here since we encountered polytopes (in particular, $(0, 1)$-polytopes and cross-polytopes) quite frequently in the previous lectures.

The idea is purely combinatorial and very simple; it was proposed by P. Erdős [15] who proved the following result.

Lemma 3.3. *Every 2-dimensional point set consisting of n points has no more than n different pairs of points realizing the diameter.*

The proof of Borsuk's conjecture for polygons relies on this lemma (see also [25]) and can be done by induction.

The situation in the 3-dimensional case is already much more complicated, and the results are far less satisfactory. J. Perkal [45] and H. G. Eggleston [12] were the first to prove Borsuk's conjecture for $d = 3$, but their proof is essentially non-elementary and provides no bound for the value

$$\alpha_3 = \sup_{\Omega \subset \mathbb{R}^3} \inf_{\Omega_1,\ldots,\Omega_4} \max_{1 \leq i \leq 4} \operatorname{diam} \Omega_i$$

better than $\alpha_3 \leq 1$. In fact, Eggleston generalized Hadwiger's method described in the previous lecture. He considered the *Gauss transformation* g of a 2-dimensional sphere S onto the boundary of an arbitrary (though not necessarily smooth) convex body in $\mathbb{R}^3$. Then he produced a partition $S = S_1 \sqcup \cdots \sqcup S_4$ such that each set $g^{-1}(g(S_i))$ did not contain antipodal points on S and showed (cf. Lecture 2) that $\operatorname{diam} g(S_i) < 1$.

A breakthrough was achieved independently by B. Grünbaum [22] and A. Heppes [27], who found universal covers in $\mathbb{R}^3$ and partitioned them into parts of smaller diameter. The result of Grünbaum was slightly better than that of Heppes: Grünbaum obtained the bound

$$\alpha_3 \leq \frac{\sqrt{6129030 - 937419\sqrt{3}}}{1518\sqrt{2}} = 0.9887\ldots$$

whereas Heppes only succeeded in showing that

$$\alpha_3 \leq \frac{\sqrt{9 + 4\sqrt{3}}}{4} = 0.99775\ldots$$

We shall discuss Grünbaum's result in some detail. As we have already mentioned, this result is based on the construction of a universal cover. The following lemma was proved by D. Gale [19].

Lemma 3.4. *Every set* $\Omega \subset \mathbb{R}^d$ *with diameter 1 can be covered by a regular octahedron in which the distance between parallel faces is equal to 1.*

Grünbaum modified the above statement a little bit. His universal cover is a polyhedron obtained from the one described in Lemma 3.4 by cutting off three pyramids that contain some three of its vertices; more precisely, one should take three pairwise orthogonal planes Π_1, Π_2, Π_3 in $\mathbb{R}^3$ that are parallel to the corresponding central sections of the octahedron and lie at a distance $\frac{1}{2}$ from its centre. If our set of diameter 1 has already been covered by an octahedron, then it lies either entirely 'below' some of the three fixed planes of the form Π_i, or entirely 'above' some of the planes Π'_i that are symmetric† to Π_i. Any choice of three planes with different indices from the set of the six planes $\Pi_i, \Pi'_i, i = 1, 2, \ldots, 6$ gives the same picture (the same polyhedron) and everything is fine.

Lemma 3.4 itself can be found in [5] and [6], for example, so we do not dwell on it. We do not describe Grünbaum's partition either, but in the next section we shall discuss its important modification.

Grünbaum's estimate for α_3 remained unimproved until very recently, when V. Makeev [40, 41] succeeded in showing that $\alpha_3 \leq 0.98$. Instead of an octahedron Makeev used a rhombic dodecahedron with the distance between parallel faces equal to 1; three pairwise orthogonal planes are taken just as was done by Grünbaum. Unlike Grünbaum, Makeev used some computer calculations (due to L. Evdokimov) in order to find a good partition of his universal cover. An important modification of this approach, leading to further improvements of the bound for α_3, will also be described below.

Recall that a ball (i.e. a segment or circle) provides the best lower bounds for α_i, $i \leq 2$. In the case $i = 3$, the best known lower bound is

---

† To be more precise: for each i we define Π'_i to be the plane consisting of all points v such that $-v \in \Pi_i$.

also given by a ball:

$$\alpha_3 \geq \sqrt{\frac{3 + \sqrt{3}}{6}} = 0.888\ldots$$

(cf. Lecture 2). D. Gale [19] even proposed a conjecture stronger than Borsuk's:

every d-dimensional set of diameter 1 can be partitioned into $d+1$ parts, none of whose diameters exceed those appearing in an optimal partition of the unit ball.

Of course this conjecture is not true for all dimensions, but for $d = 3$ the question is still open.

To complete the discussion of the situation in $\mathbb{R}^3$, note that the special case of 3-dimensional polytopes was also considered by A. Heppes and P. Révész [28] who proved the conjecture for this case. Their method used the Euler formula for 3-dimensional polytopes.

As for $d = 4$, until recently there used to be only one result, due to M. Lassak [36]: $f(d) \leq 9$. Certainly 9 is very far from the expected value (even if the conjecture fails in four dimensions). No satisfactory results are known even for the special case of polytopes.

However, very recently the author and Yu. Kalnishkan have developed a new approach to investigating Borsuk's problem in dimensions 3 and 4 (see [54]). We shall describe this approach and discuss its possible far-reaching consequences in the last part of this lecture, where, apart from everything else, we introduce the notion of a *universal covering system* generalizing that of a universal cover.

Further comments on universal covers can be found in [6, 39, 61].

3.2 A modification of Grünbaum's and Makeev's universal covers

As we mentioned in the previous section, the 3-dimensional universal cover found by Gale and Grünbaum can be slightly modified. In this part of the lecture we present this modification and estimate the diameters of the corresponding parts.

Let $\mathcal{O}$ denote Gale's octahedron from Lemma 3.4. Fix a rectangular coordinate system $Oxyz$ in $\mathbb{R}^3$. Let

$$A = \left(\tfrac{\sqrt{3}}{2}, 0, 0\right), \qquad B = \left(0, \tfrac{\sqrt{3}}{2}, 0\right), \qquad C = \left(0, 0, \tfrac{\sqrt{3}}{2}\right),$$
$$A' = \left(-\tfrac{\sqrt{3}}{2}, 0, 0\right), \qquad B' = \left(0, -\tfrac{\sqrt{3}}{2}, 0\right), \qquad C' = \left(0, 0, -\tfrac{\sqrt{3}}{2}\right)$$

be the vertices of $\mathcal{O}$. (It is easy to check that the distance between parallel faces of the octahedron with vertices A, B, C, A', B', C' equals 1.) Thus $\mathcal{O} = \text{conv}\{A, B, C, A', B', C'\}$, where 'conv' denotes the convex hull of a point set.

Let us construct polyhedra Ω^1 and Ω^2 by cutting off some parts of $\mathcal{O}$. To construct Ω^1, take three planes Π_1^1, Π_2^1 and Π_3^1, defined as follows:

- Π_1^1 is parallel to the plane containing the points B, B', C, C', i.e. to the plane Oyz, and lies at distance 0.5 from the origin O on the same side as the vertex A';

- Π_2^1 is parallel to the plane containing the points B, B', A, A', i.e. to the plane Oxy, and lies at distance 0.5 from the origin on the same side as C'; and

- Π_3^1 is parallel to the remaining coordinate plane and lies at distance 0.475 from it on the same side as B'.

The polyhedron Ω^1 is obtained from $\mathcal{O}$ by cutting off the rectangular pyramids with the vertices A', B', and C' by the planes Π_1^1, Π_2^1, and Π_3^1. In order to construct Ω^2 we need six planes, whose definitions are very close to those of Π_i^1: they are also parallel to coordinate planes, but now they go in pairs, so it suffices to specify the distances from O and to specify the vertices that will be cut off by respective planes. Let Π_1^2, Π_2^2, Π_3^2 correspond to the vertices A', B', C' respectively, each with distance 0.5 from O; and let $\Pi_4^2, \Pi_5^2, \Pi_6^2$ correspond to A, B, C respectively, each with distance 0.525.

Lemma 3.5. *Every set $\Omega \subset \mathbb{R}^3$ of diameter 1 can be inscribed either into a rigid copy of Ω^1 or into a rigid copy of Ω^2.*

Note that, strictly speaking, this lemma provides us with two sets such that any other set can be covered at least by one of them rather than with a single universal cover. This is the simplest example of a *universal covering system* (see the next part of the lecture for the exact definition). The proof of the lemma is by generalizing the arguments discussed just after Lemma 3.4.

In order to show that Lemma 3.5 leads to an improvement of Grünbaum's result ($\alpha_3 \leq 0.9887\ldots$), it remains to construct partitions of Ω^1 and Ω^2.

Remark. In the rest of this lecture, for convenience we shall write 0.866 instead of $\sqrt{3}/2$; in a similar way we shall use approximate coordinates of points. This is justified by Remark 3.6 below.

A partition for Ω^1. By cutting off the vertices A', B' and C' from the octahedron $\mathcal{O}$ we created new vertices. Let $A_1, \ldots, A_4$ correspond to A' (i.e. they are obtained by cutting off the rectangular pyramid with vertex A); let $B_1, \ldots, B_4$ correspond to B' and $C_1, \ldots, C_4$ correspond to C'. The coordinates of these new vertices can be easily computed:

$$
\begin{array}{llll}
A_1 = (-0.5, 0, 0.366) & \in A'C, & A_2 = (-0.5, 0.366, 0) & \in A'B, \\
A_3 = (-0.5, 0, -0.366) & \in A'C', & A_4 = (-0.5, -0.366, 0) & \in A'B', \\
B_1 = (0.391, -0.475, 0) & \in B'A, & B_2 = (0, -0.475, 0.391) & \in B'C, \\
B_3 = (-0.391, -0.475, 0) & \in B'A', & B_4 = (0, -0.475, -0.391) & \in B'C', \\
C_1 = (0, 0.366, -0.5) & \in C'B, & C_2 = (0.366, 0, -0.5) & \in C'A, \\
C_3 = (0, -0.366, -0.5) & \in C'B', & C_4 = (-0.366, 0, -0.5) & \in C'A'.
\end{array}
$$

Consider some more points on the boundary of Ω^1:

$$
\begin{array}{ll}
H_1 = (0, 0.458, 0.408) & \in BC, \\
H_2 = (0.433, 0, 0.433) & \in AC, \\
H_3 = (0.408, 0.458, 0) & \in AB, \\
K_1 = (-0.5, 0.105, 0.138) & \in A_1 A_2 A_3 A_4, \\
K_2 = (0.148, -0.475, 0.148) & \in B_1 B_2 B_3 B_4, \\
K_3 = (0.138, 0.105, -0.5) & \in C_1 C_2 C_3 C_4;
\end{array}
$$

and

$$
\begin{array}{ll}
I_1 = (-0.5, 0.173, 0.193) & \in A_1 A_2, \\
I_2 = (0.195, -0.475, 0.196) & \in B_1 B_2, \\
I_3 = (0.193, 0.173, -0.5) & \in C_1 C_2; \\
L_1 = (-0.5, -0.174, 0.192) & \in A_1 A_4, \\
L_2 = (0.2, -0.475, -0.191) & \in B_1 B_4, \\
L_3 = (-0.174, 0.192, -0.5) & \in C_1 C_4; \\
L'_1 = (-0.5, 0.192, -0.174) & \in A_2 A_3, \\
L'_2 = (-0.191, -0.475, 0.2) & \in B_2 B_3, \\
L'_3 = (0.192, -0.174, -0.5) & \in C_2 C_3;
\end{array}
$$

and $G = (0.2887, 0.2887, 0.2887) \in ABC$. (Here $X_1 X_2 \ldots X_k$ is short for $\mathrm{conv}\{X_1, X_2, \ldots, X_k\}$.) Finally, we get $\Omega^1 = \Omega^1_1 \cup \cdots \cup \Omega^1_4$, where the polyhedra Ω^1_i are given by

$$
\begin{aligned}
\Omega^1_1 &= OB_3 B_4 C_3 C_4 A_3 A_4 L'_1 K_1 L_1 L'_2 K_2 L_2 L'_3 K_3 L_3, \\
\Omega^1_2 &= OAB_1 C_2 GH_2 I_2 K_2 L_2 L'_3 K_3 I_3 H_3, \\
\Omega^1_3 &= OBA_2 C_1 GH_3 I_3 K_3 L_3 L'_1 K_1 I_1 H_1, \\
\Omega^1_4 &= OCB_2 A_1 GH_1 I_1 K_1 L_1 L'_2 K_2 I_2 H_2.
\end{aligned}
$$

The diameter of a polytope is attained on its vertices. It thus remains to calculate distances between all vertices of each (fixed) Ω_i^1. The largest does not exceed $0.9842 < 0.9887$ and the first partition is thereby 'good'.

A partition for Ω^2. Recall that in the case of Ω^2 we cut off six vertices of the octahedron $\mathcal{O}$. Therefore we get 24 new vertices $A_1, \ldots, A_4$, $B_1, \ldots, B_4$, $C_1, \ldots, C_4$, $A_1', \ldots, A_4'$, $\ldots$, $C_1', \ldots, C_4'$ that correspond to the vertices of $\mathcal{O}$ denoted using the same letters. The coordinates of the new vertices are

$A_1 = (-0.5, 0, 0.366)$	$\in A'C$,	$A_2 = (-0.5, 0.366, 0)$	$\in A'B$,
$A_3 = (-0.5, 0, -0.366)$	$\in A'C'$,	$A_4 = (-0.5, -0.366, 0)$	$\in A'B'$,
$B_1 = (0.366, -0.5, 0)$	$\in B'A$,	$B_2 = (0, -0.5, 0.366)$	$\in B'C$,
$B_3 = (-0.366, -0.5, 0)$	$\in B'A'$,	$B_4 = (0, -0.5, -0.366)$	$\in B'C'$,
$C_1 = (0, 0.366, -0.5)$	$\in C'B$,	$C_2 = (0.366, 0, -0.5)$	$\in C'A$,
$C_3 = (0, -0.366, -0.5)$	$\in C'B'$,	$C_4 = (-0.366, 0, -0.5)$	$\in C'A'$,
$A_1' = (0.525, 0, 0.341)$	$\in AC$,	$A_2' = (0.525, 0.341, 0)$	$\in AB$,
$A_3' = (0.525, 0, -0.341)$	$\in AC_2$,	$A_4' = (0.525, -0.341, 0)$	$\in AB_1$,
$B_1' = (-0.341, 0.525, 0)$	$\in A_2 B$,	$B_2' = (0, 0.525, 0.341)$	$\in BC$,
$B_3' = (0.341, 0.525, 0)$	$\in AB$,	$B_4' = (0, 0.525, -0.341)$	$\in BC_1$,
$C_1' = (0, 0.341, 0.525)$	$\in BC$,	$C_2' = (0.341, 0, 0.525)$	$\in AC$,
$C_3' = (0, -0.341, 0.525)$	$\in B_2 C$,	$C_4' = (-0.341, 0, 0.525)$	$\in A_1 C$.

We need some more points from the boundary of the set Ω^2. Let A_{23}, B_{23}, C_{23} denote the midpoints of the line segments $A_2 A_3$, $B_2 B_3$, $C_2 C_3$ respectively; the points A_{14}, B_{14}, C_{14}, A_{12}', A_{34}', B_{14}', B_{23}', C_{12}' and C_{34}' are defined similarly. Finally, consider the points

$$\begin{aligned} P_1 &= (-0.30175, 0.2625, -0.30175) & &\in A_2 A_3 C_4 C_1 B_1' B_4' \,, \\ P_2 &= (0.2887, 0.2887, 02887) & &\in C_1' C_2' B_2' B_3' A_1' A_2' \,, \\ P_3 &= (0.2625, -0.30175, -0.30175) & &\in A_3' A_4' B_1 B_4 C_3 C_2 \,, \\ P_4 &= (-0.30175, -0.30175, 0.2625) & &\in B_2 B_3 A_4 A_1 C_3' C_4' \,; \end{aligned}$$

these points belong to the appropriate hexagonal faces of Ω^2. We now get the partition $\Omega^2 = \Omega_1^2 \cup \cdots \cup \Omega_4^2$, where

$$\begin{aligned} \Omega_1^2 &= OC_3 C_4 A_3 A_4 B_3 B_4 C_{23} C_{14} P_1 A_{23} A_{14} P_4 B_{23} B_{14} P_3, \\ \Omega_2^2 &= OC_1 C_2 A_3' A_2' B_3' B_4' C_{23} C_{14} P_1 B_{14}' B_{23}' P_2 A_{12}' A_{34}' P_3, \\ \Omega_3^2 &= OB_1' B_2' C_1' C_4' A_1 A_2 A_{23} A_{14} P_4 C_{34}' C_{12}' P_2 B_{23}' B_{14}' P_1, \\ \Omega_4^2 &= OC_2' C_3' B_2 B_1 A_4' A_1' B_{23} B_{14} P_3 A_{34}' A_{12}' P_2 C_{12}' C_{34}' P_4. \end{aligned}$$

One can easily check that none of the diameters diam Ω_i^2 exceeds the value $0.9836 < 0.9842 < 0.9887$. This completes the construction of a 'good' partition for Ω^2.

Thus we obtain the bound $\alpha_3 < 0.9842$.

Remark 3.6. Although some coordinates in the above constructions were specified approximately, the overall error is of the order 10^{-4}; this does not affect our bound for α_3.

Remark 3.7. Of course, our choice of parameters was not optimal. At the same time, it should be possible to generalize our approach. The upper bound for α_3 can definitely be improved. On the other hand, Makeev's method (octahedron $\longrightarrow$ rhombic dodecahedron) can be modified too. This would imply estimates even better than $\alpha_3 < 0.98$. The author did not perform the corresponding calculations on a computer, but this direction of research looks really promising; it is natural to expect an estimate like $\alpha_3 < 0.975$ or even $\alpha_3 < 0.97$.

3.3 An approach due to the author and Yu. A. Kalnishkan.

We start by introducing a notion of a d-dimensional universal covering system. As usual, only sets $\Omega \subset \mathbb{R}^d$ of diameter 1 are considered.

Definition 3.8. A (possibly infinite, even uncountable) collection of sets $\mathfrak{U} = \{U\}$, $U \subset \mathbb{R}^d$, is said to be a *universal covering system* if for any set $\Omega \subset \mathbb{R}^d$ of diameter 1 there exists a (not necessarily unique) set U such that $\Omega \subset U'$, where U' is a rigid copy of U.

A trivial example of a universal covering system is provided by the collection of all sets of unit diameter. On the other hand, we have already encountered universal covering systems consisting of a single set U. We also constructed a nontrivial example of a two-element covering system. Clearly, the main problem is in finding universal covering systems $\mathfrak{U}$, which are as 'good' as possible, such that every $U \in \mathfrak{U}$ admits a partition into parts of smaller diameter.

Before proceeding to the construction of $\mathfrak{U}$ proposed by the author in his joint work with Yuri Kalnishkan, note that the diameters of the sets U from a universal covering system can be substantially greater than 1 (see, e.g. the systems from the previous parts of the lecture).

We start by constructing a partition for an arbitrary d, and then concentrate on the cases $d = 3$ and 4 where the construction is most effective. Let $r_d = \sqrt{d/2(d+1)}$. Fix r such that $\frac{1}{2} \leq r \leq r_d$ and let $B_r = B_r(O)$ be the ball of radius r with the centre at the origin O. Note that $\frac{1}{2} = r_1 < r_2 < r_3 < \cdots < r_d \to \frac{\sqrt{2}}{2}$, $d \to \infty$.

Consider the following inductive procedure. First take an arbitrary point on the sphere corresponding to the ball B_r: $\mathbf{x}_1 \in \partial B_r$. Secondly, consider the segment $O\mathbf{x}_1$ and the plane Π_1 which is orthogonal to this segment and contains the origin: the plane Π_1 divides the sphere ∂B_r into two (closed) hemispheres. Thirdly, take an arbitrary point $\mathbf{x}_2 \in \partial B_r$ lying on the (closed) hemisphere that does not contain $\mathbf{x}_1$ and such that $|\mathbf{x}_2 - \mathbf{x}_1| \leq 1$. This is the base of the inductive procedure.

Now suppose that the points $\mathbf{x}_1, \ldots, \mathbf{x}_{k-1} \in \partial B_r$ have already been constructed. Let Π_{k-1} be the plane containing the origin and parallel to the plane Π'_{k-1} passing through $\mathbf{x}_1, \ldots, \mathbf{x}_{k-1}$. Two situations may occur: either $\Pi_{k-1} = \Pi'_{k-1}$ or not. In the first case we terminate the procedure and claim that a set of points $\mathbf{x}_1, \ldots, \mathbf{x}_{k-1}$ has been constructed. Otherwise take an arbitrary point $\mathbf{x}_k \in \partial B_r$ satisfying

$$\max_{i=1,\ldots,k-1} |\mathbf{x}_i - \mathbf{x}_k| \leq 1,$$

and lying on the (closed) hemisphere that is obtained by intersecting B_r with Π_{k-1} and which does not contain any of the points $\mathbf{x}_i$, $i = 1, \ldots, k-1$. The procedure can terminate on any $k \leq d+1$, but, of course, it cannot pass through $k = d$ since $\Pi_{d+1} = \Pi'_{d+1} = \mathbb{R}^d$.

Consider an arbitrary set of points $\mathbf{x}_1, \ldots, \mathbf{x}_k$, $k \leq d+1$ constructed according to the procedure above and let $B_1(\mathbf{x}_i)$, $i = 1, \ldots, k$ be the balls of radii 1 with centres at the points $\mathbf{x}_i$. Finally, put

$$\Omega_r(\mathbf{x}_1, \ldots, \mathbf{x}_k) = B_r \cap B_1(\mathbf{x}_1) \cap \cdots \cap B_1(\mathbf{x}_k)$$

and let $\mathfrak{U} = \{\Omega_r(\mathbf{x}_1, \ldots, \mathbf{x}_k)\}$, where r ranges from $\frac{1}{2}$ to r_d and the $\mathbf{x}_i$ range over all possible outcomes of the inductive procedure (note that k also varies; see Remark 3.10 below).

Lemma 3.9. *The collection of sets $\mathfrak{U}$ is a universal covering system.*

Proof. We must show that for every set $\Omega \subset \mathbb{R}^d$ of diameter 1 there is a set $U \in \mathfrak{U}$ such that $\Omega \subset U$. In this proof $B(\Omega)$ is the ball of smallest radius circumscribed around Ω and $r(B(\Omega))$ is the radius of this ball.

Fix an arbitrary Ω. Then by Jung's theorem [30], $\frac{1}{2} \leq r(B(\Omega)) \leq r_d$. Put $r = r(B(\Omega))$ and $B_r = B(\Omega)$ (we may move the coordinate system appropriately). Since B_r is the minimal ball circumscribed around Ω, the well-known Helly theorem implies that there is a simplex $R \subset \Omega$ of an intermediate dimension $l \leq d$ such that B_r is minimal for this simplex as well (see [11] for details concerning Helly's theorem). Therefore R contains the origin O, and some m of the vertices of R lie on the sphere ∂B_r. In other words, for any plane passing through the origin, there exist vertices of R in each of the two corresponding (closed) hemispheres.

Hence if we denote the vertices of R that lie on the sphere ∂B_r by $\mathbf{x}_1, \ldots, \mathbf{x}_m$ and recall the inductive procedure, then we obtain

$$\Omega \subset U = \Omega_r(\mathbf{x}_1, \ldots, \mathbf{x}_k),$$

where k does not exceed m and the distances among the vectors $\mathbf{x}_1, \ldots, \mathbf{x}_k$ are less than or equal to 1, since $\operatorname{diam} R \leq \operatorname{diam} \Omega = 1$. Moreover, $\Omega \subset B_1(\mathbf{x}_i)$, $i = 1, \ldots, k$, for $\mathbf{x}_i \in R \subset \Omega$ and $\operatorname{diam} \Omega = 1$. The lemma follows.   □

Remark 3.10. The main advantage of the procedure is that k is large enough provided r is large enough. For example, if $r > r_{d-1}$ then $k \geq d$. The distances between the points $\mathbf{x}_1, \ldots, \mathbf{x}_k$ constructed in the procedure are also large enough. This is very important, since the 'quality' of the resulting partition depends directly on the number of intersecting balls and on the pairwise distances between their centres.

The universal covering system $\mathfrak{U}$ from Lemma 3.9 is uncountable and is therefore rather difficult to work with (i.e. to partition *every* set U that belongs to it). However the author has proposed a method of making the corresponding inductive procedure discrete, at least for $d = 3$ and 4. It is possible to choose the points not from the whole boundary of B_r but, roughly speaking, from some ε-nets defined on it for every fixed $\varepsilon > 0$, and to finally obtain some new finite universal systems $\mathfrak{U}_\varepsilon = \{U_\varepsilon\}$. The author has also proposed some refinements of the inductive procedure and 'good' partitions of the corresponding sets U_ε. Yu. Kalnishkan wrote a program calculating the diameters. It appears unreasonable to include technical details of this joint work, since they are rather cumbersome and do not reflect the essence of the approach. We discuss only some implications of the method, which are arguably the most interesting.

Theorem 3.11. *If $\Omega \subset \mathbb{R}^3$ is such that $r(B(\Omega)) \in [\frac{1}{2}, \delta_1] \cup [\delta_2, r_3]$, where $\delta_1 < \delta_2$, then Ω can be partitioned into four parts of diameter $\leq \alpha_3(\delta_1, \delta_2)$ and*

$$\lim_{\delta_1 \to \frac{1}{2}, \delta_2 \to r_3} \alpha_3(\delta_1, \delta_2) = \sqrt{\frac{3 + \sqrt{3}}{6}} = 0.888\ldots,$$

Here $\alpha_3(\delta_1, \delta_2)$ is defined similar to the quantity α_3 except for the first supremum in the definition, which is taken over all possible sets whose smallest covering balls have radii from the aforementioned union of intervals. In particular, if $\delta_1 \leq 0.53$ and $\delta_2 \geq 0.6$ then $\alpha_3(\delta_1, \delta_2) \leq 0.96$. (Compare this with Gale's conjecture from the end of §3.1.)

Remark 3.12. The result of Theorem 3.11 seems rather astonishing. Indeed, a substantial improvement of the bound $\alpha_3 \leq 0.98$ has been obtained for the cases when the radius of the minimal ball for Ω is either small enough or large enough.

However if we take into account Remark 3.10, this becomes less strange. If r is large, then k is also large, so that all the corresponding sets can be 'well-partitioned' (in our example $r \geq 0.6 > r_2 = \frac{\sqrt{3}}{2} = 0.577\ldots$ and Remark 3.10 implies that $k \geq 3$). If r is small, then even the minimal ball B_r can be 'well-partitioned'. Note that all these properties only hold provided our method is used.

Remark 3.13. Note that our method is in fact based on the use of the so-called (*generalized*) *Reuleaux polytopes* (see [55] for definitions). For instance, the method allows us to partition the ordinary 3-dimensional Reuleaux polytope into four parts of diameter roughly ≤ 0.81. Thus once again Helly's theorem implies that every $\Omega \subset \mathbb{R}^3$ with $r(B(\Omega)) = r_3$ can be partitioned into four parts of diameter ≤ 0.81. On the one hand, this means that one could even replace the limit $\delta_2 \to r_3$ in Theorem 3.11 by a bound $c < \delta_2 \leq r_3$ (in our method, the bounds are monotone in $r > c$ for some $c < r_3$). On the other hand, this leads to a partial strengthening of Gale's conjecture:

if an arbitrary $\Omega \subset \mathbb{R}^3$ satisfies the inequalities $c < r(B(\Omega)) \leq r_3$, then it can be partitioned into four parts of diameters strictly less than the diameters of the parts from optimal partition of a three-sphere.

Remark 3.14. In the case of $d = 4$, one can improve the bound $f(d) \leq 9$ due to M. Lassak [36]. Unfortunately the method does not lead to the equality $f(4) = 5$, but on some intervals similar to those appearing in Theorem 3.11 one can actually bound the number of parts of smaller diameter for any appropriate set Ω by 5.

We would like to finish this lecture by emphasizing that we strongly believe that Borsuk's conjecture is true for $\mathbb{R}^4$.

Acknowledgments.

This work was supported by the Russian Foundation for Basic Research through the grants 02-01-00912 and 00-15-96109. It is a great pleasure for me to point out how grateful I am to V. G. Boltyanski and H. Martini for their invaluable help, particularly in obtaining a copy of their wonderful book [6]. I am also very grateful to G. M. Ziegler for his constant support of my work. Furthermore, I would like to thank E. Makai for drawing my attention to [40] and [41], and L. Danzer for drawing my attention to [33]. Finally, I am especially grateful to N. G. Moshchevitin, N. P. Dolbilin and S. V. Konyagin for their constant attention to my research and for numerous fruitful discussions, and to I. Bárány for his constant help as well as for his very useful comments and suggestions concerning this article.

Bibliography

[1] M. Aigner and G. M. Ziegler. *Proofs from The Book*. Springer-Verlag, Berlin, 1999. Corrected reprint of the 1998 original.

[2] N. Alon, L. Babai, and H. Suzuki. Multilinear polynomials and Frankl-Ray-Chaudhuri-Wilson type intersection theorems. *J. Combin. Theory Ser. A*, 58(2):165–180, 1991.

[3] N. Alon and J. H. Spencer. *The probabilistic method*. Wiley-Interscience [John Wiley & Sons], New York, second edition, 2000.

[4] L. Babai and P. Frankl. *Linear algebra methods in combinatorics*. Dept. of Computer Science, University of Chicago. Preliminary version #2, 1992.

[5] V. G. Boltyanski and I. T. Gohberg. *Decomposition of figures into smaller parts*. Popular Lectures in Mathematics. Univ. of Chicago Press, Chicago-London, 1980. Translated from the Russian edition by Henry Christoffers and Thomas P. Branson.

[6] V. G. Boltyanski, H. Martini, and P. S. Soltan. *Excursions into combinatorial geometry*. Universitext. Springer-Verlag, Berlin, 1997.

[7] K. Borsuk. Über die Zerlegung einer euklidischen n-dimensionalen Vollkugel in n mengen. In *Verh. Internat Math. Kongr., Zürich*, volume 2, pages 192–193, 1932.

[8] K. Borsuk. Drei Sätze über die n-dimensionale euklidische Sphäre. *Fundamenta Math.*, 20:177–190, 1933.

[9] J. Bourgain and J. Lindenstrauss. On covering a set in $\mathbb{R}^N$ by balls of the same diameter. In *Geometric aspects of functional analysis (1989–90)*, volume 1469 of *Lecture Notes in Math.*, pages 138–144. Springer, Berlin, 1991.

[10] J. H. Conway and N. J. A. Sloane. *Sphere packings, lattices and groups*, volume 290 of *Grundlehren der Mathematischen Wissenschaften*. Springer-Verlag, New York, third edition, 1999.

[11] L. Danzer, B. Grünbaum, and V. Klee. Helly's theorem and its relatives. In *Proc. Sympos. Pure Math., Vol. VII*, pages 101–180. Amer. Math. Soc., Providence, R.I., 1963.

[12] H. G. Eggleston. Covering a three-dimensional set with sets of smaller diameter. *J. London Math. Soc.*, 30:11–24, 1955.

[13] H. G. Eggleston. *Convexity*, volume 47 of *Cambridge Tracts in Mathematics*. Cambridge University Press, New York, 1958.

[14] P. Erdős. My Scottish book 'problems'. In *The Scottish Book: Mathematics from the Scottish Café*, pages 35–43. Birkhäuser Boston, Mass., 1981.

[15] P. Erdős. On sets of distances of n points. *Amer. Math. Monthly*, 53:248–250, 1946.

[16] P. Erdős, C. Ko, and R. Rado. Intersection theorems for systems of finite sets. *Quart. J. Math. Oxford Ser. (2)*, 12:313–320, 1961.

[17] P. Erdős and J. Spencer. *Probabilistic methods in combinatorics*. Academic Press, New York–London, 1974. Probability and Mathematical Statistics, Vol. 17.

[18] P. Frankl and R. M. Wilson. Intersection theorems with geometric consequences. *Combinatorica*, 1(4):357–368, 1981.

[19] D. Gale. On inscribing n-dimensional sets in a regular n-simplex. *Proc. Amer. Math. Soc.*, 4:222–225, 1953.

[20] M. L. Gerver. On partitioning sets into parts of smaller diameter: theorems and counterexamples. *Mat. Prosveshcheniye, Ser. 3*, (3), 1999. In Russian.

[21] J. Grey and B. Weissbach. Ein weiteres Gegenbeispiel zur Borsukschen Vermutung. Univ. Magdeburg, Fakultät für Mathematik, Preprint 25, 1997.

[22] B. Grünbaum. A simple proof of Borsuk's conjecture in three dimensions. *Proc. Cambridge Philos. Soc.*, 53:776–778, 1957.

[23] B. Grünbaum. Borsuk's problem and related questions. In *Proc. Sympos. Pure Math., Vol.* VII, pages 271–284. Amer. Math. Soc., Providence, R.I., 1963.

[24] H. Hadwiger. Überdeckung einer Menge durch Mengen kleineren Durchmessers. *Comment. Math. Helv.*, 18:73–75, 1945. Cf. *ibid.* 19:72–73, 1946.

[25] H. Hadwiger and H. Debrunner. *Combinatorial geometry in the plane.* Holt, Rinehart and Winston, New York, 1964. Translated by Victor Klee. With a new chapter and other additional material supplied by the translator.

[26] F. Harary. *Graph theory.* Addison-Wesley Publishing Co., Reading, Mass.-Menlo Park, Calif.-London, 1969.

[27] A. Heppes. On the partitioning of three-dimensional point-sets into sets of smaller diameter. *Magyar Tud. Akad. Mat. Fiz. Oszt. Közl.*, 7:413–416, 1957.

[28] A. Heppes and P. Révész. Zum Borsukschen Zerteilungsproblem. *Acta Math. Acad. Sci. Hung.*, 7:159–162, 1956.

[29] A. Hinrichs. Spherical codes and Borsuk's conjecture. Preprint, 2001.

[30] H. W. E. Jung. Über die kleinste Kugel, die eine räumliche Figur einschliesst. *J. Reine Angew. Math.*, 123:241–257, 1901.

[31] J. Kahn and G. Kalai. A counterexample to Borsuk's conjecture. *Bull. Amer. Math. Soc. (N.S.)*, 29(1):60–62, 1993.

[32] G. Katona, T. Nemetz, and M. Simonovits. On a problem of Turán in the theory of graphs. *Mat. Lapok*, 15:228–238, 1964.

[33] P. Katzarowa-Karanowa. Über ein euklidisch-geometrisches Problem von B. Grünbaum. *Arch. Math. (Basel)*, 18:663–672, 1967.

[34] N. N. Kuzjurin. Asymptotic investigation of the set covering problem. *Problemy Kibernet.*, 37:19–56, 1980. In Russian.

[35] D. Larman. Open problem 6. In M. Rosenfeld and J. Zaks, editors, *Convexity and graph theory (Jerusalem, March 16–20, 1981)*, volume 20 of *Annals of Discrete Mathematics*, pages 335–339. North-Holland Publishing Co., Amsterdam, 1984.

[36] M. Lassak. An estimate concerning Borsuk partition problem. *Bull. Acad. Polon. Sci. Sér. Sci. Math.*, 30:449–451, 1982.

[37] H. Lenz. Zur Zerlegung von Punktmengen in solche kleineren Durchmessers. *Arch. Math.*, 6:413–416, 1955.

[38] N. Linial, R. Meshulam, and M. Tarsi. Matroidal bijections between graphs. *J. Combin. Theory Ser. B*, 45(1):31–44, 1988.

[39] V. V. Makeev. Universal coverings and projections of bodies of constant width. *Ukrain. Geom. Sb.*, 32:84–88, 1989. Translation in *J. Soviet Math.*, 59(2):750–752, 1992.

[40] V. V. Makeev. Affinely-inscribed and affinely-circumscribed polygons and polyhedra. *Zap. Nauchn. Sem. (POMI)*, 231:286–298, 1995. Translation in *J. Math. Sci. (New York)*, 91(6):3518–3525, 1998.

[41] V. V. Makeev. On affine images of a rhombic dodecahedron circumscribed about a three-dimensional convex body. *Zap. Nauchn. Sem. (POMI)*, 246:191–195, 1997. Translation in *J. Math. Sci. (New York)*, 100(3):2307–2309, 2000.

[42] A. Nilli. On Borsuk's problem. In *Jerusalem Combinatorics '93*, volume 178 of *Contemp. Math.*, pages 209–210. Amer. Math. Soc., Providence, RI, 1994.

[43] J. Pál. Über ein elementares Variationsproblem. *Danske Videnskab. Selskab. (Math.-Fys. Medd.)*, 3(2), 1920.

[44] C. Payan. On the chromatic number of cube-like graphsr. *Discrete Math.*, 103:271–277, 1992.

[45] J. Perkal. Sur la subdivision des ensembles en parties de diamètre inférieur. *Colloq. Math.*, 1:45, 1947.

[46] J. Petersen. Färbung von Borsuk-Graphen in niedriger Dimension. Diplomarbeit, TU Berlin, 1998.

[47] K. Prachar. *Primzahlverteilung*. Springer-Verlag, Berlin, 1957.

[48] A. M. Raĭgorodskiĭ. On dimensionality in the Borsuk problem. *Uspekhi Mat. Nauk*, 318:181–182, 1997. Translation in *Russian Math. Surveys*, 52(6):1324–1325, 1997.

[49] A. M. Raĭgorodskiĭ. On a bound in the Borsuk problem. *Uspekhi Mat. Nauk*, 326:185–186, 1999. Translation in *Russian Math. Surveys*, 54(2):453–454, 1999.

[50] A. M. Raĭgorodskiĭ. On two combinatorial problems (Borsuk's partition problem, and the defect of good admissible sets in a lattice). In *Paul Erdős and his mathematics (Budapest, 1999)*, pages 210–213. János Bolyai Math. Soc., Budapest, 1999.

[51] A. M. Raĭgorodskiĭ. Systems of common representatives. *Fundam. Prikl. Mat.*, 5(3):851–860, 1999. In Russian.

[52] A. M. Raĭgorodskiĭ. The Borsuk problem for (0, 1)-polyhedra and cross polytopes. *Dokl. Akad. Nauk*, 371(5):600–603, 2000. In Russian.

[53] A. M. Raĭgorodskiĭ. The Borsuk problem and the chromatic numbers of some metric spaces. *Uspekhi Mat. Nauk*, 337:107–146, 2001. Translation in *Russian Math. Surveys*, 56(1):103–139, 2001.

[54] A. M. Raĭgorodskiĭ and Yu. A. Kalnishkan. On Borsuk's problem, IV. Preprint.

[55] F. Reuleaux. *The kinematics of machinery*. Dover, New York, 1963. Reprint of 1876 translation.

[56] A. S. Riesling. Borsuk's problem in three-dimensional spaces of constant curvature. *Ukr. Geom. Sbornik*, 11:78–83, 1971. In Russian.

[57] C. A. Rogers. Symmetrical sets of constant width and their partitions. *Mathematika*, 18:105–111, 1971.

[58] F. Schiller. Zur Berechnung und Abschätzung von Färbungszahlen und der ϑ-Funktion von Graphen. Diplomarbeit, TU Berlin, 1999.

[59] O. Schramm. Illuminating sets of constant width. *Mathematika*, 35:180–189, 1988.

[60] P. Turán. Egy grafelmeletiszelsoertek feladatrol. *Mat. Fiz. Lapok*, 48:436–452, 1941.

[61] B. Weissbach. Polyhedral covers. In *Intuitive geometry (Siófok, 1985)*, volume 48 of *Colloq. Math. Soc. János Bolyai*, pages 639–646. North-Holland Publishing Co., Amsterdam, 1987.

[62] B. Weißbach. Sets with large Borsuk number. *Beiträge Algebra Geom.*, 41(2):417–423, 2000.

[63] G. M. Ziegler. Lectures on 0/1-polytopes. In *Polytopes–combinatorics and computation (Oberwolfach, 1997)*, volume 29 of *DMV Sem.*, pages 1–41. Birkhäuser, Basel, 2000.

[64] G. M. Ziegler. Coloring Hamming graphs, optimal binary codes, and the 0/1-Borsuk problem in low dimensions. In *Computational discrete mathematics*, volume 2122 of *Lecture Notes in Comput. Sci.*, pages 159–171. Springer, Berlin, 2001. Cf. TU Berlin preprint (June 2000).

Department of Mathematical Statistics and Random Processes
Faculty of Mechanics and Mathematics
Moscow State University
Leninskie Gory
Moscow 119992
Russia
MRaigor@yandex.ru

Embedding and knotting of manifolds in Euclidean spaces

Arkadiy B. Skopenkov

Introduction

Acknowledgements

This survey is based on lectures the author has given at various times at the Independent University of Moscow, Moscow State University, the Steklov Mathematical Institute (Moscow and St. Petersburg branches), the Technical University of Berlin, the Ruhr University of Bochum, the Lorand Eötvos University of Budapest, the University of Geneva, the University of Heidelberg, the University of Ljubljana, the University of Siegen, the University of Uppsala, the University of Warsaw and the University of Zagreb. He gratefully acknowledges the support provided by INTAS grant no. YSF-2002-393, by the Russian Foundation for Basic Research, grants nos. 05-01-00993, 04-01-00682 and 06-02-72551-NCNILa, President of Russian Federation grants MD-3938.2005.1, MD-4729.2007.1 and NSH-4578.2006.1, and by the Pierre Deligne fund based on his 2004 Balzan prize in mathematics.

The preliminary version was prepared in January 2002 after a series of lectures at the Universities of Aberdeen, Cambridge, Edinburgh and Manchester, sponsored by the London Mathematical Society via the programme 'Invitation of young Russian mathematicians'. The author would like to acknowledge all these institutions for their hospitality and personally thank P. M. Akhmetiev, V. M. Buchstaber, A. V. Chernavskiy, Y. Choi, P. Eccles, K. E. Feldman, A. T. Fomenko, U. Koschorke, M. Kreck, W. B. Lickorish, A. Haefliger, R. Levy, S. Mardesic, A. S. Mischenko, N. Yu. Netsvetaev, V. M. Nezhinskiy, M. M. Postnikov, E. Rees, D. Repovs, E. V. Schepin, Yu. P. Solovyov, A. Szücs, V. A. Vassiliev, O. Ya. Viro, C. Weber, M. Weiss, G. Ziegler and H. Zieschang for their invitations and useful discussions. It is a pleasure to express special gratitude to P. Eccles for his many remarks on the preliminary version of this paper.

Embedding and knotting problems

Many theorems in mathematics state that an arbitrary object of a given abstractly defined class is a subobject of a certain 'standard' object of this class. Examples include:

- the Cayley theorem on embedding finite groups into the symmetric groups;
- the representation of any compact Lie group as a subgroup of $GL(V)$ for a certain linear space V;
- the Urysohn theorem on embedding of normal spaces with countable basis into Hilbert space;
- the general position theorem on embedding of finite polyhedra into $\mathbb{R}^m$;
- the Menger–Nöbeling–Pontryagin theorem on embedding finite-dimensional compact spaces into $\mathbb{R}^m$;
- the Whitney theorem on embedding smooth manifolds into $\mathbb{R}^m$;
- the Nash theorem on embedding Riemannian manifolds into $\mathbb{R}^m$;
- the Gromov theorem on embedding symplectic manifolds into $\mathbb{R}^{2n}$.

The solution of the 13th Hilbert problem by Kolmogorov and Arnold can also be formulated in terms of embeddings.

Although interesting in themselves, these embeddability theorems also prove to be powerful tools for solving other problems. Subtler problems about the embeddability of a given space into $\mathbb{R}^m$ for a *given m*, as well as about counting such embeddings, are among the most important classical problems of topology.

According to Zeeman, the classical problems of topology are the following.

1) *The homeomorphism problem*: when are two given spaces homeomorphic?
2) *The embedding problem*: when does a given space embed into $\mathbb{R}^m$?
3) *The knotting problem*: when are two given embeddings isotopic?

The definitions of *embedding* and *isotopy* are recalled in the next subsection.

The embedding and knotting problems have played an outstanding role in the development of topology. Various methods for the investigation of these problems were created by such classical figures as G. Alexander, H. Hopf, E. van Kampen, K. Kuratowski, S. MacLane, L. S. Pontryagin, R. Thom, H. Whitney, W. Browder, A. Haefliger, M. Hirsch, J. F. P. Hudson, M. Irwin, J. Levine, S. Novikov, R. Penrose, J. H. C. Whitehead, E. C. Zeeman and others. For surveys see [214, Introduction] and [55, 1, 151, 153]. Nowadays interest in this subject is reviving.

There are only a few cases in which a concrete answer to the embedding and knotting problems can be given. For the best known specific case of the knotting problem, namely the theory of codimension 2 embeddings (in particular, for the classical theory of knots in $\mathbb{R}^3$), a complete concrete classification is neither known nor expected. A concrete complete description of a (nonempty) set of embeddings of a given manifold up to isotopy is *only* known either just below the stable range, or for highly-connected manifolds, or for links and knotted tori. (A concrete, complete answer to the embedding problem has been obtained additionally for projective spaces [55], products of low-dimensional manifolds or of graphs [6, 184] and certain twisted products [148, 156].)

Therefore the knotting problem is one of the hardest problems in topology. The embedding problem is also hard for similar reasons. However, the statements (but not the proofs!) are simple and accessible to non-specialists. One of the purposes of this survey is to list such statements. They are presented in Sections 2 and 3. Statements analogous to those presented (e.g. for non-closed manifolds) are often omitted.

Another purpose of this survey is to outline a general approach useful for obtaining such concrete complete results. There are interesting approaches giving nice *theoretical* results.† The application of surgery [113, 15, 68, 71, 73, 63, 24] gives good concrete results for simplest manifolds. The advantage of the surgery approach (compared with the deleted product approach, see below) is that it works in the presence of smooth knots $S^n \to \mathbb{R}^m$. The disadvantage is that it uses the homotopy type of the complement and the description of possible normal bundles, and so faces computational difficulties even for relatively simple manifolds like tori, see Section 3. (For a successful attempt to overcome this problem see [107, 175]). According to Wall [206], surgery only reduces geometric problems on embeddings to algebraic problems which are even harder to solve.

---

† The author is grateful to M. Weiss for indicating that the approach of [59, 209] also gives concrete results on homotopy type of the space of embeddings $S^1 \to \mathbb{R}^n$.

The method of the *Haefliger-Wu invariant* (or the *deleted product method*) gives many concrete results. We introduce this invariant in Section 5. In particular, most of the results from Sections 2, 3 and 4 can be deduced using the deleted product method (although originally some of them were proved directly, sometimes in a weaker form). The deleted product method is a demonstration of a general mathematical idea of 'complements of diagonals' and the 'Gauss mapping' which appeared in works of Borsuk and Lefschetz around 1930. The Haefliger-Wu invariant generalizes the linking coefficient, the Whitney obstruction and the van Kampen obstruction. The deleted product method in the theory of embeddings was developed by van Kampen (1932), Shapiro (1957), Wu (1957-59), Haefliger (1962), Weber (1967), Harris (1969) and others. The Van-Kampen-Shapiro-Wu approach works for embeddings of *polyhedra*, but is closely related to embeddings of manifolds and so is presented in Section 4. The classical Haefliger-Weber theorem (Theorem 5.5) asserts the bijectivity of the Haefliger-Wu invariant for embeddings of n-dimensional polyhedra and manifolds into $\mathbb{R}^m$ under the 'metastable' dimension restriction

$$2m \geq 3n + 4.$$

Other embedding invariants may be obtained using *p-fold deleted products* (see the end of Section 5) or using the complement together with the normal bundle [15, 24], but these are hard to compute. So the investigation of embeddings for $2m < 3n + 4$ naturally leads to the problem of finding conditions under which the Haefliger-Wu invariant is complete without the metastable dimension assumption. There have been many examples showing that for embeddings of manifolds the metastable dimension restriction is sharp in many senses (Boechat, Freedman, Haefliger, Hsiang, Krushkal, Levine, Mardesic, Segal, Skopenkov, Spież, Szarba, Teichner, Zeeman; see Section 5). So it is surprising (Theorem 5.6) that in the *piecewise linear* category, for d-connected n-dimensional manifolds, the metastable restriction in the Haefliger-Weber Theorem can be weakened to

$$2m \geq 3n + 3 - d.$$

Examples in this paper

We present many beautiful examples motivated by the embedding and isotopy problems. In particular:

- in Section 2 we present a construction of the Hudson torus (which is simpler than the original one);

- in Section 3 we construct examples illustrating the distinction between piecewise linear and smooth embeddings;

- in Section 6 we prove some results on the deleted product of the 'torus' $S^p \times S^q$ and on the Haefliger-Wu invariant of knotted tori.

In Sections 6 and 7 we construct most of the examples of the incompleteness of the van Kampen obstruction and the Haefliger-Wu invariant, announced in Sections 4 and 5. The construction of these examples is based on knotted tori (see Section 6) or on (higher-dimensional) Casson finger moves (see Section 7). For some other examples we only give references.

The Haefliger-Weber theorem and its analogue below the metastable case were obtained by a combination of and the improvement of methods and results from various parts of topology: the theory of immersions, homotopy theory, engulfing, the Whitney trick, van Kampen finger moves, the Freedman-Krushkal-Teichner trick and their generalizations. The most important method is the *disjunction method* (see the end of Section 4, and Section 8). These methods are also applied in other areas. In Section 8 we prove the surjectivity of the Haefliger-Wu invariant in the piecewise linear case. For the reader's convenience, we take a historical approach to the exposition: the disjunction method is applied in its complete generality only after illustration in simpler particular cases. We also prove the analogue of the Haefliger-Weber theorem below the metastable range for the simplest case. We do not prove many other results of Sections 2–5 but give references and sometimes sketch the proofs.

Sections 6, 7 and 8 depend on Section 5; otherwise the sections are independent of each other, except for minor details that can well be omitted during the first reading.

Further references

Let us give a (by no means complete) list of references for closely related problems in geometric topology (the references inside the papers listed here could also be useful for a reader). In the problems of embeddability and isotopy the space $\mathbb{R}^m$ can be replaced by an *arbitrary* space Y. The case when Y is a manifold has been studied most extensively; for the case when Y is a product of trees see [193, Theorem 4.6 and Remark] and [54, 53, 219, 110].

- For embeddings *up to cobordism* see [16, 116]; for embeddings *up to homotopy* see [33, 191], [206, §11], [92, 31, 62].
- For the classification of *link maps* see [131, 122, 104, 105, 125, 64, 180].
- For embeddings of polyhedra in *some* manifolds see [204, 138, 115, 150].
- For the problem of *embeddability of compacta* and the closely-related problem of *approximability by embeddings* see† [26, 126, 173, 169, 98, 34, 3, 2], [151, §9], [134, 152, 5, 155, 7, 127, 156, 185, 128].
- For the problem of *intersection of compacta* see [39, 189].
- For *basic embeddings* see [193, 177] [153, §5], [110].
- For *immersions* see [55, 1], cf. [181].

1 Preliminaries

1.1 Definitions and notation

A smooth *embedding* is a smooth injective map $f : N \to \mathbb{R}^m$ such that df is a monomorphism at each point.

By a polyhedron we shall understand a *compact* polyhedron. A map $f : N \to \mathbb{R}^m$ of a polyhedron N is *piecewise-smooth* if it is smooth on each simplex of some smooth triangulation of N. We denote the piecewise-smooth category by PL. This is the usual notation for the piecewise-linear category but the classification of piecewise-smooth embeddings (or immersions) coincides with the classification of piecewise linear embeddings (or immersions) [74]. A PL *embedding* is a PL injective map $f : N \to \mathbb{R}^m$.

We write CAT for DIFF or PL. We often omit CAT if a statement holds in both PL and DIFF categories.

Two embeddings $f, g : N \to \mathbb{R}^m$ are said to be *(ambient) isotopic* (Figure 1.1), if there exists a homeomorphism onto (an *ambient isotopy*) $F : \mathbb{R}^m \times I \to \mathbb{R}^m \times I$ such that

- $F(y, 0) = (y, 0)$ for each $y \in \mathbb{R}^m$,
- $F(f(x), 1) = (g(x), 1)$ for each $x \in N$, and
- $F(\mathbb{R}^m \times \{t\}) = \mathbb{R}^m \times \{t\}$ for each $t \in I$.

---

† The author is grateful to P. Akhmetiev for indicating that the paper [3] contains a mistake for $n = 3, 7$ and that the paper [2] contains a 'preliminary version' of the proof, the complete version being submitted elsewhere.

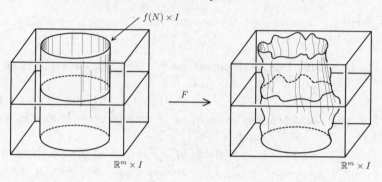

Fig. 1.1. Isotopic embeddings

An ambient isotopy is also a homotopy $\mathbb{R}^m \times I \to \mathbb{R}^m$ or a family of maps $F_t : \mathbb{R}^m \to \mathbb{R}^m$, generated by the map F in the obvious manner. Evidently, ambient isotopy is an equivalence relation on the set of embeddings of N into $\mathbb{R}^m$. An embedding is said to be *trivial* if it is isotopic to the standard embedding (the latter is evidently defined from the context).

We use the notation of [161]. An equality between sets denotes a one-to-one correspondence. Denote by $V_{m,n}$ the Stiefel manifold of n-frames in $\mathbb{R}^m$. Let $\mathbb{Z}_{(k)}$ be $\mathbb{Z}$ for k even and $\mathbb{Z}_2$ for k odd. Note that $\mathbb{Z}_{(k)} = \pi_k(V_{n,n-k})$ for $1 < k < n$. If the coefficients are omitted from the notation of (co)homology groups, then $\mathbb{Z}$-coefficients are assumed. For a manifold or a polyhedron N we denote its dimension by $n = \dim N$. Denote by $\mathrm{Emb}^m_{CAT}(N)$ the set of CAT embeddings $N \to \mathbb{R}^m$ up to ambient CAT isotopy. If $|\mathrm{Emb}^m_{CAT}(N)| = 1$, we shall say that N *CAT unknots* in $\mathbb{R}^m$. For a map $f : N \to \mathbb{R}^m$ we denote by $\Sigma(f) = \mathrm{Cl}\{x \in N : |f^{-1}fx| \geq 2\}$ its *self-intersection set*.

A closed manifold N is called *homologically k-connected* if N is connected and $H_i(N) = 0$ for each $i = 1, \ldots, k$. This condition is equivalent to $\tilde{H}_i(N) = 0$ for each $i = 0, \ldots, k$, where the $\tilde{H}_i$ are reduced homology groups. A pair $(N, \partial N)$ is called *homologically k-connected* if $H_i(N, \partial N) = 0$ for each $i = 0, \ldots, k$. Note that if $H_0(N, \partial N) = 0$, then the manifold N has no closed connected components. We use the following conventions: 0-connectedness is equivalent to homological 0-connectedness and to connectedness; k-connectedness for $k < 0$ is an empty condition. We put $\pi^S_l = 0$ for $l < 0$.

1.2 Other equivalence relations and problems

Ambient isotopy is a stronger equivalence relation than non-ambient isotopy, isoposition, concordance, bordism, etc. Two embeddings $f, g : N \to \mathbb{R}^m$ are called *(non-ambient) isotopic*, if there exists an embedding $F : N \times I \to \mathbb{R}^m \times I$ such that

- $F(x, 0) = (f(x), 0)$,
- $F(x, 1) = (g(x), 1)$ for each $x \in N$ and
- $F(N \times \{t\}) \subset \mathbb{R}^m \times \{t\}$ for each $t \in I$.

In the DIFF category or for $m - n \geq 3$ in the PL (or TOP) category *isotopy implies ambient isotopy* [95, 88, 8], [40, §7]. For $m - n \leq 2$ this is not so: for example, any knot $S^1 \to S^3$ is non-ambiently PL isotopic to the trivial one, but not necessarily ambiently PL isotopic to it.

Two embeddings $f, g : N \to \mathbb{R}^m$ are said to be (orientation preserving) *isopositioned*, if there is an (orientation preserving) homeomorphism $h : \mathbb{R}^m \to \mathbb{R}^m$ such that $h \circ f = g$. For embeddings into $\mathbb{R}^m$, orientation preserving isoposition is *equivalent to isotopy* (this is the *Alexander-Guggenheim theorem*) [161, 3.22].

Two embeddings $f, g : N \to \mathbb{R}^m$ are said to be *ambiently concordant* if there is a homeomorphism (onto) $F : \mathbb{R}^m \times I \to \mathbb{R}^m \times I$ (which is called a *concordance*) such that

- $F(y, 0) = (y, 0)$ for each $y \in \mathbb{R}^m$ and
- $F(f(x), 1) = (g(x), 1)$ for each $x \in N$.

The definition of *non-ambient concordance* is analogously obtained from that of non-ambient isotopy by dropping the last condition of level-preservation. In the DIFF category or for $m - n \geq 3$ in the PL (or TOP) category *concordance implies ambient concordance and isotopy* [114, 91, 94] (this is not so in codimension 2). This result allows a reduction of the problem of isotopy to the relativized problem of embeddability.

2 The simplest-to-state results on embeddings

2.1 Embeddings just below the stable range

Theorem 2.1 (General position theorem). *Every n-polyhedron or n-manifold embeds into $\mathbb{R}^{2n+1}$ and unknots in $\mathbb{R}^m$ for $m \geq 2n + 2$.*

The restriction $m \geq 2n + 2$ in Theorem 2.1 is sharp, as the Hopf linking $S^n \sqcup S^n \to \mathbb{R}^{2n+1}$ shows (Figure 2.1(a)). The number $2n + 1$ in the theorem is the minimal possible for polyhedra.

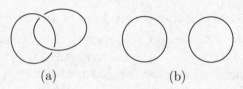

Fig. 2.1.

Example 2.2. For each n there exists an n-polyhedron, non-embeddable in $\mathbb{R}^{2n}$.

In Example 2.2 one can take the n-th power of a nonplanar graph [184], the n-skeleton of a $(2n+2)$-simplex [197, 45] or the $(n+1)$-th join power of the three-point set (see the proof for this case in Section 5).

Note that $N \times I$ embeds into $\mathbb{R}^{2n+1}$ for each n-polyhedron N [157], and unknots in $\mathbb{R}^{2n+2}$ for each n-polyhedron N. (Let us sketch a proof of the second statement, as follows: by general position every two embeddings $N \times I \to \mathbb{R}^{2n+2}$ are regular homotopic, and their restrictions to the spine $N \times \{\frac{1}{2}\}$ are isotopic).

Theorem 2.3 ([197, 211]). *Every n-manifold embeds into $\mathbb{R}^{2n}$.*

Theorem 2.3 (as well as Theorem 2.5 below) is proved using general position and the Whitney trick; the proofs in the smooth and PL case are sketched in Section 4 and in [161], [153, §8], respectively.

The dimension $2n$ in Theorem 2.3 is the best possible when $n = 2^k$ because $\mathbb{R}P^{2^k}$ does not embed into $\mathbb{R}^{2^{k+1}-1}$ (this is proved using the mod 2 Whitney obstruction defined below [133, 154, 174]). However it is not the best possible for other n, by Theorem 2.4 below.

A celebrated and difficult conjecture is that every closed n-manifold embeds into $\mathbb{R}^{2n+1-\alpha(n)}$, where $\alpha(n)$ is the number of units in the dyadic expansion of n. The analogous conjecture for immersions was proved in [112, 29]. Note that if $n = 2^{k_1} + \cdots + 2^{k_{\alpha(n)}}$ and $k_1 < \cdots < k_{\alpha(n)}$, then the n-manifold $N = \mathbb{R}P^{2^{k_1}} \times \cdots \times \mathbb{R}P^{2^{k_{\alpha(n)}}}$ does not embed into $\mathbb{R}^{2n-\alpha(n)}$ (this is proved again using the mod 2 Whitney obstruction).

Theorem 2.4.

(a) *Every n-manifold (except that if $n = 2^k$ and the manifold is closed, we need to assume that it is orientable) embeds into $\mathbb{R}^{2n-1}$.*

(b) *Suppose that N is a closed n-manifold, where n is even, $n \neq 2^k(2^h+1)$*

for integers $k, h \geq 2$, and $H_1(N) = 0$. Then N embeds into $\mathbb{R}^{2n-2}$, provided $n \geq 6$ in the PL category or $n \geq 8$ in the smooth category.

(c) *Suppose that N is a closed n-manifold, where $\alpha(n) \geq 5$ and*

  – *either $n = 0, 1(4)$ and N is orientable, or*
  – *$n = 2, 3(4)$ and N is non-orientable.*

  Then N embeds into $\mathbb{R}^{2n-2}$.

The classical results of Theorem 2.4 are much more complicated to prove than Theorem 2.3: one needs a generalization of the Whitney trick and calculation of characteristic classes, both nontrivial. Theorems 2.4(a) and (b) follow† from the analogue for $d = 0$ of Theorem 2.12 below (which is true for orientable manifolds) and [123, Theorem 1.c and Corollary 2] and [124, Theorem 1], except that Theorem 2.4(a) for $n = 3, 4$ has to be proved separately [78, 79, 159, 202, 14, 37, 41], cf. [139, 42]. Theorem 2.4(c) follows from the Haefliger-Weber theorem (Theorem 5.5 below) and [10, Theorem 45].

The condition '$\alpha(n) \geq 5$' in Theorem 2.4(c) can be weakened to '$n \geq 7$ and $\bar{w}_{n-i}(N) = 0$ for each $i \leq 4$' (the classes $\bar{w}_i(N)$ are defined in Section 2.3 below).

Theorem 2.5. *Every connected n-manifold unknots in $\mathbb{R}^{2n+1}$ for $n > 1$* [212].

Here for each n the connectedness assumption is indeed necessary and the dimension $2n+1$ is the best possible, as shown by the examples of the Hopf linking (see above) and the Hudson torus (see below). The Hopf linking is distinguished from the standard linking using the simplest ($\mathbb{Z}$-valued or $\mathbb{Z}_2$-valued) *linking coefficient* (whose definition is obtained from the definition of the Whitney invariant by setting $m = 2n + 1$, see Section 2.3 below).

Theorem 2.6. *Suppose that $n \geq 2$ and a compact n-manifold N has s closed orientable connected components and t closed non-orientable connected components (and, possibly, some non-closed components). Then the set of pairwise linking coefficients defines a one-to-one correspondence*

$$\mathrm{Emb}^{2n+1}(N) \to \mathbb{Z}^{\frac{s(s-1)}{2}} \oplus \mathbb{Z}_2^{st+\frac{t(t-1)}{2}}.$$

Note that every n-polyhedron N such that $H^n(N) = 0$

---

† From the proof of [123, p. 100] it follows that the restriction in [123, Theorem 1.c] should be stated as "the number of h_i's which are equal to $h_q + 1$ is even", cf. [78].

(a) PL (if $n = 2$, only TOP) embeds into $\mathbb{R}^{2n}$;

(b) PL unknots in $\mathbb{R}^{2n+1}$.

Assertion (a) for $n \geq 3$ is deduced from Theorems 4.2 or 5.5 below [212, 208], see also [82]. For $n = 2$ it was proved independently [102] and for $n = 1$ it is trivial. Assertion (b) for $n \geq 2$ is deduced from Theorems 4.6 or 5.5 below [147] and for $n = 1$ it is trivial. In these assertions for each n the dimensions are the best possible and the $H^n(N) = 0$ assumption is indeed necessary.

2.2 Embedding and unknotting of highly-connected manifolds

Theorem 2.7. *The sphere S^n, or even any homology n-sphere,*

(a) *PL unknots in $\mathbb{R}^m$ for $m - n \geq 3$ [216, 190, 56, 167];*

(b) *DIFF unknots in $\mathbb{R}^m$ for $m \geq \frac{3n}{2} + 2$ [66, 69], [1, §7];*

(c) *PL (if $n = 3$, only TOP) embeds into $\mathbb{R}^{n+1}$ (this follows from [101]);*

(d) *DIFF embeds into $\mathbb{R}^{[3n/2]+2}$ [66, 69], [1, §7].*

Theorem 2.7(a) is also true [163, 167] in the TOP locally flat category (see Section 3 for the definition). Here the local flatness assumption is indeed necessary.

Knots in codimension 2 and the trefoil knot (Example 3.4 below) show that the dimension restrictions are sharp (even for standard spheres) in Theorems 2.7(a) and 2.7(b) respectively. By [113, 83], cf. [118, pp. 407–408] the dimension restriction in Theorem 2.7(d) is indeed necessary (and conjecturally almost sharp) even for *homotopy spheres*. However, from [13], it follows that any $4k$-dimensional homotopy sphere embeds into $\mathbb{R}^{6k+1}$.

Theorems 2.5 and 2.7 may be generalized as follows.

Theorem 2.8 (The Haefliger-Zeeman embedding and isotopy theorems). *For $n \geq 2d+2$, every closed homologically d-connected n-manifold*

(a) *embeds into $\mathbb{R}^{2n-d}$ ($n \neq 2d + 2$ in the DIFF case), and*

(b) *unknots in $\mathbb{R}^{2n-d+1}$.*

Theorem 2.8 was proved directly in [141, 66, 217, 96, 90] for *homotopically d-connected* manifolds. The proofs in [77, 199, 69, 207] and [1, §7] work for *homologically d-connected* manifolds; such proofs are based either on embedding the complement of a ball or on the *deleted product method* (Section 5 below), cf. [178, 181]. Theorem 2.8 follows from Theorems 2.12 and 2.13 below.

Note that if $n \leq 2d + 1$, then every closed homologically d-connected n-manifold is a homology sphere, so the PL case of Theorem 2.8 gives nothing more than Theorem 2.7.

For generalizations of Theorem 2.8 see Theorems 2.12, 2.13 below or [89, 65, 93, 60, 97]. We shall use of them (the simplest case $2m \geq 3n + 3$ of) the following relative version of Theorem 2.8(a).

Theorem 2.9 (The Penrose-Whitehead-Zeeman-Irwin embedding theorem [141, 96]). *If $m - n \geq 3$, then any proper map from a $(2n - m)$-connected PL n-manifold with boundary to a $(2n - m + 1)$-connected PL m-manifold with boundary, whose restriction to the boundary is an embedding, is homotopic (relatively to the boundary) to a PL embedding.*

The dimension assumption in Theorem 2.8(b) is sharp, as shown by the following example.

Example 2.10 (The Hudson torus example). For each $p \leq q$ there exists a nontrivial embedding $S^p \times S^q \to \mathbb{R}^{p+2q+1}$ [87].

We sketch a simplified construction (see [175]).† Take the standard embedding $2D^{p+q+1} \times S^q \subset \mathbb{R}^{p+2q+1}$. The *Hudson torus* is the (linked!) connected sum of the $(p + q)$-sphere $2\partial D^{p+q+1} \times x$ with the standard embedding $\partial D^{p+1} \times S^q \subset D^{p+q+1} \times S^q \subset \mathbb{R}^{p+2q+1}$.

Remark. The Hudson torus can be distinguished from the standard embedding using the *Whitney invariant* defined in the subsection under the same name below [175] or the *Haefliger-Wu invariant* defined in Section 5 [181].

The rest of this subsection can be omitted on a first reading.

A simplified construction of the Hudson torus [181, 183]

Define a map $S^0 \times S^q \to D^{q+1}$ to be the constant $0 \in D^{q+1}$ on one component and the standard embedding φ on the other component. This map gives an *embedding*

$$S^0 \times S^q \to D^{q+1} \times S^q \subset D^{q+2} \times S^q \subset \mathbb{R}^{2q+2}$$

(Figure 2.2).

Each disc $D^{q+2} \times x$ intersects the image of this embedding at two points lying in $D^{q+1} \times x$. For each $x \in S^q$, extend this embedding $S^0 \to D^{q+1} \times x$ to an embedding $S^1 \to D^{q+2} \times x$ (Figure 2.3).

---

† This construction works even for $p = q = 1$, in which case the existence of a knotting $S^1 \times S^1 \to \mathbb{R}^4$ is easy in contrast to the higher-dimensional case.

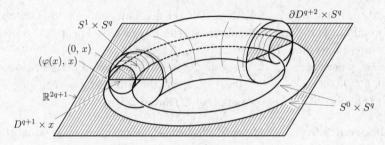

Fig. 2.2.

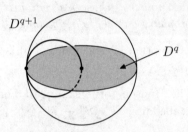

Fig. 2.3.

We thus obtain a nonstandard embedding

$$S^1 \times S^q \to D^{q+2} \times S^q \subset \mathbb{R}^{2q+2}.$$

Taking spheres of dimensions $p \le q$ we obtain analogously an embedding

$$S^p \times S^q \to D^{q+p+1} \times S^q \subset \mathbb{R}^{2q+p+1}.$$

(Taking as φ above an arbitrary CAT-map $S^q \to \partial D^{m-q-p}$ we obtain analogously a CAT embedding $S^p \times S^q \xrightarrow{\varphi \times \mathrm{pr}_2} D^{m-q} \times S^q \to \mathbb{R}^m$.)

It would be interesting to know whether the smooth case of Theorem 2.8(a) holds for $n = 2d + 2$, i.e. for d-connected $2(d+1)$-manifolds.

An *almost smooth* embedding is a PL embedding which is a smooth embedding outside a point.

Theorem 2.11. *Let N be a closed smooth $(l-1)$-connected $2l$-manifold.*

(a) *If $l \equiv 3, 5, 7 \mod 8$ and $l \ne 2^s - 1$, then N almost smoothly embeds into $\mathbb{R}^{2l+1}$.*

(b) *For l even the manifold N almost smoothly embeds into $\mathbb{R}^{2l+1}$ if and only if N is stably parallelizable.*

(c) *If N is almost parallelizable, then N almost smoothly embeds into $\mathbb{R}^{2l+2}$ [119].*

(d) *For each even l there exists a closed smooth $(l-1)$-connected (even almost parallelizable) $2l$-manifold which does not smoothly embed into $\mathbb{R}^{3l-1}$ [119].*

Theorems 2.11(a) and 2.11(b) are proved in [135, Corollary 1, Theorem 2 and Addendum (1)], cf. [35]. By Theorem 2.11(c) the manifold from 2.11(d) almost smoothly embeds into $\mathbb{R}^{2l+2}$.

Proof of 2.11(c). If $l = 2$, then the result holds by [120, Corollary 10.11], [20]. So assume that $l \neq 2$. Let N_0 be a complement in N to some open $2l$-ball. Then N_0 is parallelizable and hence there is an immersion $f : N_0 \to \mathbb{R}^{2l+1}$. Since N_0 is $(l-1)$-connected and $l \neq 2$, it follows that it has an l-dimensional spine [201, 81]. By general position the restriction of f to this spine is an embedding. Hence the restriction of f to some neighbourhood of this spine is an embedding. But this neighbourhood is homeomorphic to N_0. So there is an embedding $g : N_0 \to \mathbb{R}^{2l+1}$. Extending the embedding $g|_{\partial N_0}$ as a cone in $\mathbb{R}^{2l+2}$ we obtain an almost smooth embedding of N into $\mathbb{R}^{2l+2}$. $\qquad\square$

Proof of 2.11(d). Take the Kervaire-Milnor closed smooth almost parallelizable $4k$-manifold N whose signature $\sigma(N)$ is nonzero [120, 132]. We can modify this by surgery [120] and assume further that it is $(l-1)$-connected. Hence the Pontryagin class $p_{l/2}(N, \mathbb{R})$ is nonzero by the Hirzebruch formula. Therefore $\bar{p}_{l/2}(N, \mathbb{R}) \neq 0$ by the duality theorem for real Pontryagin classes ([214], cf. [133]). Hence N does not smoothly embed into $\mathbb{R}^{3l-1}$ (it does not even immerse in $\mathbb{R}^{3l-1}$) [142, 214]. $\qquad\square$

The dimension $2n-d$ in Theorem 2.8(a) can be decreased by 1 for some pairs (n, d), as Theorem 2.4 shows. However, we conjecture that the dimension $2n-d$ in Theorem 2.8(a) cannot be significantly decreased for some (n, d). This is so for $d = 0$ (as the example $N = \mathbb{R}P^{a_1} \times \cdots \times \mathbb{R}P^{a_s}$ shows) and for $n = 2, 4, 8, d = \frac{n}{2} - 1$ (take $N = \mathbb{R}P^2, \mathbb{C}P^2, \mathbb{H}P^2$ or apply Theorem 2.11(d)). Examples of highly-connected but badly embeddable manifolds were also exhibited in [84, 35].

2.3 The Whitney obstruction

Let N be a closed manifold. We present the definition in the piecewise linear case; in the smooth case, there are both an analogous definition

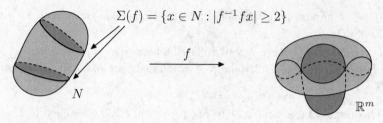

$$\Sigma(f) = \{x \in N : |f^{-1}fx| \geq 2\}$$

Fig. 2.4. The self-intersection set $\Sigma(f)$

and a simpler definition [175]. Take any general position map $f : N \to \mathbb{R}^m$. Recall the definition of the self-intersection set $\Sigma(f)$ from Section 1.1 (see Figure 2.4).

Take a triangulation T of N such that f is linear on simplices of T. Then the self-intersection set $\Sigma(f)$ is a subcomplex of T. Denote by $[\Sigma(f)] \in C_{2n-m}(N; \mathbb{Z}_2)$ the sum of the top-dimensional simplices of $\Sigma(f)$ (Figure 2.4). Then $[\Sigma(f)]$ is a cycle [90, Lemma 11.4], [92, Lemma 1].

Sketch of the proof. It suffices to show that each $(2n-m-1)$-simplex η of T is in the boundary of an *even* number of $(2n-m)$-simplices $\sigma \subset \Sigma(f)$. We can restrict without loss of generality to the case $\eta \subset \Sigma(f)$.

By general position, $f^{-1}f\eta$ consists of simplices $\eta = \eta_1, \ldots, \eta_k$. The link of each $\mathrm{lk}_T \eta_i$ is a sphere of dimension $n - (2n - m - 1) - 1 = m - n$, while the link $\mathrm{lk}_{\mathbb{R}^m} f\eta$ is a sphere of dimension $m - (2n - m - 1) - 1 = 2(m-n)$. The intersection of two f-images $f(\mathrm{lk}_T \eta_i)$ of $(m-n)$-spheres in the $2(m-n)$-sphere $\mathrm{lk}_{\mathbb{R}^m} f\eta$ consists of an even number of points. These intersection points are in 1–1 correspondence with $(2n - m)$-simplices $\sigma \subset \Sigma(f)$ containing η in their boundaries. $\square$

The *modulo 2 Whitney obstruction* is the homology class

$$\bar{w}_{m-n}(N) := [\Sigma(f)] \in H_{2n-m}(N; \mathbb{Z}_2).$$

The class $\bar{w}_i$ is called *the normal Stiefel-Whitney class*. This definition of the normal Stiefel-Whitney classes is equivalent to other definitions [133], up to Poincaré duality.

The independence of $\bar{w}_{m-n}(N)$ from our choice of f follows from the equality

$$[\Sigma(f_0)] - [\Sigma(f_1)] = \partial[\Sigma(F)]$$

for a general position homotopy $F : N \times I \to \mathbb{R}^m \times I$ between general position maps $f_0, f_1 : N \to \mathbb{R}^m$.

Hence these classes are obstructions to the embeddability of N into $\mathbb{R}^m$: if N embeds into $\mathbb{R}^m$ then $\bar{w}_i(N) = 0$ for $i \geq m - n$ [210].

The case of N orientable and $m - n$ odd. Fix in advance any orientation of N and of $\mathbb{R}^m$. The definition of the Whitney obstruction is analogous to the one above, but here $[\Sigma(f)]$ is the sum of *oriented* simplices σ with $\pm$ signs defined as follows. (For $m - n$ even, the signs can also be defined but are not used because $[\Sigma(f)]$ is not necessarily a cycle with integer coefficients).

By general position there is a unique simplex τ of T such that $f(\sigma) = f(\tau)$. The orientation on σ induces an orientation on $f\sigma$ and then on τ. The orientations on σ and τ induce orientations on normal spaces in N to these simplices. These two orientations (in this order) together with the orientation on $f\sigma$ induce an orientation on $\mathbb{R}^m$. If this orientation agrees with the fixed orientation of $\mathbb{R}^m$, then the coefficient of σ is $+1$, otherwise -1. Clearly, the change of orientation of σ changes the sign of σ in $[\Sigma(F)]$, so the sign is well-defined.

Remark. An equivalent definition of the signs in $[\Sigma(f)]$ is as follows. The orientation on σ induces an orientation on $f\sigma$ and then on τ, hence it induces an orientation on their links. Consider the oriented sphere $\mathrm{lk}_{\mathbb{R}^m} f\sigma$, that is the link of $f\sigma$ in a certain triangulation of $\mathbb{R}^m$ 'compatible' with T. The dimension of this sphere is $m - 1 - (2n - m) = 2(m - n) - 1$. This sphere contains disjoint oriented $(m - n - 1)$-spheres $f(\mathrm{lk}_T \sigma)$ and $f(\mathrm{lk}_T \tau)$. The coefficient of σ in $[\Sigma(f)]$ is their linking coefficient, which equals ± 1.

The *Whitney obstruction* is the homology class

$$\bar{W}_{m-n}(N) := [\Sigma(f)] \in H_{2n-m}(N; \mathbb{Z}).$$

If N embeds into $\mathbb{R}^{n+2i}$ then $\bar{w}_i(N) = 0$ [210]. (Pontryagin introduced for each closed orientable n-manifold N the *Pontryagin classes* $\bar{p}_i \in H_{n-4i}(N; \mathbb{Z})$, which obstruct embeddability into $\mathbb{R}^{n+2i-1}$ [142].)

By definition $\bar{w}_i(N)$ is a modulo 2 reduction of $\bar{W}_i(N)$. Thus the classes $\bar{w}_i(N)$ are easier to compute; however, they are possibly weaker than $\bar{W}_i$.

Recall the definition of $\mathbb{Z}_{(k)}$ from Section 1.1. For a closed orientable n-manifold N denote by $\bar{W}_{m-n}(N) \in H_{2n-m}(N, \mathbb{Z}_{(m-n-1)})$ the class $\bar{W}_{m-n}(N)$ for $m - n$ odd and the class $\bar{w}_{m-n}(N)$ for $m - n$ even.

Theorem 2.12. *Let N be a closed d-connected n-manifold, $d \geq 1$. The manifold N embeds into $\mathbb{R}^{2n-d-1}$ if $\bar{W}_{n-d-1}(N) = 0$, provided $n \geq d+4$ or $n \geq 2d+5$ in the PL or DIFF cases respectively.*

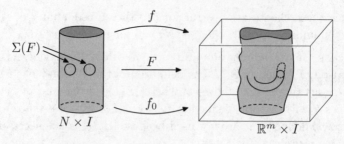

Fig. 2.5.

See the references listed after Theorem 2.13.

In Theorem 2.12 the d-connectedness assumption can be weakened to *homological* d-connectedness, except when $n = 2d + 2$ in the PL case. The PL case of Theorem 2.12 gives nothing but Theorem 2.7(c) for $d + 4 \leq n \leq 2d + 1$. The smooth case of Theorem 2.12 is true if d is even and $n = 2d + 3$ [181, Corollary 1.7].

2.4 The Whitney invariant

Let N be a closed connected orientable n-manifold. Let $f_0 : N \to \mathbb{R}^m$ be a certain fixed ('standard') embedding and let $f : N \to \mathbb{R}^m$ be an arbitrary embedding. Take a general position homotopy $F : N \times I \to \mathbb{R}^m \times I$ between f and f_0 (Figure 2.5).

Analogously to the above, the self-intersection set $\Sigma(F)$ supports a $(2n - m + 1)$-cycle $[\Sigma(F)]$ in $N \times I \simeq N$ with the coefficients $\mathbb{Z}_{(m-n-1)}$. The *Whitney invariant* of f is the homology class of this cycle:

$$W(f) := [\Sigma(F)] \in H_{2n-m+1}(N, \mathbb{Z}_{(m-n-1)}).$$

Again, $W(f)$ depends only on f and f_0 but not on the choice of F [90, §11], cf. [77, 199, 175].

Theorem 2.13. *Let N be a closed orientable homologically d-connected n-manifold, $d \geq 0$. Then the Whitney invariant*

$$W : \mathrm{Emb}^{2n-d}(N) \to H_{d+1}(N, \mathbb{Z}_{(n-d-1)})$$

is a bijection, provided $n \geq d + 3$ or $n \geq 2d + 4$, in the PL or DIFF cases respectively.

Theorems 2.12 and 2.13 were proved† in [77], [90, §11], [14, 13, 199] for *homotopically* d-connected manifolds (except the PL case of Theorem 2.12). The proof works for *homologically* d-connected manifolds.

By Theorem 2.13 we obtain that the Whitney invariant

$$W : \mathrm{Emb}^{p+2q+1}(S^p \times S^q) \to \mathbb{Z}_{(q)}$$

is bijective for $1 \le p \le q - 2$, cf. Theorem 3.8 below. The generator is the Hudson torus.

The PL case of Theorem 2.13 gives nothing but Theorem 2.7(a) for $d + 3 \le n \le 2d + 1$.

Analogously to Theorem 2.13 it may be proved that if N is a closed connected non-orientable n-manifold, then

$$\mathrm{Emb}^{2n}(N) = \begin{cases} H_1(N, \mathbb{Z}_2) & n \text{ odd,} \\ \mathbb{Z} \oplus \mathbb{Z}_2^{s-1} & n \text{ even and } H_1(N, \mathbb{Z}_2) \cong \mathbb{Z}_2^s, \end{cases}$$

provided $n \ge 3$ or $n \ge 4$, in the PL or DIFF case respectively [10, 199]. (There is a mistake in the calculation for the non-orientable case in [69], [207, Theorem B].)

Because of the existence of knots the analogues of Theorem 2.13 for $n = d+2$ in the PL case, and for (most) $n \le 2d+3$ in the smooth case are false. So for the smooth category and $n \le 2d+3$ a classification is much harder: until recently the *only* known concrete complete classification results were for spheres and their disjoint unions. Recently the following two results were obtained using the Kreck modification of surgery theory.

Theorem 2.14 ([175]). *Let N be a closed homologically $(2k-2)$-conn-ected $(4k-1)$-manifold. Then the Whitney invariant*

$$W : \mathrm{Emb}^{6k}_{DIFF}(N) \to H_{2k-1}(N)$$

is surjective, and for each $u \in H_{2k-1}(N)$ there is a one-to-one correspon-dence $\eta_u : W^{-1}u \to \mathbb{Z}_{d(u)}$, where $d(u)$ is the divisibility of the projection of u to the free part of $H_1(N)$.

(Recall that the divisibility of zero is zero and the divisibility of $x \in G - \{0\}$ is $\max\{d \in \mathbb{Z} \mid \text{there exists } x_1 \in G \text{ with } x = dx_1\}$.)

Theorem 2.14 implies that the Whitney invariant

$$W : \mathrm{Emb}^{6k}(S^{2k-1} \times S^{2k}) \to \mathbb{Z}$$

---

† The author is grateful to J. Boechat for indicating that [13, Theorem 4.2] needs a correction [175, §5]; this does not affect the main result of [13].

is surjective, and that for each $u \in \mathbb{Z}$ there is a one-to-one correspondence $W^{-1}u \to \mathbb{Z}_u$.

Theorem 2.15 ([107]).

(a) *Let N be a closed connected smooth 4-manifold such that $H_1(N) = 0$ and the signature $\sigma(N)$ of N is free of squares (i.e. is not divisible by a square of an integer $s \geq 2$). Then the Whitney invariant $W : \mathrm{Emb}^7_{DIFF}(N) \to H_2(N)$ is injective. There exists $x_0 \in H_2(N)$ such that $x_0^2 = \sigma(N)$ and $x_0 \mod 2 = w_2(N)$, and moreover*

$$\mathrm{im}\, W = \{y \in H_2(N) \mid y^2 + y \cap x_0 = 0\}.$$

(b) *Let N be a closed simply-connected smooth 4-manifold embeddable into S^6. Take a composition $f : N \to S^6 \subset S^7$ of an embedding and the inclusion. Then $\#W^{-1}W(f) = 12$.*

Corollary 2.16.

(a) *There is a unique embedding $f : \mathbb{C}P^2 \to \mathbb{R}^7$ up to isoposition (i.e. for each two embeddings $f, f' : \mathbb{C}P^2 \to \mathbb{R}^7$ there is a diffeomorphism $h : \mathbb{R}^7 \to \mathbb{R}^7$ such that $f' = h \circ f$).*

(b) *For each embedding $f : \mathbb{C}P^2 \to \mathbb{R}^7$ and each nontrivial knot $g : S^4 \to \mathbb{R}^7$ the embedding $f \# g$ is isotopic to f.*

Conjecture. *Every smooth embedding $S^1 \times S^1 \to \mathbb{R}^4$ is PL isotopic to a connected sum of a knot $S^2 \to S^4$ either with the standard embedding, or with the right Hudson torus, or with the left Hudson torus, or with the composition of Dehn twist along the parallel and the right Hudson torus.*

A similar conjecture can be stated for arbitrary closed 2-manifolds, see [43, 44].

2.5 Low-dimensional manifolds

For relatively low-dimensional manifolds there are the following results not covered by Theorems 2.3, 2.5, 2.4, 2.8(a) and 2.12. (We need not specify whether PL or DIFF manifolds are under consideration because every PL manifold of dimension at most 7 is smoothable.)

Theorem 2.17.

(a) *A closed orientable 4-manifold N PL embeds into $\mathbb{R}^6$ if and only if $\overline{w}_2(N) = 0$ [120, Corollary 10.11], [20].*

(b) *A closed orientable 4-manifold smoothly embeds into $\mathbb{R}^6$ if and only if $\overline{w}_2(N) = 0$ and $p_1(N) = 0$ [120, Corollary 10.11], [20, 162].*

(c) *Every 2-connected closed 6-manifold is a connected sum of $S^3 \times S^3$ [187, Theorem B] and therefore embeds into $\mathbb{R}^7$.*

(d) *Every closed non-orientable 6-manifold N such that $\overline{w}_2(N) = 0$ and $\overline{w}_3(N) = 0$ PL embeds into $\mathbb{R}^{10}$ [181].*

(e) *Let N be a closed simply-connected 6-manifold whose homology groups are torsion-free, and with $\overline{w}_2(N) = 0$. Then N*

 - *embeds into $\mathbb{R}^7$ if and only if it is a connected sum of copies of $S^2 \times S^4$ and $S^3 \times S^3$;*
 - *smoothly (or PL locally flat) embeds into $\mathbb{R}^8$ if and only if $p_1(N) = 0$;*
 - *smoothly embeds into $\mathbb{R}^{10}$ [205, Theorems 12 and 13].*

(f) *Every closed homologically 2-connected 7-manifold PL embeds into $\mathbb{R}^{11}$ [178, 181].*

We note that in Theorem 2.17(e), the embeddability in $\mathbb{R}^{10}$ is true also in the PL case, but this is covered by Theorem 2.4(b).

Remark. Take the Dold 5-manifold N such that

$$\overline{w}_{2,3}(N) := \overline{w}_2(N)\overline{w}_3(N) \neq 0$$

and make a surgery killing $\pi_1(N)$. We obtain a simply connected 5-manifold N' with $\overline{w}_{2,3}(N') \neq 0$, therefore $\overline{w}_3(N') \neq 0$ and hence N' does not embed into $\mathbb{R}^8$. This remark of Akhmetiev shows that the dimension $2n - d$ is the minimal in Theorem 2.8(a) for $n = 5$ and $d = 1$.

We conjecture that there exists a 1-connected 6-manifold N with normal Stiefel-Whitney class $\overline{W}_3(N) \neq 0$ so that N does not embed into $\mathbb{R}^9$, see [205, 220, 221].

3 Links and knotted tori

3.1 The linking coefficient

Definition of the linking coefficient

Fix orientations of S^p, S^q, S^m and D^{m-p}. Assume that $m \geq q + 3$ and $f : S^p \sqcup S^q \to S^m$ is an embedding. Take an embedding $g : D^{m-q} \to S^m$ such that gD^{m-q} intersects fS^q transversally at exactly one point with positive sign (Figure 3.1). Then the restriction of g to ∂D^{m-q} is an orientation preserving homotopy equivalence

$$h : S^{m-q-1} \to S^m - fS^q. \tag{3.1}$$

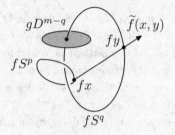

Fig. 3.1.

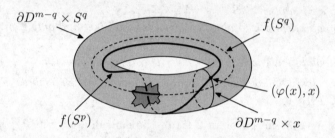

Fig. 3.2.

The induced isomorphism of homotopy groups does not depend on g. The *linking coefficient* is

$$\lambda_{12}(f) = \left[S^p \xrightarrow{f|_{S^p}} S^m - fS^q \xrightarrow{h} S^{m-q-1} \right] \in \pi_p(S^{m-q-1}).$$

Clearly, $\lambda_{12}(f)$ is indeed independent of h.

Analogously we may define $\lambda_{21}(f) \in \pi_q(S^{m-p-1})$ for $m \geq p+3$. The definition works for $m = q+2$ if the restriction of f to S^q is PL unknotted (this is always so for $m \geq q+3$ by Theorem 2.7(a)). For $m = p+q+1$ there is a simpler alternative definition.

Construction of a link with prescribed linking coefficient for
$$p \leq q \leq m-2$$

Define f on S^q to be the standard embedding into $\mathbb{R}^m$. Take any CAT map $\varphi : S^p \to \partial D^{m-q}$. Define the CAT embedding f on S^p by

$$S^p \xrightarrow{\varphi \times i} \partial D^{m-q} \times S^q \subset D^{m-q} \times S^q \subset \mathbb{R}^m,$$

where $i : S^p \to S^q$ is the equatorial inclusion and the latter inclusion is the standard one. See Figure 3.2. Clearly, $\lambda_{12}(f) = \varphi$.

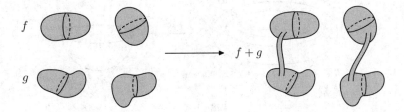

f

g

$f + g$

Fig. 3.3. Connected sum

If $m \geq \frac{p}{2} + q + 2$, then $\Sigma^{\infty} : \pi_p(S^{m-q-1}) \to \pi^S_{p+q+1-m}$ is an isomorphism.

Consider the commutative group structure on $\mathrm{Emb}^m(S^p \sqcup S^q)$, defined by 'connected sum' (Figure 3.3) in [71, 73] for $m - 3 \geq p, q$.

Theorem 3.1 (The Haefliger-Zeeman theorem). *If $1 \leq p \leq q$, then the map*

$$\Sigma^{\infty} \lambda_{12} : \mathrm{Emb}^m(S^p \sqcup S^q) \to \pi^S_{p+q+1-m}$$

is an isomorphism for $m \geq \frac{p}{2} + q + 2$ and for $m \geq \frac{3q}{2} + 2$, in the PL and DIFF cases respectively.

The surjectivity of λ_{12} is proved above. The injectivity is proved in [67, 217], or follows from the Haefliger-Weber theorem (Theorem 5.5) and the deleted product lemma 5.3(a) below.

By Theorem 3.1 we have the following table for $m \geq \frac{3q}{2} + 2$.

m	$2q+2$	$2q+1$	$2q$	$2q-1$	$2q-2$	$2q-3$	$2q-4$
$\mathrm{Emb}^m(S^q \sqcup S^q)$	0	$\mathbb{Z}$	$\mathbb{Z}_2$	$\mathbb{Z}_2$	$\mathbb{Z}_{24}$	0	0

The stable suspension of the linking coefficient can be described alternatively as follows. For an embedding $f : S^p \sqcup S^q \to S^m$ define a map $\widetilde{f} : S^p \times S^q \to S^{m-1}$ by

$$\widetilde{f}(x, y) = \frac{fx - fy}{|fx - fy|}.$$

For $p \leq q \leq m - 2$ define the α-invariant by

$$\alpha(f) = [\widetilde{f}] \in [S^p \times S^q, S^{m-1}] \xrightarrow[v^*]{\cong} \pi_{p+q}(S^{m-1}) \cong \pi^S_{p+q+1-m}.$$

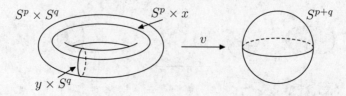

$S^p \times S^q$ $\qquad$ $S^p \times x$ $\qquad$ S^{p+q}

v

$y \times S^q$

Fig. 3.4.

The second isomorphism in the formula for $\alpha(f)$ is given by the Freudenthal suspension theorem. The map $v : S^p \times S^q \to \frac{S^p \times S^q}{S^p \vee S^q} \cong S^{p+q}$ is the quotient map (Figure 3.4). The map v^* is an isomorphism for $m \geq q + 2$.

(For $m \geq q+3$ this follows by general position and for $m = q+2$ by the cofibration Barratt-Puppe exact sequence of the pair $(S^p \times S^q, S^p \vee S^q)$ and by the existence of a retraction $\Sigma(S^p \times S^q) \to \Sigma(S^p \vee S^q)$, cf. [122, §3].)

By [100, Lemma 5.1] we have $\alpha = \pm\Sigma^\infty \lambda_{12}$. Note that the α-invariant can be defined in more general situations [104].

3.2 Borromean rings, the Whitehead link and the trefoil knot

An analogue of the Haefliger-Zeeman theorem holds for links with many components. However, the collection of pairwise α-invariants (or even linking coefficients) is not injective for $2m < 3n + 4$ and n-dimensional links with more than two components in $\mathbb{R}^m$. This is implied by the following example.

Example 3.2 (The Borromean rings). The *Borromean rings*

$$S^{2l-1} \sqcup S^{2l-1} \sqcup S^{2l-1} \to \mathbb{R}^{3l}$$

form a nontrivial embedding whose restrictions to 2-componented sub-links are trivial. ([68, 4.1], [67], cf. [125])

Consider the space $\mathbb{R}^{3l}$ with coordinates

$$(x, y, z) = (x_1, \dots, x_l, y_1, \dots, y_l, z_1, \dots, z_l).$$

The (higher-dimensional) Borromean rings are three embedded spheres (Figures 3.5 and 3.6(b)) given by the equations

$$\begin{cases} x = 0 \\ y^2 + 2z^2 = 1 \end{cases}, \qquad \begin{cases} y = 0 \\ z^2 + 2x^2 = 1 \end{cases} \quad \text{and} \quad \begin{cases} z = 0 \\ x^2 + 2y^2 = 1 \end{cases}.$$

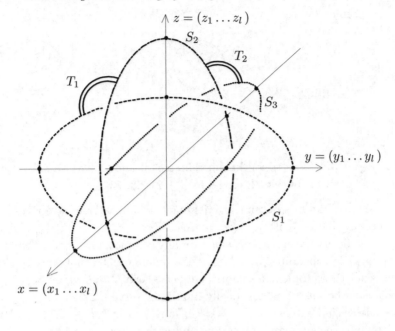

Fig. 3.5.

The following classical example shows that the invariant

$$\alpha = \pm \Sigma^\infty \lambda_{12} : \mathrm{Emb}^m(S^p \sqcup S^q) \to \pi^S_{p+q+1-m}$$

can be *incomplete for $m < \frac{p}{2} + q + 2$ and links with two components*, i.e. that the dimension restriction in the Haefliger-Zeeman theorem is sharp.

Example 3.3 (The Whitehead link). The *Whitehead link*

$$w : S^{2l-1} \sqcup S^{2l-1} \to \mathbb{R}^{3l}$$

is nontrivial, although $\alpha(w) = \Sigma^\infty \lambda_{12}(w) = 0$.

The Whitehead link is obtained from Borromean rings by joining two components with a tube (Figure 3.6(w)). We have

$$\alpha(w) = 0 \quad \text{but} \quad \lambda_{12}(w) = [\iota_l, \iota_l] \neq 0 \quad \text{for} \quad l \neq 1, 3, 7.$$

Cf. [67, §3]. Note that for $l = 1, 3, 7$ the Whitehead link is still nontrivial, although $\lambda_{12}(w) = \lambda_{21}(w) = 0$ (again, see [67, §3]).

(b)

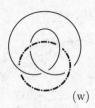
(w)

(t)

Fig. 3.6.

Example 3.4 (The trefoil knot). The trefoil knot $S^{2l-1} \to \mathbb{R}^{3l}$ is not smoothly trivial (but is PL trivial for $l \ge 2$) [68, 73].

The trefoil knot is obtained by joining the three Borromean rings by two tubes (Figure 3.6(t)).

If we take a cone or a suspension over any codimension 2 knot, then we obtain a PL embedding f of a ball or of a sphere which is not *smoothable*, i.e. is not PL isotopic to a smooth (not necessarily standard) embedding. This is so because f is not locally flat. † Observe that for $m \ge n + 3$ the suspension extension $S^n \to \mathbb{R}^m$ of any knot $S^{n-1} \to \mathbb{R}^{m-1}$ is PL isotopic to the standard embedding and is therefore smoothable.

Example 3.5 (The Haefliger torus). There is a PL embedding $S^{2k} \times S^{2k} \to \mathbb{R}^{6k+1}$ which is (locally flat but) not PL isotopic to a smooth embedding [68], [14, p. 165], [13, 6.2].

In order to construct the Haefliger torus take the above trefoil knot $S^{4k-1} \to \mathbb{R}^{6k}$. Extend this knot to a conical embedding $D^{4k} \to \mathbb{R}^{6k+1}_{-}$. By [68], the trefoil knot also extends to a smooth embedding $S^{2k} \times S^{2k} - \mathring{D}^{4k} \to \mathbb{R}^{6k+1}_{+}$ (Figure 3.7). These two extensions together form the Haefliger torus (Figure 3.8).

3.3 A classification of knots and links below the metastable range

Let $C_q^{m-q} := \mathrm{Emb}^m_{DIFF}(S^q)$. The 'connected sum' commutative group structure on C_q^{m-q} was defined for $m \ge q + 3$ in [71], cf. [73]. Theorem 2.7(b) states that $C_q^{m-q} = 0$ for $2m \ge 3q + 4$. It is known [71, 129, 107] that

$$C_{4k-1}^{2k+1} \cong \mathbb{Z}, \quad C_{4k+1}^{2k+2} \cong \mathbb{Z}_2, \quad C_4^3 \cong \mathbb{Z}_{12}, \quad C_{4k-2}^{2k} = 0,$$

---

† Recall that an embedding $N \subset \mathbb{R}^m$ of a PL n-manifold N is *locally flat* if each point $x \in N$ has a closed neighbourhood U such that $(U, U \cap N) \cong (D^m, D^n)$.

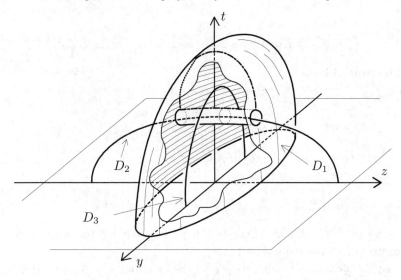

Fig. 3.7.

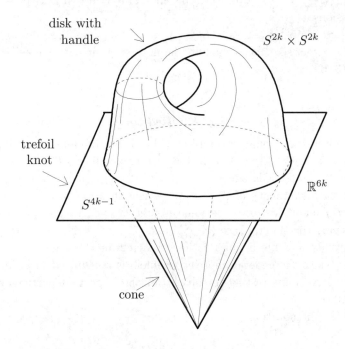

Fig. 3.8.

while

$$C_{8s+4}^{4s+3} \cong \mathbb{Z}_4 \quad \text{for} \quad s > 0 \quad \text{and} \quad C_{8s}^{4s+1} \cong \mathbb{Z}_2 \oplus \mathbb{Z}_2.$$

Theorem 3.6.

(a) *[73], cf. [63] If $p, q \leq m - 3$, then*

$$\mathrm{Emb}_{DIFF}^m(S^p \sqcup S^q) \cong \mathrm{Emb}_{PL}^m(S^p \sqcup S^q) \oplus C_q^{m-q} \oplus C_p^{m-p}.$$

(b) *[73, Theorem 10.7], [186] If $p \leq q \leq m - 3$ and $3m \geq 2p + 2q + 6$, then for large enough M*

$$\mathrm{Emb}_{PL}^m(S^p \sqcup S^q) \cong \pi_p(S^{m-q-1}) \oplus \pi_{p+q+2-m}(V_{M+m-p-1,M}).$$

The isomorphism in Theorem 3.6(b) is given by the sum of λ_{12}-invariant and the β-*invariant* [186].

By Theorem 3.6(b) (and its proof) the invariant $\lambda_{12} \oplus \lambda_{21}$ is injective for $l \geq 2$ and PL embeddings $S^{2l-1} \sqcup S^{2l-1} \to \mathbb{R}^{3l}$; its range is isomorphic to $\pi_{2l-1}(S^l) \oplus \mathbb{Z}_{(l)}$. However, this invariant is not injective in other dimensions.

The set $\mathrm{Emb}^m(S^{n_1} \sqcup \cdots \sqcup S^{n_s})$ for $m \geq n_i + 3$ has been described in terms of exact sequences involving homotopy groups of spheres [71, 73], cf. [113, 63].

3.4 Knotted tori

Many interesting counterexamples in the theory of embeddings [9, 106, 87, 203, 195, 14, 13, 130, 181, 183] are embeddings $S^p \times S^q \to \mathbb{R}^m$, i.e. *knotted tori*. Classification of knotted tori is a natural next step (after the link theory of [73]) towards classification of embeddings of *arbitrary* manifolds: it gives some insight or even precise information concerning the general case (cf. [87] and [77], [90, §12], [199], see [176]) and reveals new interesting relations to algebraic topology.

Also, since the general knotting problem is recognized to be unsolvable, it is very interesting to solve it for the important particular case of knotted tori.

We use the notation

$$KT_{p,q,CAT}^m := \mathrm{Emb}_{CAT}^m(S^p \times S^q).$$

Notice the change in the role of p in this subsection compared with the previous ones.

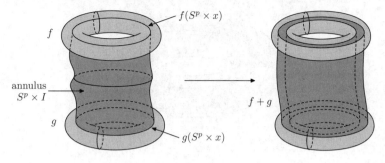

Fig. 3.9.

From the Haefliger-Zeeman isotopy theorem (Theorem 2.8(b)) it follows that $KT_{p,q}^m = 0$ for $p \leq q$ and $m \geq p + 2q + 2$. The dimension restriction in this result is sharp by Example 2.10 (the Hudson torus).

Theorem 3.7 (Group structure theorem). *The set $KT_{p,q}^m$ has a commutative group structure for $m \geq 2p + q + 3$ in the smooth case and $m \geq \max\{2p + q + 2, q + 3\}$ in the PL case [183].*

Idea of the proof. See Figure 3.9. By [183] under the dimension assumptions for any embedding $f : S^p \times S^q \to \mathbb{R}^m$ there is a *web*, i.e. an embedding $u : D^{p+1} \to \mathbb{R}^m$ such that

$$u(D^{p+1}) \cap f(S^p \times S^q) = u(\partial D^{p+1}) = f(S^p \times 1).$$

Moreover, a web is unique up to isotopy.

Now take two embeddings $f_0, f_1 : S^p \times S^q \to \mathbb{R}^m$ and their webs D_0^{p+1} and D_1^{p+1}. Join the centres of D_0^{p+1} and D_1^{p+1} by an arc. Construct an embedding $\partial D^{p+1} \times I \to \mathbb{R}^m$ 'along this arc' so that

$$\partial D^{p+1} \times I \cap f_i(S^p \times S^q) = \partial D^{p+1} \times i = f_i(S^p \times 1) \quad \text{for} \quad i = 0, 1.$$

Take a 'connected sum' of f_0 and f_1 'along $\partial D^{p+1} \times I$'. The resulting embedding $S^p \times S^q \to \mathbb{R}^m$ is the sum of f_0 and f_1. $\qquad\square$

Theorem 3.8 ([77, 87, 199]).

$$KT_{p,q,PL}^{p+2q+1} \cong \begin{cases} \mathbb{Z}_{(q)} & 1 \leq p < q \\ \mathbb{Z}_{(q)} \oplus \mathbb{Z}_{(q)} & 2 \leq p = q \end{cases}$$

and

$$KT_{p,q,DIFF}^{p+2q+1} \cong \mathbb{Z}_{(q)} \quad for \quad 1 \leq p \leq q - 2.$$

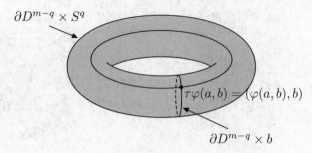

Fig. 3.10.

The theorem follows from Theorem 2.13 (as well as from Theorem 3.9 below). In the PL case of Theorem 3.8 for $p = q$ we only have a one-to-one correspondence of sets (because Theorem 3.7 does not give a group structure for such dimensions). A description of $KT^{6k}_{2k-1,2k,DIFF}$ is given after Theorem 2.14.

This result can be generalized as follows.

Theorem 3.9 ([181, Corollary 1.5.a]). *If $2m \geq 3q + 2p + 4$ or $2m \geq 3q + 3p + 4$, in the PL or DIFF cases respectively, then*

$$KT^m_{p,q} \cong \pi_q(V_{m-q,p+1}) \oplus \pi_p(V_{m-p,q+1}).$$

Note that $\pi_p(V_{m-p,q+1}) = 0$ for $m \geq 2p + q + 2$ (which is automatic for $p \leq q$ and $2m \geq 3p + 3q + 4$). Theorem 3.9 follows from Theorems 5.5, 5.6 and the deleted product lemma 5.3(b) below; it was proved for $2m \geq 3p + 3q + 4$ and $m \geq 2q + 3$ in [11]. For $m \geq 2p + q + 2$ there is an alternative direct proof [183], but for $m < 2p + q + 2$ (when no group structure exists) no proof is known of Theorem 3.9 that does not refer to the deleted product method. For an application see [23].

Let us construct a map $\tau : \pi_q(V_{m-q,p+1}) \to KT^m_{p,q}$ giving one summand in Theorem 3.9. Recall that $\pi_q(V_{m-q,p+1})$ is isomorphic to a group of CAT maps $S^q \to V_{m-q,p+1}$ up to CAT homotopy. The latter maps can be considered as CAT maps $\varphi : S^q \times S^p \to \partial D^{m-q}$. Define the CAT embedding $\tau(\varphi)$ (Figure 3.10) as the composition

$$S^p \times S^q \xrightarrow{\varphi \times \mathrm{pr}_2} \partial D^{m-q} \times S^q \subset D^{m-q} \times S^q \subset \mathbb{R}^m.$$

Let us present some calculations based on Theorem 3.9 and the calculation of $\pi_q(V_{a,b})$ [140, 36]. Recall that

$$\pi_q(V_{m-q,2}) \cong \pi_q(S^{m-q-1}) \oplus \pi_q(S^{m-q-2}) \qquad \text{for } m - q \text{ even}$$

(and $\pi_q(V_{m-q,2}) \cong \pi^S_{2q+1-m} \oplus \pi^S_{2q+2-m}$ for $m-q$ even and $2m \geq 3q+6$) because the sphere S^{m-q-1} has a nonzero vector field. In all tables of this subsection u^v means $(\mathbb{Z}_u)^v$.

The Haefliger-Zeeman theorem suggests the following question: *how can we describe* $KT^m_{1,q}$? We have the following table for $2m \geq 3q+6$ and for $2m \geq 3q+7$, in the PL and DIFF cases respectively.

m	$2q+2$	$2q+1$	$2q$	$2q-1$	$2q-2$	$2q-3$
$KT^m_{1,q}$, q even	$\mathbb{Z}$	2	2^2	2^2	24	0
$KT^m_{1,q}$, q odd	2	$\mathbb{Z} \oplus 2$	4	$2 \oplus 24$	2	0

Theorem 3.8 and [130] suggest the following problem: describe $KT^m_{p,q}$ for $m \leq 2q+p$. We have the following table for $q \geq 4$ or $q \geq p+4$, in the PL or DIFF cases respectively.

p	1	$2 \leq p \leq q-2$	$q-1$	q
$KT^{p+2q}_{p,q}$, $q = 4s$	2	0	2	0
$KT^{p+2q}_{p,q}$, $q = 4s+2$	2	2	2^2	2^2
$KT^{p+2q}_{p,q}$, $q = 4s+1$	$\mathbb{Z} \oplus 2$	2^2	$\mathbb{Z} \oplus 2^2$	2^4
$KT^{p+2q}_{p,q}$, $q = 4s-1$	$\mathbb{Z} \oplus 2$	4	$\mathbb{Z} \oplus 4$	4^2

Classification of *smooth* embeddings $S^p \times S^q \to \mathbb{R}^m$ for $2m \leq 3p + 3q+3$, as well as PL embeddings for $2m \leq 2p+3q+3$, is much harder (because of the existence of smooth knots and the incompleteness of the Haefliger-Wu invariant). However, the *statements* are simple.

Theorem 3.10.

(a) *[183] If $p \leq q$, $m \geq 2p+q+3$ and $2m \geq 3q+2p+4$, then*

$$KT^m_{p,q,DIFF} \cong KT^m_{p,q} \oplus C^{m-p-q}_{p+q} \cong \pi_q(V_{m-q,p+1}) \oplus C^{m-p-q}_{p+q}.$$

(b) *[175] If $1 \leq p \leq 2k-2$, then*

$$KT^{6k}_{p,4k-1-p} \cong \pi_{4k-1-p}(V_{2k+p+1,p+1}) \oplus \mathbb{Z}.$$

Theorem 3.11 (cf. [183]).

(a)

$$KT^{6k+1}_{1,4k-1,DIFF} \cong \pi^S_{2k-2} \oplus \pi^S_{2k-1} \oplus \mathbb{Z} \oplus G_k, \qquad \text{(DIFF)}$$

where $k > 1$ and G_k is an abelian group of order 1, 2 or 4, and

$$KT^{6k+1}_{1,4k-1,PL} \cong \pi^S_{2k-2} \oplus \pi^S_{2k-1} \oplus \mathbb{Z}. \qquad \text{(PL)}$$

(b) *For each $k > 0$ we have*

$$KT^{6k+4}_{1,4k+1,PL} \cong KT^{6k+4}_{1,4k+1,DIFF} \cong \mathbb{Z}^a_2 \oplus \mathbb{Z}^b_4$$

for some integers $a = a(k)$, $b = b(k)$ such that

$$a + 2b - \mathrm{rk}(\pi^S_{2k} \otimes \mathbb{Z}_2) - \mathrm{rk}(\pi^S_{2k-1} \otimes \mathbb{Z}_2) = \begin{cases} 0 & k \in \{1,3\} \\ 1 & k+1 \text{ is not} \\ & \text{a power of } 2 \\ 1 \text{ or } 0 & k+1 \geq 8 \text{ is} \\ & \text{a power of } 2. \end{cases}$$

For a generalization of this theorem and its relation to homotopy groups of Stiefel manifolds see [183, 136].

By the theorem just given and [140, 196, 36] (see the details in [183]) we have the following table.

l	2	3	4	5	
$KT^{3l+1}_{1,2l-1,PL}$	$\mathbb{Z}^2 \oplus 2$	4	$\mathbb{Z} \oplus 24 \oplus 2$	2^2	
l	6	7	8	9	10
$KT^{3l+1}_{1,2l-1,PL}$	$\mathbb{Z}$	2	$\mathbb{Z} \oplus 240 \oplus 2$	$2\tilde{\times}2^2$	$\mathbb{Z} \oplus 2^5$

The following strong result was proved using a clever generalization of methods from [183].

Theorem 3.12 ([25]). *Assume that*

$$p \leq q, \quad p + \frac{4}{3}q + 2 < m < p + \frac{3}{2}q + 2 \quad \text{and} \quad m > 2p + q + 2.$$

Then the group $KT^m_{p,q,DIFF}$ is infinite if and only if either $q + 1$ or $p + q + 1$ is divisible by 4.

We conclude this subsection with some open problems.

- Find $\widehat{C^{m-p-q}_{p+q}}$, at least for particular cases.
- Describe $KT^{6k+2}_{2k,2k+1,DIFF}$. Note that $\#KT^{6k+2}_{2k,2k+1,DIFF} \in \{2,3,4\}$ [175], cf. Theorem 3.9(b).
- Find $KT^{6k+3}_{2k,2k+1,DIFF}$. In this case the Whitney invariant is a surjection onto $\mathbb{Z}_2$, and both preimages consist of 1 or 2 elements.
- Find $KT^{3q+1}_{q,q,DIFF}$ for $q \geq 2$. In this case the image of the Whitney invariant is $\mathbb{Z} \vee \mathbb{Z}$ for q even, and is either $\mathbb{Z}_2 \vee \mathbb{Z}_2$ or $\mathbb{Z}_2 \oplus \mathbb{Z}_2$ for q odd. (Here, if G is a group, $G \vee G$ is defined by

$$G \vee G = \{(x,y) \in G \oplus G \mid \text{either } x = 0 \text{ or } y = 0\}.)$$

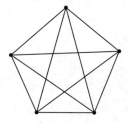

Fig. 4.1.

The nonempty preimages of the Whitney invariant consist of 1 element for q odd, of 1, 2, 4 elements for q even ≥ 4, and of 1, 2, 3, 4, 6, 12 elements for $q = 2$.

It would be interesting to find an action of the group of CAT auto-homeomorphisms of $S^p \times S^q$ on $KT^m_{p,q,CAT}$ (E. Rees). The above classification of knotted tori could perhaps be applied to solve for knotted tori the Hirsch problem about the description of possible normal bundles for embeddings of manifolds into $\mathbb{R}^m$, cf. [130]. The same remark holds for the following *Hirsch-Rourke-Sanderson problem* [80, 160], cf. [195, 200]: *which embeddings $N \to \mathbb{R}^{m+1}$ are isotopic to embeddings $N \to \mathbb{R}^m$?*

4 The van Kampen obstruction

4.1 The embeddability of n-complexes in $\mathbb{R}^{2n}$

By the general position theorem (Theorem 2.1) the first nontrivial case of the embedding problem is the investigation of embeddability of n-polyhedra in Euclidean space $\mathbb{R}^{2n}$, cf. Example 2.2. For $n = 1$ this problem was solved by the Kuratowski criterion [109], see also [151, §2], [182] and references there. However for $n > 1$, such a simple criterion does not exist [165]. (Note that there are infinitely many closed non-orientable 2-surfaces that do not embed into $\mathbb{R}^3$, and these do not contain a common subspace non-embeddable into $\mathbb{R}^3$.)

In [197] an obstruction to the embeddability of n-polyhedra in $\mathbb{R}^{2n}$ was constructed for arbitrary n (see also a historical remark at the end of Section 5).

To explain the idea of van Kampen, we sketch a proof of the nonplanarity of K_5 (i.e. of the complete graph with 5 vertices, see Figure 4.1). Take any general position map $f : K_5 \to \mathbb{R}^2$. For each two edges σ, τ the intersection $f\sigma \cap f\tau$ consists of a finite number of points.

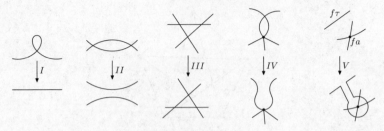

Fig. 4.2. 'Reidermeister moves'

Let v_f be the sum mod 2 of the numbers $|f\sigma \cap f\tau|$, over all non-ordered pairs $\{\sigma, \tau\}$ of disjoint edges of K_5.

For the map f shown in Figure 4.1, $v_f = 1$. Every general position map $f : K_5 \to \mathbb{R}^2$ can be transformed to any other such map through isotopies of $\mathbb{R}^2$ and 'the Reidemeister moves' for graphs in the plane from Figure 4.2.

This assertion is proved analogously to the Reidemeister theorem for knots. We will not prove it, since it is needed only for this sketch proof and not for the rigorous proof. For each edge of K_5 with vertices a, b, the graph $K_5 - \{a, b\}$, obtained by deleting from K_5 the vertices a, b and the interiors of the edges adjacent to a, b, is a circle (this is the very property of K_5 we need for the proof). Therefore v_f is invariant under the 'Reidemeister moves'. Hence $v_f = 1$ for *each* general position map $f : K_5 \to \mathbb{R}^2$. So K_5 is nonplanar. (For a proof without use of the assertion on the Reidemeister moves see below or [174].)

Similarly, one can prove that the graph $K_{3,3}$ (three houses and three wells) is not embeddable into $\mathbb{R}^2$ and that the 2-skeleton of the 6-simplex is not embeddable into $\mathbb{R}^4$ (again, compare this with Example 2.2).

4.2 Ramsey link theory

Now let us discuss some generalizations of the above proof, which are interesting in themselves and are used in Section 7. From that proof one actually gets a stronger assertion. Let e be an edge of K_5 and Σ^1 the cycle in K_5, formed by the edges of K_5 disjoint with e. Then $K_5 - \mathring{e}$ embeds into $\mathbb{R}^2$ (Figure 4.3) and for each embedding $g : K_5 - \mathring{e} \to \mathbb{R}^2$ the g-images of the ends of e (which form a 0-sphere $g\partial e = g\Sigma^0$) lie on different sides of $g\Sigma^1$.

Moreover, let e be a 2-simplex of the 2-skeleton Δ_6^2 of the 6-simplex

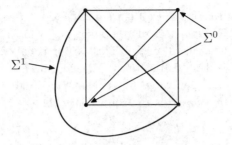

Fig. 4.3. Embedding of $K_5 - \mathring{e}$

and $P = \Delta_6^2 - \mathring{e}$. Then P embeds into $\mathbb{R}^4$. Let $\Sigma^1 = \partial e$ and Σ^2 be the sphere formed by 2-faces disjoint with e. Then for each embedding $P \to \mathbb{R}^4$ the images of these spheres link with a nonzero (more precisely, with an odd) linking number [45].

Using the above idea one can prove the following.

Theorem 4.1 ([164, 32]). *For any embedding $K_6 \to \mathbb{R}^3$ there are two cycles in K_6 whose images are linked with an odd linking coefficient.*

For generalizations see [158, 137, 12, 184, 144].

4.3 The van Kampen obstruction mod 2

Fix a triangulation T of a polyhedron N. The space

$$\widetilde{T} = \cup \{\sigma \times \tau \in T \times T \mid \sigma \cap \tau = \emptyset\}$$

is called the *simplicial deleted product* of N. By the simplicial deleted product lemma (Lemma 5.4 below) the equivariant homotopy type of $\widetilde{T}$ depends only on N, so we write $\widetilde{N}$ instead of $\widetilde{T}$ in this section. Consider the involution $t : \widetilde{N} \to \widetilde{N}$ defined by $t(x, y) = (y, x)$. Let $N^* = \widetilde{N}/t$.

For any general position PL map $f : N \to \mathbb{R}^2$ and disjoint edges σ, τ of T, the intersection $f(\sigma) \cap f(\tau)$ consists of a finite number of points. Let

$$v_f(\sigma, \tau) = |f(\sigma) \cap f(\tau)| \mod 2.$$

Then v_f is an element of the group $C^2(N^*; \mathbb{Z}_2)$ of vectors (which are called *cochains*) with components from $\mathbb{Z}_2$ indexed by 2-simplices of N^*, i.e. by non-ordered products of disjoint edges of N.

This cochain v_f is invariant under isotopy of $\mathbb{R}^2$ and the first four 'Reidemeister moves' (Figure 4.2, I–IV). The fifth 'Reidemeister move' (passing of $f\tau$ through fa, see Figure 4.2,V) adds to v_f the vector that assumes value 1 on the class of 2-simplex $\sigma \times \tau$ for $a \in \sigma$, and value 0 on the other 2-simplices of N^*. This vector is called an *elementary coboundary* and is denoted by $\delta[a \times \tau]$. Denote by $B^2(N^*;\mathbb{Z}_2)$ the subgroup of $C^2(N^*;\mathbb{Z}_2)$ generated by elementary coboundaries. The *van Kampen obstruction mod 2* is the equivalence class

$$v(N) := [v_f] \in H^2(N^*;\mathbb{Z}_2) = C^2(N^*;\mathbb{Z}_2)/B^2(N^*;\mathbb{Z}_2).$$

(Since $\dim N^* = 2$, it follows that $C^2 = Z^2$.)

Sketch of a proof† that $v(N)$ is independent of f. We follow [172, Lemma 3.5]. Consider an arbitrary general position homotopy $F : N \times I \to \mathbb{R}^2 \times I$ between general position maps $f_0, f_1 : N \to \mathbb{R}^2$. Colour a pair $a \times \tau$ in red if $F(a \times I)$ intersects $F(\tau \times I)$ in an odd number of points. Then $v_{f_0} - v_{f_1}$ is the sum of $\sum \delta(a \times \tau)$ over all red pairs $a \times \tau$. Hence v_f is independent on f.   □

Remark. Clearly, $v(N) = 0$ for all planar graphs N.

Analogously, one can define the mod 2 van Kampen obstruction

$$v(N) \in H^{2n}(N^*;\mathbb{Z}_2)$$

which obstructs embeddability of an n-polyhedron N into $\mathbb{R}^{2n}$.

4.4   The integral van Kampen obstruction

Fix a triangulation of N and define $\widetilde{N}$, t and N^* as above. Choose an orientation of $\mathbb{R}^{2n}$ and orientations of the n-simplices of $\widetilde{N}$. The latter give orientations on $2n$-simplices of $\widetilde{N}$. Clearly, $t(\sigma \times \tau) = (-1)^n(\tau \times \sigma)$ (the case $n = 1$ helps to check the sign) see [172, p .257].

For any general position map $f : N \to \mathbb{R}^{2n}$ and any two disjoint n-simplices σ, τ of N the intersection $f(\sigma) \cap f(\tau)$ consists of a finite number of points. Define the *intersection cochain* $V_f \in C^{2n}(\widetilde{N})$ by the formula

$$V_f(\sigma, \tau) = f\sigma \cdot f\tau := \sum_{P \in f\sigma \cap f\tau} \operatorname{sign} P.$$

Here $\operatorname{sign} P = +1$, if for the positive n-bases $s_1, \ldots, s_n$ and $t_1, \ldots, t_n$ of σ and τ, respectively, we have that $fs_1, \ldots, fs_n, ft_1, \ldots, ft_n$ is a positive $2n$-base of $\mathbb{R}^{2n}$; and $\operatorname{sign} P = -1$ otherwise. Clearly [172],

$$V_f(\sigma \times \tau) = (-1)^n V_f(\tau \times \sigma) = V_f(t(\sigma \times \tau)).$$

---

† This argument does not use the assertion about 'the Reidemeister moves', which is not proved here.

So V_f induces a cochain in the group $C^{2n}(N^*)$. Denote this new cochain by the same notation V_f (we use the old V_f only in the proof of Lemma 4.4, so no confusion will arise).

The *van Kampen obstruction* is the equivalence class

$$V(N) := [V_f] \in H^{2n}(N^*) = C^{2n}(N^*)/B^{2n}(N^*).$$

This class is independent of f (the proof is analogous to that for $v(N)$).

Clearly, $V(N)$ is an obstruction to the embeddability of N into $\mathbb{R}^{2n}$. One can easily show that $V(N)$ depends on the choice of orientations of $\mathbb{R}^{2n}$ and of the n-simplices of N only up to an automorphism of the group $H^{2n}(N^*)$.

Remark. The author is grateful to S. Melikhov for indicating that in [49, 108, 12] the signs are not accurate, so that the van Kampen obstruction for odd n erroneously assumes its values in in $H^{2n}_{\mathbb{Z}_2}(\widetilde{N}; \mathbb{Z})$ (where the involution acts on $\widetilde{N}$ by exchanging factors and on $\mathbb{Z}$ by $(-1)^n$).

4.5 The van Kampen-Shapiro-Wu theorem

Theorem 4.2. *If an n-polyhedron N embeds into $\mathbb{R}^{2n}$, then $V(N) = 0$. For $n \neq 2$ the converse is true, whereas for $n = 2$ it is not [197, 172, 212, 49].*

The necessity of $V(N) = 0$ in Theorem 4.2 was actually proved in the construction of the van Kampen obstruction. Sufficiency in Theorem 4.2 for $n \geq 3$ follows from Lemmas 4.4 and 4.5 below, and for $n = 1$ is obtained using the Kuratowski graph planarity criterion. A counterexample to the completeness of the van Kampen obstruction for $n = 2$ is presented in Section 7.

Definition 4.3. A map $g : N \to \mathbb{R}^m$ is of a polyhedron N is called a *nondegenerate almost embedding* if there exists a triangulation T of N such that $g|_\alpha$ is an embedding for each $\alpha \in T$ and $g\alpha \cap g\beta = \emptyset$ for each $\alpha \times \beta \subset \widetilde{T}$.

Lemma 4.4. *If N is an n-polyhedron and $V(N) = 0$, then there exists a general position nondegenerate almost embedding $g : N \to \mathbb{R}^{2n}$ cf. [49, Lemmas 2 and 4].*

Sketch of the proof. Let T be a triangulation of N. Let $\varphi : N \to \mathbb{R}^{2n}$ be a map linear on the simplices of T. Then φ is nondegenerate. The condition $V(N) = 0$ implies that $V_\varphi \in C^{2n}(\widetilde{N})$ is a symmetric coboundary. Hence V_φ is the sum of some 'elementary' symmetric coboundaries $\delta(\sigma^n \times \nu^{n-1}) + \delta t(\sigma^n \times \nu^{n-1})$ over some $\sigma^n \times \nu^{n-1} \in \widetilde{T}$. Applying the van Kampen finger moves (higher-dimensional analogues of Figure 4.2.V) for all pairs $\sigma^n \times \nu^{n-1}$ from this sum we obtain a general position nondegenerate map $f : N \to \mathbb{R}^{2n}$ such that $f\alpha \cdot f\beta = 0$ for each $\alpha, \beta \in \widetilde{T}$.

Then by induction on pairs of n-simplices of $\widetilde{T}$ and using the Whitney trick (see below) in the inductive step we obtain the required map g. See the details in [49].   □

Let us illustrate the application of the Whitney trick by the following argument.

Sketch proof of Theorem 2.3 in the smooth case. We must show that every smooth n-manifold N smoothly embeds into $\mathbb{R}^{2n}$. For $n \leq 2$ the proof is trivial, so assume that $n \geq 3$.

Using the higher-dimensional analogue of the first Reidemeister move (Figure 4.2.I), any smooth general position map $f : N \to \mathbb{R}^{2n}$ can be modified so that a single self-intersection point with a prescribed sign will be added. Hence there exists a general position map $f : N \to \mathbb{R}^{2n}$ whose self-intersections consist of an even number of isolated points, with algebraic sum zero.

In order to conclude the proof, we 'kill' these double points in pairs. This procedure is analogous to the second Reidemeister move (Figure 4.2.II) and is called the *Whitney trick*. More precisely, take two double points of opposite sign:

$$x_1, y_1, x_2, y_2 \in N \quad \text{so that} \quad f(x_1) = f(x_2), \quad f(y_1) = f(y_2).$$

Join x_1 to y_1 and x_2 to y_2 by arcs l_1 and l_2 so that these double points have 'opposite signs' along these arcs (Figure 4.4).

By general position ($n \geq 2$), we may assume that the restrictions $f|_{l_1}$ and $f|_{l_2}$ are embeddings and that l_1 and l_2 do not contain other double points of f. Since $n \geq 3$, by general position we can embed a 2-disc C into $\mathbb{R}^{2n}$ so that

$$\partial C = f(l_1) \cup f(l_2) \quad \text{and} \quad C \cap f(N) = \partial C$$

(such a disc D is called *Whitney's disc*). We can move the f-image of a regular neighbourhood of l_1 in N 'along' C so that we 'cancel' the double points $f(x_1) = f(x_2)$ and $f(y_1) = f(y_2)$. For details see [1, 145].   □

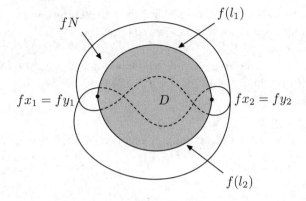

Fig. 4.4. The Whitney trick

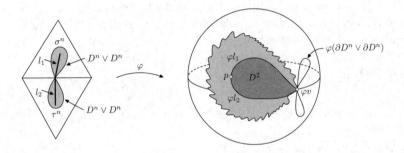

Fig. 4.5. A nondegenerate almost embedding

Lemma 4.5 (The Freedman-Krushkal-Teichner lemma). *If there is a nondegenerate almost embedding* $\varphi : N \to \mathbb{R}^{2n}$ *of an n-polyhedron* N *and* $n \geq 3$, *then there is an embedding* $f : N \to \mathbb{R}^{2n}$ *[49, Lemma 5], cf. [207, §6], [179].*

Proof. Take a triangulation T as in Definition 4.3. We may assume by induction that $\varphi\alpha \cap \varphi\beta = \varphi(\alpha \cap \beta)$ for each $\alpha, \beta \in T$ such that $\alpha \cap \beta \neq \emptyset$ except for $(\alpha, \beta) = (\sigma^n, \tau^n)$. In addition, we may assume that $\varphi\mathring{\sigma}^n \cap \varphi\mathring{\tau}^n$ is a point, p say. Let v be a point of $\sigma^n \cap \tau^n$. Take PL arcs $l_1 \subset v \cup \mathring{\sigma}^p$ and $l_2 \subset v \cup \mathring{\tau}^q$ joining v to $\varphi^{-1}p$ and containing no self-intersection points of φ in their interiors (Figure 4.5).

Then $\varphi(l_1 \cup l_2)$ is a circle. Since $n \geq 3$, this circle bounds a PL embedded 2-disc $D \subset \mathbb{R}^{2n}$.

We have by general position $(n+2 < 2n)$ that $\mathring{D} \cap \varphi N = \emptyset$. There is a small neighbourhood D^{2n} of D in $\mathbb{R}^{2n}$ rel $\varphi(v)$ that is PL homeomorphic to the $2n$-ball and such that $\varphi^{-1}D^{2n}$ is a neighbourhood of $l_1 \cup l_2$ in N rel v and is homeomorphic to the wedge $D^n \vee D^n$. By the "unknotting wedges theorem" [114], the restriction $\varphi : \partial\varphi^{-1}D^{2n} \to \partial D^{2n}$ is unknotted. Hence it can be extended to an embedding $h : \partial\varphi^{-1}D^{2n} \to D^{2n}$. In order to conclude the proof, set f equal to φ on $N - \partial\varphi^{-1}D^{2n}$ and to h on $\partial\varphi^{-1}D^{2n}$.   □

Observe that for the embedding f constructed above we have $\widetilde{f} \simeq_{eq} \widetilde{\varphi}$ on $\widetilde{T}$.

4.6 Generalizations of the van Kampen obstruction

The idea of the van Kampen obstruction can be applied to calculate the minimal m such that a polyhedron, which is a product of graphs, embeds into $\mathbb{R}^m$ [184], cf. [52, 6].

Analogously, one can construct the *van Kampen-Wu invariant* $U(f) \in H^{2n}(N^*)$ of an embedding $f : N \to \mathbb{R}^{2n+1}$.

Theorem 4.6. *If embeddings* $f, g : N \to \mathbb{R}^{2n+1}$ *of a finite n-polyhedron N are isotopic, then* $U(f) = U(g)$. *For $n \geq 2$ the converse is true, whereas for $n = 1$ it is not [214].*

Note that embeddings $f, g : N \to \mathbb{R}^3$ of a graph N such that $U(f) = U(g)$ are *homologous* [194].

We shall not present a proof of this theorem in this paper. For $n \geq 2$ it is proved analogously to Theorem 4.2 using the ideas of [153, §12], and for $n = 1$ it is trivial.

As was pointed out by Shapiro, when $V(N) = 0$ (and hence N is embeddable in $\mathbb{R}^{2n}$ for $n \geq 3$), one can construct the 'second obstruction' to the embeddability of N in $\mathbb{R}^{2n-1}$, etc, cf. Section 5.

For a subpolyhedron A of a polyhedron N one can analogously define the obstruction to extending a given embedding $A \subset \partial B^m$ to an embedding $N \to B^m$ [49]. This *relative* van Kampen obstruction is complete for $n \neq 2$ (for $n \geq 3$ see [214] and for $n = 1$ this follows from a relative version of the Kuratowski criterion [182]) and is incomplete for $n = 2$ (see Section 7).

Remark. For the van Kampen obstruction for *approximability by embeddings*, see [22, §4], [152], [7, §4] [156, 127, 185].

5 The Haefliger-Wu invariant

5.1 Basic idea

The ideas of the 'complement of the diagonal' and the Gauss map play a great role in different branches of mathematics [57, 198]. The Haefliger-Wu invariant is a manifestation of these ideas in the theory of embeddings.

The idea of taking the complement of the diagonal originated from two celebrated theorems: the Lefschetz fixed point theorem and the Borsuk-Ulam antipodes theorem. In order to state the latter denote the antipode of a point $x \in S^n$ by $-x$ and recall that a map $f : S^n \to S^m$ between spheres is *equivariant* (or *odd*) if $f(-x) = -f(x)$ for each $x \in S^n$.

Theorem 5.1 (The Borsuk-Ulam theorem).

(a) *For any map $f : S^n \to \mathbb{R}^n$ there exists $x \in S^n$ such that $f(x) = f(-x)$.*
(b) *There are no equivariant maps $S^n \to S^{n-1}$.*
(c) *Every equivariant map $S^n \to S^n$ is not homotopic to the constant map.*

Part (c) is nontrivial, see e.g. the proof in [146, 8.8], and we do not prove it here. We show briefly how part (c) implies the other assertions.

Sketch of a deduction of (a) *and* (b) *from* (c). Part (b) follows from (c) because if $\varphi : S^n \to S^{n-1}$ is an equivariant map, then the restriction $\varphi|_{S^{n-1}}$ extends to S^n and therefore is null-homotopic.

In order to present the idea of the Gauss map in the simplest case, let us deduce (a) from (b). Suppose (a) does not hold, i.e. there exists a map $f : S^n \to \mathbb{R}^n$ which does not identify any antipodes. Then a map

$$\widetilde{f} : S^n \to S^{n-1} \quad \text{is well-defined by} \quad \widetilde{f}(x) = \frac{f(x) - f(-x)}{|f(x) - f(-x)|}.$$

Evidently $\widetilde{f}$ is equivariant, which contradicts (b). $\square$

Construction of Example 2.2. We present a simplified construction invented by Schepin [168, Appendix] and D. Repovs together with the author [155] (and possibly others). Let T be a triod, i.e. the graph with four vertices O, A, B, C and three edges OA, OB and OC. The product T^{n+1} is a cone over some n-polyhedron N.

In order to prove that N does not embed into $\mathbb{R}^{2n}$ it suffices to prove that T^{n+1} does not embed in $\mathbb{R}^{2n+1}$. Suppose to the contrary that there is an embedding $f : T^{n+1} \to \mathbb{R}^{2n+1}$. Let $p : D^2 \to T$ be a map which

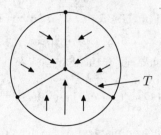

Fig. 5.1. The map $p : D^2 \to T$

Fig. 5.2. The deleted product

does not identify any antipodes of $S^1 = \partial D^2$ (e.g. the map from Figure 5.1). It is easy to check that the map $p^{n+1}|_{\partial D^{2n+2}} : \partial D^{2n+2} \to T^{n+1}$ also does not identify any antipodes. Then the composition of p^{n+1} and f again does not identify antipodes. This contradicts Theorem 5.1(a). □

5.2 Definition of the Haefliger-Wu invariant

The *deleted product* $\widetilde{N}$ of a topological space N is the product of N with itself, minus the diagonal:

$$\widetilde{N} = \{(x, y) \in N \times N \mid x \neq y\}.$$

This is the configuration space of ordered pairs of distinct points of N.

Now suppose that $f : N \to \mathbb{R}^m$ is an embedding. Then the map $\widetilde{f} : \widetilde{N} \to S^{m-1}$ is well-defined by the Gauss formula

$$\widetilde{f}(x, y) = \frac{f(x) - f(y)}{|f(x) - f(y)|}.$$

This map is equivariant with respect to the involution $t(x, y) = (y, x)$ on N and the antipodal involution a_{m-1} on S^{m-1}. Thus the existence of at least one equivariant map $\widetilde{N} \to S^{m-1}$ is a necessary condition for the embeddability of N in $\mathbb{R}^m$.

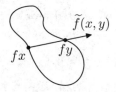

Fig. 5.3. The Gauss map

Consider isotopic embeddings $f_0, f_1 : N \to \mathbb{R}^m$ and an isotopy $f_t : N \times I \to \mathbb{R}^m$ between them. The homotopy $\widetilde{f_t}$ is an equivariant homotopy between $\widetilde{f_0}$ and $\widetilde{f_1}$. Hence the existence of an equivariant homotopy between $\widetilde{f_0}$ and $\widetilde{f_1}$ is necessary for embeddings $f_0, f_1 : N \to \mathbb{R}^m$ to be isotopic.

Definition 5.2 ([213, 66]). The *Haefliger-Wu invariant* $\alpha(f)$ of the embedding f is the equivariant homotopy class of the map $\widetilde{f}$.

(This is an application of the 'h-principle' idea [61, 2.1.E].)

Let $\pi^{m-1}_{eq}(\widetilde{N}) = [\widetilde{N}; S^{m-1}]_{eq}$ be the set of equivariant maps $\widetilde{N} \to S^{m-1}$ up to equivariant homotopy. Thus the Haefliger-Wu invariant is a mapping

$$\alpha = \alpha^m_{CAT}(N) : \mathrm{Emb}^m_{CAT}(N) \to \pi^{m-1}_{eq}(\widetilde{N})$$

defined by $\alpha(f) = [\widetilde{f}] \in \pi^{m-1}_{eq}(\widetilde{N})$.

5.3 Calculations of the Haefliger-Wu invariant

It is important that the set $\pi^{m-1}_{eq}(\widetilde{N})$ can be explicitly calculated in many cases using methods of algebraic topology. Let us give several examples (see also [51], [30, start of §2], [69],[70, 1.7.1] [10, 11], [1, 7.1] [153, 181]. We denote by '$\cong$' a bijection between sets.

Theorem 5.3 (The deleted product lemma).

(a) $\pi^{m-1}(\widetilde{S^p \sqcup S^q}) \cong \pi^S_{p+q+1-m}$ *for* $m \geq q + 2$.

(b) $\pi^{m-1}(\widetilde{S^p \times S^q}) \cong \pi_q(V_{m-q,p+1}) \oplus \pi_p(V_{m-p,q+1})$ *for* $2m \geq 3q + 2p + 4$.

The deleted product lemma will be proved in Section 6; part (b) follows from the torus lemma (Lemma 6.1). From the proof of part (a) it follows that the α-invariants of Sections 3 and 5 do indeed coincide for $N = S^p \sqcup S^q$.

Lemma 5.4 (The simplicial deleted product lemma). *Fix a triangulation T of a polyhedron N. The simplicial deleted product*

$$\widetilde{T} = \bigcup \{\sigma \times \tau \in T \times T \mid \sigma \cap \tau = \emptyset\}$$

is an equivariant strong deformation retract of $\widetilde{N}$ [172, Lemma 2.1], [86, §4].

Sketch of a proof. Let $E_{\sigma\tau} := \bigcup (U_\sigma \times U_\tau)$, where the union is taken over all mutually disjoint U_σ, U_τ that are nonempty faces of σ, τ respectively. Then

$$\sigma \times \tau \cong E_{\sigma\tau} * \operatorname{diag}(\sigma \cap \tau).$$

So for $\sigma \cap \tau \neq \emptyset$ there is an equivariant strong deformation retraction $\sigma \times \tau - \operatorname{diag}(\sigma \cap \tau) \to E_{\sigma\tau}$. These retractions agree on intersections, so together they form an equivariant strong deformation retraction $\widetilde{N} \to \widetilde{T}$. □

Now we sketch how to deduce in a purely algebraic way all the necessary conditions for embeddability and isotopy presented in Sections 2 and 4 from the 'deleted product necessary conditions'.

Let N be a polyhedron (in particular, a smooth manifold). By Lemma 5.4 there exists an equivariant map $\widetilde{N} \to S^{m-1}$ if and only if there exists an equivariant map $\widetilde{T} \to S^{m-1}$. Define an S^{m-1}-bundle

$$\gamma : \frac{\widetilde{T} \times S^{m-1}}{(x,y,s) \sim (y,x,-s)} \xrightarrow{\;S^{m-1}\;} \frac{\widetilde{T}}{(x,y) \sim (y,x)}$$

by $\gamma[(x,y,s)] := [(x,y)]$. The existence of an equivariant map $\widetilde{T} \to S^{m-1}$ is equivalent to the existence of a cross-section of γ; the existence of an equivariant homotopy between $\widetilde{f}_0$ and $\widetilde{f}_1$ is equivalent to the equivalence of the corresponding cross-sections of the bundle γ. Thus the existence of either can be checked using methods of obstruction theory.

In particular, the Whitney and van Kampen obstructions (Sections 2 and 4) are the first obstructions to the existence of a cross-section of the bundle γ [214]; the Whitney and van Kampen-Wu invariants (Sections 2 and 4) are the first obstructions to an equivariant homotopy of $\widetilde{f}_0$ and $\widetilde{f}_1$ [214].

5.4 The completeness of the Haefliger-Wu invariant

However trivial the 'deleted product necessary conditions' may seem, the material above shows that they are very useful. Thus it is very interesting to find out for which cases they are also *sufficient* for embeddability and isotopy, i.e. for which cases the following assertions hold:

(e) If there exists an equivariant map $\Phi : \widetilde{N} \to S^{m-1}$, then N embeds into $\mathbb{R}^m$.

(s) If there exists an equivariant map $\Phi : \widetilde{N} \to S^{m-1}$, then there exists an embedding $f : N \to \mathbb{R}^m$ such that $\widetilde{f} \simeq_{eq} \Phi$.

(i) If $f_0, f_1 : N \to \mathbb{R}^m$ are two embeddings and $\widetilde{f}_0 \simeq_{eq} \widetilde{f}_1$, then f_0 and f_1 are isotopic.

Clearly, (s) and (i) assert the surjectivity and injectivity, respectively, of α. Obviously, (s) implies (e).

Remark. From (e) it follows that if N TOP embeds into $\mathbb{R}^m$, then N PL or DIFF embeds into $\mathbb{R}^m$: in particular, the PL or DIFF embeddability of N into $\mathbb{R}^m$ does not depend on the PL or DIFF structure on N. Condition (i) has an analogous corollary.

Thus the surjectivity and injectivity of α are directly related to the embedding and knotting problems.

Theorem 5.5 (The Haefliger-Weber theorem; cf. [70, Theorem 1'], [207, Theorems 1, 1']). *For embeddings $N \to \mathbb{R}^m$ of either an n-polyhedron or a smooth n-manifold N the Haefliger-Wu invariant is*

- *bijective if $2m \geq 3n + 4$;*
- *surjective if $2m = 3n + 3$.*

The *metastable dimension restrictions* in the Haefliger-Weber theorem are sharp in the smooth case, as shown by the trefoil knot (Example 3.4) and other examples of smooth knots ([71]), and by Theorems 2.11(c), (d) (because the PL embedability $N \to \mathbb{R}^{2l+2}$ implies the existence of an equivariant map $\widetilde{N} \to S^{2l+1} \subset S^{3l-2}$). Such dimension restrictions appeared also in the PL cases of the classical theorems on embeddings of highly-connected manifolds and of Poincaré complexes (see Theorems 2.7 and 2.8 and [163, 18, 19]), but were later weakened to $m \geq n + 3$.

Since the 1960's it has been conjectured by Viro, Dranishnikov, Koschorke, Szücs, Schepin and others that in Theorem 5.5 the metastable dimension restrictions can also be weakened to $m \geq n+3$ for the PL case and connected N (possibly at the price of adding the *p-fold Haefliger-Wu invariant*, see Section 5.6 below). This turned out to be false not only for polyhedra (Examples 5.8(d) and 5.10(c) below) but even for PL manifolds (Examples 5.8(b), 5.8(c) and 5.9(b) below). Surprisingly, the metastable dimension restrictions *can* be weakened to $m \geq n +$

3 for *highly connected* PL n-manifolds (less highly connected than in Theorems 2.8, 2.12 and 2.13).

Theorem 5.6 ([181]). *For embeddings $N \to \mathbb{R}^m$ of a closed d-connected PL n-manifold N, with $m \geq n+3$, the Haefliger-Wu invariant is:*

- *bijective if $2m \geq 3n+3-d$;*
- *surjective if $2m = 3n+2-d$.*

For $d = 1$ we need only *homological* 1-connectedness in Theorem 5.6.

Observe that Theorem 5.6 is not quite the result expected in the 1960s, and that its proof cannot be obtained by direct generalization of the Haefliger-Weber proof without the invention of new ideas. This follows from the preceding discussion and the following two remarks.

Remark. The PL case of the Haefliger-Weber Theorem 5.5 holds for polyhedra, but Theorem 5.6 holds only for highly enough connected PL manifolds.

Remark. The same $(3n - 2m + 2)$-connectedness assumption as in the surjectivity part of Theorem 5.6 appeared in the Hudson PL version of the Browder-Haefliger-Casson-Sullivan-Wall embedding theorem for closed manifolds (roughly speaking, it states that a homotopy equivalence between PL manifolds is homotopic to a PL embedding, and it was proved by engulfing) [89].

This assumption was soon proved to be superfluous (by surgery) [75, 92]. It was therefore natural to expect that the $(3n-2m+3)$-connectedness assumption in Theorem 5.6 is superfluous (Theorem 5.6 is proved using a generalization of the engulfing approach). However, the examples of non-injectivity of the next subsection (Examples 5.8(b), (c)) show that this assumption is *sharp*.

In Section 8 we sketch the proof of surjectivity in Theorem 5.5 in the PL case and present the idea of the proof of Theorem 5.6. Most of the results of Sections 2, 3 and 4 are corollaries of Theorems 5.5 and 5.6, although some of them were originally proved independently (sometimes in a weaker form).

Corollary 5.7. *If N is a homologically 1-connected closed smooth n-manifold, then $\alpha_{DIFF}^m(N)$ is injective for $2m = 3n+2$, $n = 4s+2$ and surjective for $2m = 3n+1$, $n = 4s+3$.*

Sketch proof of the corollary. Theorem 5.6 and smoothing theory [74, 1.6], [72, 11.1] imply the following.

Let N be a closed d-connected (for $d = 1$, just homologically 1-connected) smooth n-manifold and $m \geq n + 3$.

- If $2m \geq 3n+2-d$, then for each $\Phi \in \pi_{eq}^{m-1}(\widetilde{N})$ there is a PL embedding $f : N \to \mathbb{R}^m$ smooth outside a point and such that $\alpha(f) = \Phi$; a complete obstruction to the smoothing of f lies in C_{n-1}^{m-n}.
- If $2m \geq 3n+3-d$, then any two smooth embeddings $f_0, f_1 : N \to \mathbb{R}^m$ such that $\alpha(f_0) = \alpha(f_1)$ can be joined by a PL isotopy, which is smooth outside a point; a complete obstruction to the smoothing of such a PL isotopy lies in C_n^{m-n}.

The corollary follows from these assertions and $C_{4k-2}^{2k} = 0$; see† [71, 8.15] and [129, Corollary C]. $\square$

In Theorem 5.6 the surjectivity is not interesting for $m < \frac{5n+6}{4}$. Indeed, if $\frac{5n+6}{4} > \frac{3n+2-d}{2}$ then $d > \frac{n}{2}-1$ and $n \geq 6$; hence N is a homotopy sphere, so $N \cong S^n$ and the surjectivity in Theorem 5.6 is trivial. But the proof is not simpler for $m \geq \frac{5n+6}{4}$; the proof can also be considered as a step towards the analogue of Theorem 5.6 for embeddings into *manifolds*, which is interesting even for $m < \frac{5n+6}{4}$. An analogous remark can be made about the injectivity in Theorem 5.6.

The Haefliger-Weber theorem (Theorem 5.5) has relative and approximative versions [70, 1.7.2], [207, Theorems 3 and 7], [152], which require that a constructed embedding or isotopy extend or are close to a given one. But Theorem 5.6 has such versions only under some additional assumptions.

Remark.

(i) An interesting corollary of [109, 27, 28] was deduced in [214] for graphs and in [179] for the general case: a Peano continuum N embeds into $\mathbb{R}^2$ if and only if there exists an equivariant map $\widetilde{N} \to S^1$.

(ii) An interesting corollary of [117] was deduced in [214]: embeddings $f, g : N \to \mathbb{R}^2$ of a Peano continuum N are isotopic if and only if $\widetilde{f} \simeq_{eq} \widetilde{g}$.

5.5   The incompleteness of the Haefliger-Wu invariant

Clearly, $\widetilde{S^n} \simeq_{eq} S^n$. Therefore the Haefliger-Wu invariant is not injective in codimension 2 (e.g. for knots in $\mathbb{R}^3$) and any smoothly nontrivial

---

† There is a misprint in [71, 8.15]: instead of "$C_{4k-2}^{3k} = 0$" it should be "$C_{8k-2}^{4k} = 0$".

knot $S^n \to \mathbb{R}^m$ demonstrates the non-injectivity of $\alpha_{DIFF}^m(S^n)$. The 'deleted product necessary conditions' for embeddability or isotopy do not reflect either the ambience of isotopy or the distinction between the DIFF and PL (or TOP) categories. The same assertions hold for generalized Haefliger-Wu invariants (see below) or 'isovariant maps invariants' [70], [1, §7].

Let us present other examples. All the examples in this subsection hold for *each* set of the parameters k, l, m, n, p satisfying the conditions in the statement.

Example 5.8 (Examples of non-injectivity). The following maps are not injective:

(a) $\alpha^{3l}(S^{2l-1} \sqcup S^{2l-1} \sqcup S^{2l-1})$, $\alpha^{3l}(S^{2l-1} \sqcup S^{2l-1})$ and $\alpha_{DIFF}^{3l}(S^{2l-1})$;
(b) $\alpha^{6k}(S^p \times S^{4k-1})$ for $p < k$ [181];
(c) $\alpha^{3l+1}(S^1 \times S^{2l-1})$ if $l + 1$ is not a power of 2 [183];
(d) $\alpha_{PL}^m((S^n \vee S^n) \sqcup S^{2m-2n-3})$ for $n + 2 \leq m \leq (3n+3)/2$ [181].

Example 5.8(a) follows from the examples of the Borromean rings, the Whitehead link and the trefoil knot as considered in Section 3. Examples 5.8(b), 5.8(c) and 5.8(d) are constructed below in Sections 6 and 7.

The construction (but not the proof [183]) of Example 5.8(c) is very simple and explicit.

Construction of Example 5.8(c) in the PL case. Add a strip to the Whitehead link $\omega_{0,PL}$ (see Section 3), i.e. extend it to an embedding

$$\omega_0' : S^0 \times S^{2l-1} \bigcup_{S^0 \times D_+^{2l-1} = \partial D_+^1 \times D_+^{2l-1}} D_+^1 \times D_+^{2l-1} \to \mathbb{R}^{3l}.$$

This embedding contains connected sum of the components of the Whitehead link. The union of ω_0' and the cone over the connected sum forms an embedding $D_+^1 \times S^{2l-1} \to \mathbb{R}_+^{3l+1}$. This latter embedding can clearly be shifted to a proper embedding. The *PL Whitehead torus* $\omega_{1,PL} : S^1 \times S^{2l-1} \to \mathbb{R}^{3l+1}$ is the union of this proper embedding and its mirror image with respect to $\mathbb{R}^{3l} \subset \mathbb{R}^{3l+1}$ (cf. the definition of μ' in Section 6).

It is easy to prove that $\alpha(\omega_{1,PL}) = \alpha(f_0)$, where f_0 is the standard embedding [183]. It is nontrivial that $\omega_{1,PL}$ is not PL isotopic to the standard embedding when $l + 1$ is not a power of 2.

Remark. Example 5.8(a) for $\alpha^{3l}(S^{2l-1} \sqcup S^{2l-1})$ is based on the linking coefficient. Example 5.8(c) is much more complicated because $S^1 \times S^{2l-1}$

is connected, so the linking coefficient cannot be defined (the linking coefficient for the restriction to $S^0 \times S^{2l-1}$ gives the weaker Example 5.8(b)). A new invariant [183] is therefore required.

Theorem 5.9 (Examples of non-surjectivity). *For $m \geq n+3$ the following maps are not surjective:*

(a) $\alpha^m(S^n \sqcup S^n)$, *if $\Sigma^\infty : \pi_n(S^{m-n-1}) \to \pi^S_{2n+1-m}$ is not epimorphic, e.g. for*

n	6	9	12	13	14	21
m	$\frac{3n+2}{2}$	$\frac{3n-1}{2}$	$\frac{3n}{2}$	$\frac{3n-1}{2}$	$\frac{3n+2}{2}$	$\frac{3n-1}{2}$
	10	13	18	19	22	31

(b) $\alpha^m(S^1 \times S^{n-1})$, *if $m - n$ is odd and $\Sigma^\infty : \pi_{n-1}(S^{m-n}) \to \pi^S_{2n-m-1}$ is not epimorphic [181], e.g. for*

n	7	10	13	14	15	22
m	$\frac{3n-1}{2}$	$\frac{3n-4}{2}$	$\frac{3n-3}{2}$	$\frac{3n-4}{2}$	$\frac{3n-1}{2}$	$\frac{3n-4}{2}$
	10	13	18	19	22	31

(c) $\alpha^{6k+1}_{DIFF}(S^{2k} \times S^{2k})$.

Here $\Sigma^\infty : \pi_{n+k}(S^n) \to \pi_{2k+2}(S^{k+2}) = \pi^S_k$ denotes the stable suspension mapping *$(n \leq k+2)$.*

Example 5.9(a) follows from the formula $\alpha = \pm\Sigma^\infty\lambda_{12}$ and the construction of a link with the prescribed linking coefficient of Section 3. Example 5.9(b) is constructed in Section 6.

Example 5.9(c) follows because $\alpha^{6k+1}_{PL}(S^{2k} \times S^{2k})$ is bijective by Theorem 5.6 (or by [13, 69]) but as shown by the Haefliger torus (Example 3.5) there exists a PL embedding $S^{2k} \times S^{2k} \to \mathbb{R}^{6k+1}$ that is not PL isotopic to a smooth embedding.

Links give many other examples of non-injectivity and non-surjectivity of α. From a link example, by gluing an arc joining connected components, we can obtain a highly connected polyhedral example.

Theorem 5.10 (Examples of non-embeddability). *There exists an equivariant map $\widetilde{N} \to S^{m-1}$, and yet N does not CAT embed into $\mathbb{R}^m$ (hence $\alpha^m_{CAT}(N)$ is not surjective), in the following cases:*

(a) *CAT=DIFF, $m = n+3$, $n \in \{8, 9, 10, 16\}$ and a certain homotopy n-sphere N;*

(b) *CAT=DIFF, $m = 6k-1$, $n = 4k$ and a certain (almost parallelizable $(2k-1)$-connected) n-manifold N;*

(c) *CAT=PL*, $\max\{3, n\} \le m \le \frac{3n+2}{2}$ *and a certain n-polyhedron* N *[121, 171, 49, 170, 58].*

Example 5.10(a) follows from the existence of a homotopy n-sphere N which is non-embeddable in codimension 3 [83, 113], cf. [149, §2], [118, pp. 407–408] (because $N \cong S^n$ topologically and so $\widetilde{N} \simeq_{eq} \widetilde{S^n}$). Example 5.10(b) follows from Theorems 2.11(c) and 2.11(d), while Example 5.10(c) is proved in Section 7 below.

Remark. In [179] it is proved that although the 3-adic solenoid Σ (i.e. the intersection of an infinite sequence of filled tori, each inscribed in the previous one with degree 3) does not embed into $\mathbb{R}^2$, there nevertheless exists an equivariant map $\widetilde{\Sigma} \to S^1$.

We conjecture that there exists a nonplanar *tree-like* continuum N, for which there are no equivariant maps $\widetilde{N} \to S^1$.

5.6 The generalized Haefliger-Wu invariant

The example of the Borromean rings (Example 3.2) suggests that one can introduce an obstruction to embeddability, analogous to the deleted product obstruction (and the van Kampen obstruction, see Section 4) but deduced from a triple, quadruple, or higher order product. Moreover, the vanishing of this obstruction should be sufficient for embeddability even when this is not so for the deleted product obstruction.

Such an obstruction can indeed be constructed as follows, cf. [108]. Suppose that G is a subgroup of the group S_p of permutations of p elements and let

$$\widetilde{N}_G = \{(x_1, \ldots, x_p) \in N^p \mid x_i \ne x_{\sigma(i)} \text{ for each } \sigma \in G \setminus \mathrm{id}, \ i = 1, \ldots, p\}$$

where id denotes the identity element of G. The space $\widetilde{N}_G$ is called the *deleted G-product* of N. The group G obviously acts on the space $\widetilde{N}_G$. For an embedding $f : N \to \mathbb{R}^m$ the map $\widetilde{f}_G : \widetilde{N}_G \to \widetilde{\mathbb{R}^m}_G$ is well-defined by the formula $\widetilde{f}_G(x_1, \ldots, x_p) = (fx_1, \ldots, fx_p)$. Clearly, the map $\widetilde{f}_G$ is G-equivariant. We can then define the *G-Haefliger-Wu invariant*

$$\alpha_G = \alpha_G^m(N) : \mathrm{Emb}^m(N) \to [\widetilde{N}_G, \widetilde{\mathbb{R}^m}_G]_G$$

by $\alpha_G(f) = [\widetilde{f}_G]$. The *deleted G-product obstruction* for the embeddability of N in $\mathbb{R}^m$ is the existence of a G-equivariant map $\Phi : \widetilde{N}_G \to \widetilde{\mathbb{R}^m}_G$.

This approach works well in link theory (the simplest example is the classification of 'higher-dimensional Borromean rings' [67, §3][125,

Proposition 8.3] by means of $\alpha_{S_3}^m$). In contrast, surprisingly, the examples of non-injectivity in Example 5.8 (except for the Borromean rings of 5.8(a)) demonstrate the non-injectivity of α_G for each G: in their formulations α can be replaced by α_G for each G. This follows by the construction of these examples. Clearly, if α is not surjective, then neither is α_G for each G. Under the conditions for non-embeddability in Example 5.10, property (e) is false even if we replace the $\mathbb{Z}_2$-equivariant map $\widetilde{N} \to S^{m-1}$ by a G-equivariant map $\widetilde{N}_G \to \widetilde{\mathbb{R}^m}_G$.

One can obtain other invariants, analogously to the construction of the generalized Haefliger-Wu invariant, using *isovariant* rather than *equivariant* maps. These invariants are possibly stronger but apparently harder to calculate (at least for manifolds more complicated than disjoint unions of spheres; for a disjoint union of spheres see Theorem 3.6 and [183, remarks after the Haefliger Theorem 1.1]. This seems to be one of the reasons why these invariants were not mentioned in [214] (whose last section considered even more complicated generalizations).

5.7 Historical remarks

A particular case of Theorem 5.5 (Theorem 4.2) was discovered by van Kampen [197]. Van Kampen's proof of sufficiency in Theorem 4.2 contained a mistake; however, he modified his argument to prove the PL case of Theorem 2.3. Based on the idea of the Whitney trick invented in [211], Shapiro and Wu completed the proof [172, 212].

Subsequently their argument was generalized by Haefliger and Weber (using some ideas of Shapiro and Zeeman) in order to prove the Haefliger-Weber Theorem 5.5. The second part of the Weber proof was simplified in [179] using the idea of the Freedman-Krushkal-Teichner lemma (Lemma 4.5).

The Whitney trick, on which the proof of sufficiency in Theorem 4.2 for $n \geq 3$ is based, cannot be performed for $n = 2$ [99, 111]. Sarkaria has found a proof† of the case $n = 1$ of Theorem 4.2 based on the 1-dimensional Whitney trick. Sarkaria also asked whether the sufficiency in Theorem 4.2 holds for the case $n = 2$. Freedman, Krushkal and Teichner have constructed an example showing that it does not [49].

The dimension restriction $2m \geq 3n+3$ in the Haefliger-Weber theorem (Theorem 5.5) comes from the use of the Freudenthal suspension theorem, the Penrose-Whitehead-Zeeman-Irwin embedding theorem (Theo-

---

† The author is grateful to K. Sarkaria and M. Skopenkov for indicating that the argument in [166] is incomplete.

rem 2.9), a relative version of the Zeeman unknotting theorem (Theorem 2.7(a)) and general position arguments (see Section 8). Toruńczyk and Spież showed that one can try to relax the restriction coming from the Zeeman unknotting theorem (using relative regular neighbourhoods) and those coming from the Freudenthal theorem (using Whitehead's 'hard part' of the Freudenthal theorem and the Whitehead higher-dimensional finger moves, see [47, §10]‡) [188, 189], see also [38, 39]. (We note that the application of higher-dimensional finger moves in this situation was first suggested by Schepin.)

In 1992 Dranishnikov therefore conjectured that surjectivity might hold in the Haefliger-Weber Theorem 5.5 for $2m = 3n + 2$. However, Segal and Spież [171] constructed a counterexample (using the same higher-dimensional finger moves), which was a weaker version of the non-embeddability example 5.10(c). Their construction used the Adams theorem on Hopf invariant 1, and therefore has exceptions corresponding to the exceptional values $1, 3, 7$. In 1995 the author suggested a way to remove the codimension-2 exceptions; subsequently, this idea was generalized independently by Segal-Spież and the author to obtain a simplification of [171] which did not use the Adams theorem, and therefore has no exceptions [170]. This simplified construction leads to the example of non-embeddability given as Example 5.10(c).

6 The deleted product of the torus

6.1 Proof of the deleted product lemma

Proof of Theorem 5.3(a). We have

$$\widetilde{S^p \sqcup S^q} \simeq_{eq} S^p \times S^q \sqcup S^p \times S^q \sqcup S^p \sqcup S^q$$

where the involution on the right-hand side exchanges antipodes in S^p and in S^q, as well as the corresponding points from the two copies of $S^p \times S^q$ (Figure 6.1). Therefore, analogously to the definition of the α-invariant in Section 3, we have

$$\pi_{eq}^{m-1}(\widetilde{S^p \sqcup S^q}) \cong [S^p \times S^q, S^{m-1}] \cong \pi_{p+q+1-m}^S \quad \text{for} \quad m - 2 \geq p, q$$

as required.   □

---

‡ A preliminary version of the material in [47] can be found in English translation as [46].

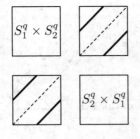

Fig. 6.1. $\widetilde{S^p \sqcup S^q}$

Recall that the *equivariant Stiefel manifold* V_{mn}^{eq} is the space of equivariant maps $S^{n-1} \to S^{m-1}$. Denote by $a_k : S^k \to S^k$ the antipodal map.

Lemma 6.1 (The torus lemma [181]). *For $p \leq q$ and $m \geq p+q+3$ there exist groups $\Pi_{p,q}^{m-1}$ and $\Pi_{q,p}^{m-1}$, a group structure on $\pi_{eq}^{m-1}(\widetilde{S^p \times S^q})$, and homomorphisms σ, γ, ρ, τ and α' forming the diagram below, in which the right-hand square commutes and the left-hand square either commutes or anticommutes.*

$$
\begin{array}{ccccc}
\pi_q(V_{m-q,p+1}) & \xrightarrow[\frac{3q}{2}+p+2]{\tau} & KT_{p,q}^m & \xrightarrow{\alpha} & \pi_{eq}^{m-1}(\widetilde{S^p \times S^q}) \\
\rho \downarrow {\scriptstyle \frac{3q}{2}+p+2} & & \downarrow \alpha' & & \gamma \downarrow {\scriptstyle p+q+3} \\
\pi_q(V_{m-q,p+1}^{eq}) & \xrightarrow[\frac{3q+p}{2}+2]{\sigma} & \Pi_{pq}^{m-1} & \xleftarrow[q+2p+2]{\mathrm{pr}_1} & \Pi_{p,q}^{m-1} \oplus \Pi_{q,p}^{m-1}
\end{array}
\qquad (6.1)
$$

Each of the homomorphisms σ, γ, ρ and pr_1 is an isomorphism under the dimension restriction $m \geq A$, where A is shown near the corresponding arrow in Diagram (6.1).

Proof. There is an equivariant deformation retraction

$$
H_1 : \widetilde{S^p \times S^q} \to \mathrm{adiag}\, S^p \times S^q \times S^q \bigcup_{\mathrm{adiag}\, S^p \times \mathrm{adiag}\, S^q} S^p \times S^p \times \mathrm{adiag}\, S^q,
$$

where adiag denotes the antidagonal (see Figure 6.2). More precisely, for non-antipodal points x and y of S^k and $t \in [0,1]$, let

$$
[x, y, t] = \frac{(1-t)x + ty}{|(1-t)x + ty|}.
$$

Define the deformation $H_t : \widetilde{S^p \times S^q} \to \widetilde{S^p \times S^q}$ by setting $H(x, y, x_1, y_1)$

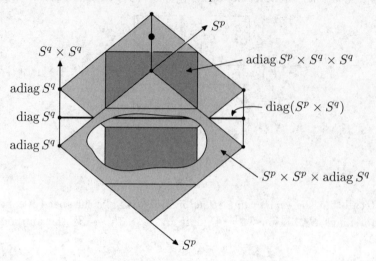

Fig. 6.2.

to be

$$([x, y, t], [y, x, t], [x_1, y_1, 2\delta t], [y_1, x_1, 2\delta t]) \quad \text{if } |x, y| \geq |x_1, y_1|$$
$$([x, y, 2(1 - \delta)t], [y, x, 2(1 - \delta)t], [x_1, y_1, t], [y_1, x_1, t]) \quad \text{if } |x_1, y_1| \geq |x, y|$$

where $\delta = \frac{|x,y|}{|x_1,y_1| + |x,y|}$.

Let

$$\Pi_{pq}^{m-1} := \pi_{eq}^{m-1}(S^p \times S^{2q})$$

where the involution on $S^p \times S^{2q}$ is $a_p \times t_q$ and $t_q : S^{2q} \to S^{2q}$ is the symmetry with respect to $S^q \subset S^{2q}$. The group structure on Π_{pq}^{m-1} is defined as follows: for equivariant maps $\varphi, \psi : S^p \times S^{2q} \to S^{m-1}$ define the map $\varphi + \psi : S^p \times S^{2q} \to S^{m-1}$ on $x \times S^{2q}$ to be the ordinary sum of the restrictions of φ and ψ to $x \times S^{2q}$; and analogously, define the identity and the inverse of φ on $x \times S^{2q}$ to be the ordinary identity and the ordinary inverse of $\varphi|_{x \times S^{2q}}$.

Let $v_q : S^q \times S^q \to \frac{S^q \times S^q}{S^q \vee S^q} \cong S^{2q}$ be the quotient map, cf. Section 3. Consider the involution $(s, x, y) \to (-s, y, x)$ on $S^p \times S^q \times S^q$. For $m \geq p + q + 3$ by general position we have a one-to-one correspondence

$$(\mathrm{id}_{S^p} \times v_q)^* : \pi_{eq}^{m-1}(S^p \times S^q \times S^q) \xrightarrow{\cong} \Pi_{pq}^{m-1}.$$

One can check that the involution on $S^q \times S^q$ exchanging factors corresponds to t_q.

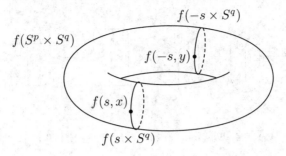

Fig. 6.3.

Consider the restrictions of an equivariant map $\widetilde{S^p \times S^q} \to S^{m-1}$ to adiag $S^p \times S^q \times S^q$ and to $S^p \times S^p \times$ adiag S^q. Define the map γ to be the direct sum of the compositions of such restrictions and the isomorphisms $(\mathrm{id}_{S^p} \times v_q)^*$ and $(\mathrm{id}_{S^q} \times v_p)^*$. If

$$\dim(\text{adiag } S^p \times \text{adiag } S^q) = p + q \le (m-1) - 2$$

then γ is a one-to-one correspondence, by general position and the Borsuk homotopy extension theorem. Take the group structure on $\pi_{eq}^{m-1}(\widetilde{S^p \times S^q})$ induced by γ. Then γ is an isomorphism.

By general position, for $2p + q \le m - 2$ we have $\Pi_{qp}^{m-1} = 0$; hence pr_1 is an isomorphism.

Given an embedding $f : S^p \times S^q \to \mathbb{R}^m$, let $\alpha'(f)$ be the map corresponding under the isomorphism $(\mathrm{id}_{S^p} \times v_q)^*$ to the map $S^p \times S^q \times S^q \to S^{m-1}$ defined by

$$(s, x, y) \mapsto \widetilde{f}((s, x), (-s, y)).$$

Clearly the right-hand square of Diagram (6.1) commutes.

The map τ was defined after Theorem 3.9. Recall also that ρ is the inclusion-induced homomorphism; by [76, 183], ρ is an isomorphism for $m \ge \frac{3q}{2} + p + 2$. We define σ as follows: an element $\varphi \in \pi_q(V_{m-q,p+1}^{eq})$ can be considered as a map $\varphi : S^p \times S^q \to S^{m-q-1}$ such that $\varphi(-x, y) = -\varphi(x, y)$ for each $x \in S^p$. Let $\sigma(\varphi)$ be the q-fold S^p-fibrewise suspension of such a map φ, i.e. $\sigma(\varphi)|_{x \times S^{2q}} = \Sigma^q(\varphi|_{x \times S^q})$. It is easy to see that σ is a homomorphism.

The (anti)commutativity of the left-hand square in Diagram (6.1) is proved for $p = 0$ using the representation $S^{m-1} \cong S^{m-q-1} * S^{q-1}$ and deforming $\alpha' \tau(\varphi)$ to the S^p-fibrewise suspension $\sigma(\varphi)$ of φ [100]. For $p > 0$ we apply this deformation for each $x \in S^p$ independently.

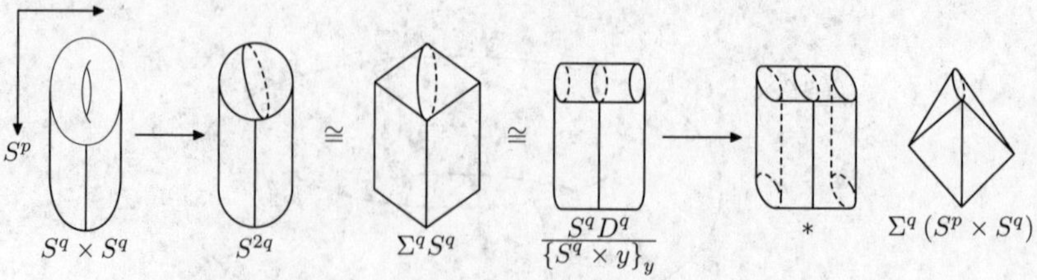

Fig. 6.4. The map $\mathrm{pr} : S^p \times \Sigma^q S^q \to \Sigma^q (S^p \times S^q)$

It remains to prove that σ is an isomorphism for $m \geq \frac{3q+p}{2} + 2 \geq p + q + 2$. For this, observe that σ is a composition

$$\pi_q(V^{eq}_{m-q,p+1}) = \pi^{m-q-1}_{eq}(S^p \times S^q) \xrightarrow{\Sigma^q} \pi^{m-1}_{eq}(\Sigma^q(S^p \times S^q)) \xrightarrow{\mathrm{pr}^*} \Pi^{m-1}_{pq}$$

Here the involution on $S^p \times S^q$ is $a_p \times \mathrm{id}_{S^q}$, the involutions on $\Sigma^q S^q$ and on $\Sigma^q(S^p \times S^q)$ are the 'suspension' involutions over id_{S^q} and $a_p \times \mathrm{id}_{S^q}$; the map

$$\mathrm{pr} : S^p \times \Sigma^q S^q = S^p \times \frac{S^q \times D^q}{S^q \times y, \ y \in \partial D^q}$$
$$\longrightarrow \frac{S^p \times S^q \times D^q}{S^p \times S^q \times y, \ y \in \partial D^q} = \Sigma^q(S^p \times S^q)$$

is a quotient map (Figure 6.4). The S^p-fibrewise group structures on $\pi^{m-q-1}_{eq}(S^p \times S^q)$ and on $\pi^{m-1}_{eq}(\Sigma^q(S^p \times S^q))$ are defined analogously to that on $\Pi^{m-1}_{p,q}$. By the equivariant version of the Freudenthal suspension theorem, it follows that the above map Σ^q is an isomorphism for $p + q \leq 2(m - q - 1) - 2$.

The nontrivial preimages of pr are $S^p \times [S^q \times y]$, $y \in \partial D^q$ and their union is homeomorphic to $S^p \times \partial D^q$. Since $\dim(S^p \times \partial D^q) = p + q - 1$, by general position it follows that pr^* is an isomorphism for $p + q - 1 \leq m - 3$. Therefore σ is an isomorphism for $m \geq \frac{3q+p}{2} + 2 \geq p + q + 2$, and we are done. $\square$

Note that the above map $(\mathrm{id}_{S^p} \times v_q)^*$ is an isomorphism also for $m = p + q + 2$. This is proved using the cofibration exact sequence of the pair $(S^p \times S^q \times S^q, S^p \times (S^q \vee S^q))$, together with a retraction

$$\Sigma(S^p \times S^q \times S^q) \to \Sigma(S^p \times (S^q \vee S^q))$$

obtained from the retraction

$$\mathrm{id}_{S^p} \times r_q : S^p \times \Sigma(S^q \times S^q) \to S^p \times \Sigma(S^q \vee S^q)$$

by shrinking the product of S^p with the vertex of the suspension to a point.

Lemma 6.2 (The generalized torus lemma [181]).

(a) *If $s \geq 3$, $p_1 \leq \cdots \leq p_s$, $n = p_1 + \cdots + p_s$ and*

$$N = S^{p_1} \times \cdots \times S^{p_s},$$

then the same assertion as in Lemma 6.1 holds for the following diagram:

$$
\begin{array}{ccccc}
\pi_{n-p_1}(V_{m-n+p_1,p_1+1}) & \xrightarrow{\ \ \tau\tau\ \ } & \mathrm{Emb}^m(N) & \xrightarrow{\ \ \alpha\ \ } & \pi_{eq}^{m-1}(\widetilde{N}) \\[2pt]
{\scriptstyle \frac{3n-p_1}{2}+2}\Big\downarrow\rho & & \Big\downarrow\alpha' & & {\scriptstyle 2n-p_1-p_2+3}\Big\downarrow\gamma \\[2pt]
\pi_{n-p_1}(V^{eq}_{m-n+p_1,p_1+1}) & \xrightarrow[{\frac{3n}{2}-p_1+2}]{\ \ \sigma\ \ } & \Pi^{m-1}_{p_1,n-p_1} & \xleftarrow[{2n-p_2+2}]{\ \ \mathrm{pr}_1\ \ } & \oplus_i \Pi^{m-1}_{p_i,n-p_i}
\end{array}
$$

$$\text{(6.2)}$$

(b) *Suppose that $s \geq 3$, $n = p_1 + \cdots + p_s$, $p_1 \leq \cdots \leq p_s$ and $m \geq 2n - p_1 - p_2 + 3$. (For $s = 3$ and $CAT=DIFF$ assume also that $m \geq \frac{3n}{2} + 2$.) Then*

$$\mathrm{Emb}^m(S^{p_1} \times \cdots \times S^{p_s}) = \bigoplus_{i=1}^{s} \pi_{n-p_i}(V_{m-n+p_i,p_i+1}).$$

Proof. Part (b) follows from part (a) and Theorems 5.5 and 5.6. The proof of (a) is analogous to the proof of the torus lemma (Lemma 6.1). We shall only define τ and σ and omit the details.

The map τ is defined as follows. An element $\varphi \in \pi_{n-p_1}(V_{m-n+p_1,p_1+1})$ is represented by a map $S^{n-p_1} \times S^{p_1} \to S^{m-n+p_1-1}$. Consider the projections

$$\mathrm{pr}_1 : N \to S^{p_1} \times S^{p_2+\cdots+p_s} = S^{p_1} \times S^{n-p_1} \quad \text{and} \quad \mathrm{pr}_2 : N \to S^{p_2} \times \cdots \times S^{p_s}.$$

Analogously to the case $s = 2$, define an embedding $\tau(\varphi)$ as the composition

$$S^{p_1} \times S^{p_2} \times \cdots \times S^{p_s} \xrightarrow{(\varphi \circ \mathrm{pr}_1) \times \mathrm{pr}_2} \partial D^{m-n+p_1} \times S^{p_2} \times \cdots \times S^{p_s} \subset \mathbb{R}^m.$$

The map σ is defined analogously to the case $s = 2$ as the S^{p_1}-fibrewise $(n - p_1)$-fold suspension.

Let $q_1 := n - p_1$. Equivalently, σ is a composition

$$\pi_{q_1}(V^{eq}_{m-q_1,p_1+1}) = \pi^{m-q_1-1}_{eq}(S^{p_1} \times S^{q_1})$$

$$\xrightarrow{\ \Sigma^{q_1}\ } \pi^{m-1}_{eq}(\Sigma^{q_1}(S^{q_1}(S^{p_1} \times S^{q_1}))) \xrightarrow{\ \mathrm{pr}^*\ } \Pi^{m-1}_{p_1,q_1}.$$

Here the maps Σ^{q_1} and pr^* are isomorphisms for $2m \geq 3n - 2p_1 + 4$ and $m \geq n + 2$, respectively. Therefore σ is an isomorphism for $m \geq 3n/2 - p_1 + 2$. $\square$

Note that under the assumptions of Lemma 6.2(b) we have $m \geq \frac{3n}{2} + 2$ for $s \geq 4$.

6.2 Proofs of non-injectivity and non-surjectivity for Examples 5.8(b) and 5.9(b)

Lemma 6.3 (The decomposition lemma [181]). *For $m \geq 2p + q + 1 \geq q + 3$ there is the following (anti)commutative diagram, in which the first and the third lines are exact. The map ν is epimorphic for $m - q$ even and $\mathrm{im}\,\nu$ is the subgroup of elements of order 2 for $m - q$ odd.*

$$
\begin{array}{ccccc}
\pi_q(V_{m-q-1,p}) & \xrightarrow{\ \mu''\ } & \pi_q(V_{m-q,p+1}) & \xrightarrow{\ \nu''\ } & \pi_q(V_{m-q,1}) \\
\downarrow{\scriptstyle \tau_{p-1}} & & \downarrow{\scriptstyle \tau} & & \downarrow{=} \\
\mathrm{Emb}_{PL}^{m-1}(S^{p-1} \times S^q) & \xrightarrow{\ \mu'\ } & \mathrm{Emb}_{PL}^m(S^p \times S^q) & \xrightarrow{\ \nu'\ } & \pi_q(S^{m-q-1}) \\
\downarrow{\scriptstyle \alpha'_{p-1}} & & \downarrow{\scriptstyle \alpha'} & & \downarrow{\scriptstyle \Sigma^\infty} \\
\Pi_{p-1,q}^{m-2} & \xrightarrow{\ \mu\ } & \Pi_{p,q}^{m-1} & \xrightarrow{\ \nu\ } & \pi_{2q+1-m}^S
\end{array}
$$

$$(6.3)$$

Only the right-hand squares and the exactness at Π_{pq}^{m-1} of Diagram (6.3) are used for the examples. The left-hand squares are interesting in themselves and are useful elsewhere [183]. (The definition of μ' here is simpler than that in [181].)

Definition of the maps from the diagram. Let ν'' and μ'' be the homomorphisms induced by the 'forgetting the first p vectors' bundle $V_{m-q,p+1} \xrightarrow{V_{m-q-1,p}} V_{m-q,1}$.

For an embedding $f : S^p \times S^q \to \mathbb{R}^m$ let $\nu'(f)$ be the linking coefficient of $f|_{x \times S^q}$ and $f|_{-x \times S^q}$ in $\mathbb{R}^m$. Define the map $\nu : \Pi_{p,q}^{m-1} \to \Pi_{0,q}^{m-1} \cong \pi_{2q-m+1}^S$ to be 'the restriction over $* \times S^{2q}$'.

In order to define μ', denote by

$$
S^p = D_+^p \bigcup_{\partial D_+^p = S^{p-1} = \partial D_-^p} D_-^p
$$

the standard decomposition of S^p. Define $\mathbb{R}_\pm^m$ and $\mathbb{R}^{m-1}$ analogously.

Add a strip to an embedding $f : S^{p-1} \times S^q \to \mathbb{R}^{m-1}$, i.e. extend it to an embedding

$$f' : S^{p-1} \times S^q \bigcup_{S^{p-1} \times D_+^q = \partial D_+^p \times D_+^q} D_+^p \times D_+^q \to \mathbb{R}^m.$$

Since $m \geq 2p+q+1$, it follows that this extension is unique up to isotopy. The union of f' with the cone over the restriction of f' to the boundary forms an embedding $D_+^p \times S^q \to \mathbb{R}_+^m$. This latter embedding can clearly be shifted to a proper embedding. Define $\mu' f : S^p \times S^q \to \mathbb{R}^m$ to be the union of this proper embedding and its mirror image with respect to $\mathbb{R}^{m-1} \subset \mathbb{R}^m$.

Let us define the map μ first for the case $p = 1$. For a map $\varphi : S^{2q} \to S^{m-2}$ define the map $\mu\varphi$ to be the equivariant extension of the composition

$$D^1 \times S^{2q} \xrightarrow{\text{pr}} \Sigma S^{2q} \xrightarrow{\Sigma\varphi} S^{m-1}.$$

In order to define the map μ for arbitrary p, replace $\Pi_{p-1,q}^{m-2}$ and Π_{pq}^{m-1} by $\pi_{eq}^{m-2}(\Sigma^q(S^{p-1} \times S^q))$ and $\pi_{eq}^{m-1}(\Sigma^q(S^p \times S^q))$ respectively (see the proof of Lemma 6.1). For an equivariant map $\varphi : \Sigma^q(S^{p-1} \times S^q) \to S^{m-2}$ let $\mu\varphi$ be the composition

$$\Sigma^q(S^p \times S^q) = \Sigma^q(\Sigma S^{p-1} \times S^q) \xrightarrow{\Sigma^q \text{pr}} \Sigma^{q+1}(S^{p-1} \times S^q) \xrightarrow{\Sigma\varphi} \Sigma S^{m-2}$$

where pr is the map from the proof of Lemma 6.1. Clearly, the definition for arbitrary p agrees with that for $p = 1$.

Remark. Note that except for the map μ', all the maps in Diagram (6.3) are defined for $m \geq p + q + 3$.

The embedding $\mu' f$ can also be defined by the Penrose-Whitehead-Zeeman-Irwin embedding theorem (Theorem 2.9) and its isotopy analogue. For $m \geq 2p + q + 2$ any embedding $f : S^{p-1} \times S^q \to S^{m-1}$ can be extended to a PL embedding $f_\pm : D_\pm^p \times S^q \to \mathbb{R}_\pm^m$, uniquely up to isotopy. Then two embeddings f_+ and f_- define an embedding $\mu'(f) : S^p \times S^q \to \mathbb{R}^m$.

Proof of Lemma 6.3. It is easy to check that both ν and μ are homomorphisms.

Clearly, the upper-left square of the diagram commutes; and clearly the upper-right square of the diagram commutes (see details in [195, Lemma 3]). The bottom-right square of the diagram (anti)commutes by [100, Lemma 5.1].

We prove the commutativity of the bottom-left square for $p = 1$; the proof is analogous for the general case. Take an embedding $f : S^0 \times S^q \to \mathbb{R}^{m-1}$. Then $\alpha\mu'f = \mu\alpha'_0 f$ on $S^0 \times S^{2q}$. Also, for each $y \in S^1 \times S^q \times S^q$ the points $(\alpha'\mu'f)y$ and $(\mu\alpha'_0 f)y$ are either both in the upper or both in the lower hemisphere of S^{m-1}. Hence $\alpha'\mu'f \simeq_{eq} \mu\alpha'_0 f$.

Let us prove exactness at Π^{m-1}_{pq}. Clearly, $\nu\mu = 0$. On the other hand, if $\Phi : S^p \times S^{2q} \to S^{m-1}$ is an equivariant map such that $\Phi|_{*\times S^{2q}}$ is null-homotopic, then by the Borsuk homotopy extension theorem Φ is equivariantly homotopic to a map which maps $* \times S^{2q}$ and $a_p(*) \times S^{2q}$ to antipodal points of S^{m-1}. By the equivariant suspension theorem the latter map is in $\operatorname{im}\mu$, since $p - 1 + 2q \le 2(m-2) - 1$. So $\ker\nu = \operatorname{im}\mu$.

Clearly $\operatorname{im}\nu$ consists of homotopy classes $\varphi \in \Pi^{m-1}_{0,q}$ extendable to a map $D^1 \times S^{2q} \to S^{m-1}$. Such maps φ, considered as maps $\varphi : S^{2q} \to S^{m-1}$, are exactly those which satisfy $a_{m-1} \circ \varphi \circ t_q \simeq \varphi$. The latter condition is equivalent to $(-1)^m\varphi = (-1)^q\varphi$ (for m odd this follows by [143, complement to lecture 6, (10), p. 264], since $h_0 : \pi_{2q}(S^{m-1}) \to \pi_{2q}(S^{2m-3})$ and $2q < 2m - 3$). So $\operatorname{im}\nu = \ker(1 - (-1)^{m-q})$. $\quad\square$

Proof of non-surjectivity in Example 5.9(b). Set $q = n - 1 \le m - 4$. Look at the bottom-right square of Diagram (6.3) and use the surjectivity of ν for $m - q$ even. The specific examples can be found using [196, §14] (set $l = m - n = m - q - 1$ and $k = 2q + 1 - m$). $\quad\square$

Proof of non-injectivity in Example 5.8(b). Since $p < k$, we have $m \ge 2p+q+2$. Look at the right squares of Diagram (6.3) and use Lemma 6.4 below. $\quad\square$

Lemma 6.4.

(a) Π^{m-1}_{pq} is finite when $p + q + 2 \le m \le 2q$.

(b) *The image of the restriction-induced homomorphism*

$$\nu''_p : \pi_{4k-1}(V_{2k+1,p+1}) \to \pi_{4k-1}(S^{2k})$$

is infinite for $p < 2k$.

Proof. Let us prove (a) by induction on p. The base of the induction is $p = 0$, when $\Pi^{m-1}_{0q} \cong \pi_{2q}(S^{m-1})$ is indeed finite. The inductive step of (a) follows by the induction hypothesis and the exactness of the bottom line from Diagram (6.3).

In order to prove (b) for $p = 0$ observe that the map ν_0 is an isomorphism and $\pi_{4k-1}(S^{2k})$ is infinite. Suppose that $p \ge 1$ and there is an infinite set $\{x_i\} \in \pi_{4k-1}(V_{2k+1,p})$ with distinct ν''_{p-1}-images. Consider the Serre fibration $S^{2k+1-p} \to V_{2k+1,p+1} \xrightarrow{\psi} V_{2k+1,p}$ and the following segment of its exact sequence:

$$\pi_{4k-1}(V_{2k+1,p+1}) \xrightarrow{\psi_*} \pi_{4k-1}(V_{2k+1,p}) \to \pi_{4k-2}(S^{2k-p}).$$

Since $\pi_{4k-2}(S^{2k-p})$ is finite, by exactness it follows that the number of congruence classes of $\pi_{4k-1}(V_{2k+1,p})$ modulo $\mathrm{im}\,\psi_*$ is finite. Therefore an infinite number of the x_i (we may assume all the x_i) lie in the same congruence class. By passing from $\{x_i\}$ to $\{x_i - x_1\}$ we may assume that this congruence class is the subgroup $\mathrm{im}\,\psi_*$ itself. Hence the inductive step follows from $\nu_p'' = \nu_{p-1}''\psi_*$. $\qquad\square$

To state our final remarks in this section we need some more notation. For a group G define $G_{[k]}$ to be G when k is even and the subgroup of G formed by elements of order 2 when k is odd.

For $m - q$ even, it follows from the existence of a section $s : \pi_q(V_{m-q,1}) \to \pi_q(V_{m-q,2})$ that ν'' and hence ν' is epimorphic.

We also have $\mathrm{im}\,\nu = \mathrm{im}\,\nu' = \mathrm{im}\,\nu'' = \pi_{2q-m+1,[m-q]}$ for $2m \geq 3q + 4$. Note that $\mathrm{im}\,\nu = \pi_{2q-m+1,[m-q]}$ even for $2m \leq 3q + 3$, but by Lemma 6.4(b) $\mathrm{im}\,\nu'' \neq \pi_q(S^{m-q-1})_{[m-q]}$ for $2m \leq 3q+3$.

We *conjecture* that if $m - q$ is even ≥ 4 and $2m \geq 3q + 4$, then $\Pi_{1,q}^{m-1} \cong \pi_{2q-m+2}^S \oplus \pi_{2q-m+1}^S$. We also conjecture that in general Π_{1q}^{m-1} is adjoint to $\pi_{2q-m+2,[m-q]}^S \oplus \pi_{2q-m+1,[m-q]}^S$, unless $m = 2q + 1$ and $q = 2l$ is even, when $\Pi_{1,2l}^{4l} \cong \mathbb{Z}_2$ (cf. the formula for $\pi_q(V_{m-q,2})$ after Theorem 3.9). Since $\mathrm{im}\,\nu = \pi_{2q-m+1,[m-q]}^S$, by the decomposition lemma (Lemma 6.3) the conjecture would follow from $\mathrm{coim}\,\mu \cong \pi_{2q-m+2,[m-q]}^S$ (recall that we identify $\Pi_{0q}^{m-2} = \pi_{2q-m+2}^S$).

7 The Borromean rings and the Haefliger-Wu invariant

All examples illustrating that the metastable dimension restrictions in embedding theorems are sharp have their origin in the Borromean rings (Example 3.2). So let us illustrate the ideas of Examples 5.8(d) and 5.10(c) by giving an *alternative* construction of three circles embedded into $\mathbb{R}^3$ so that every pair of them is unlinked but all three are linked together. Our construction is based on the fact that the fundamental group is not always commutative.

Take two unknotted circles Σ and $\bar{\Sigma}$ in $\mathbb{R}^3$ far away from each other. Embed in $\mathbb{R}^3 - (\Sigma \sqcup \bar{\Sigma})$ the *figure eight*, i.e. the wedge C of two oriented circles, so that the inclusion $C \subset \mathbb{R}^3 - (\Sigma \sqcup \bar{\Sigma})$ induces an isomorphism of the fundamental groups. Take generators a and $\bar{a}$ of $\pi_1(C) \cong$

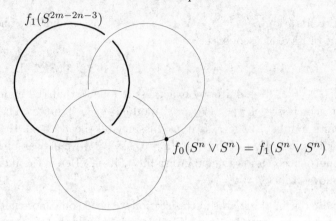

Fig. 7.1. Another construction of the Borromean rings

$\pi_1(\mathbb{R}^3 - (\Sigma \sqcup \bar{\Sigma}))$ represented by the two arbitrarily oriented circles of the figure eight. Consider a map $S^1 \to C \subset \mathbb{R}^3$ representing the element $a\bar{a}a^{-1}\bar{a}^{-1}$. By general position, there is an embedding $f : S^1 \to \mathbb{R}^3$ very close to this map. It is easy to choose f so that Σ and $f(S^1)$, $\bar{\Sigma}$ and $f(S^1)$ are unlinked (because f induces the zero homomorphism of the 1-dimensional homology groups). Then Σ, $\bar{\Sigma}$ and $f(S^1)$ are as required (see Figure 7.1, where $m = 3$, $n = 1$, $f = f_1$ and $C = f_0(S^n \vee S^n)$.) Indeed, Σ and $\bar{\Sigma}$ are unlinked by their definition. But f induces a nonzero homomorphism of the fundamental groups. Therefore the three circles Σ, $\bar{\Sigma}$ and $f(S^1)$ are linked together.

Remark. Higher-dimensional Boromean rings can be constructed analogously, using *Whitehead products* instead of commutators.

Sketch of a counterexample. The following construction provides a counterexample to the relative versions of Theorem 4.2 for $n = 2$ and to the surjectivity in the Haefliger-Weber Theorem 5.5 for $m = 2n = 4$.
    Let

$$N = D^2 \sqcup D^2 \sqcup D^2 \quad \text{and} \quad A = \partial D^2 \sqcup \partial D^2 \sqcup \partial D^2.$$

Let $A \subset S^3 \cong \partial D^4$ be (generalized) Borromean rings.
    Since all three rings are linked, it follows that the embedding $A \to \partial D^4$ cannot be extended to an embedding $N \to D^4$. But since each pair of Borromean rings is unlinked, it follows that the corresponding relative Haefliger-Wu or van Kampen obstruction to this extension vanishes. This is so because both the Haefliger-Wu obstruction and the van Kam-

pen obstruction involve 2-fold products and double intersections, but involve neither 3-fold products nor triple intersections.

Proof of non-injectivity in Example 5.8(d). The reader is recommended to read this proof first for $n = 1$ and $m = 3$. Let $N = S^n \vee S^n \sqcup S^{2m-2n-3}$ and take the standard embedding $f_0 : N \to S^m$. Then

$$S^m - f_0(S^n \vee S^n) \simeq S^{m-n-1} \vee S^{m-n-1}.$$

Take a map $\varphi : S^{2m-2n-3} \to S^m - f_0(S^n \vee S^n)$ representing the Whitehead product (for $m - n = 2$, a commutator) of generators. If $n = 1$ and $m = 3$, then φ is homotopic to an embedding by general position. If $n > 1$, then $2m \le 3n+3$ implies that $m \le 2n$, i.e. $m-(2m-2n-3) \ge 3$. Since also $2(2m-2n-3)-m+1 \le m-n-2$ by the Penrose-Whitehead-Zeeman-Irwin embedding theorem (Theorem 2.9), it follows that φ is homotopic to an embedding. Define $f_1 : N \to S^m$ on $S^n \vee S^n$ as f_0 and on $S^{2m-2n-3}$ as such an embedding (Figure 7.1). Since the homotopy class of φ is nontrivial, it follows that $f_1 : N \to S^m$ is not isotopic to the standard embedding $f_0 : N \to S^m$.

Let us prove that $\alpha(f_0) = \alpha(f_1)$. Using 'finger moves' (analogously to the construction of Example 7.1 below) we construct a map $F : N \times I \to \mathbb{R}^m \times I$ such that

$$F(x,0) = (f_0(x),0), \quad F(x,1) = (f_1(x),1) \quad \text{and}$$
$$F((S^n \vee S^n) \times I) \cap F(S^{2m-2n-3} \times I) = \emptyset.$$

Then there is a triangulation T of N such that no images of disjoint simplices intersect throughout F_t. Then the map $\widetilde{F}_t$ is well-defined on the simplicial deleted product $\widetilde{T}$. Since $\widetilde{T}$ is an equivariant deformation retract of $\widetilde{N}$, it follows that $\alpha(f_0) = \alpha(f_1)$.

(One can also check that in general $\alpha_G^m(N)f_0 = \alpha_G^m(N)f_1$ for each G, so $\alpha_G^m(N)$ is not injective.) $\square$

The above gives us a 3-*dimensional visualization* of the celebrated *Casson finger moves*. Combine the homotopy F constructed in the above proof for $n = 1$ and $m = 3$ with the 'reverse' homotopy. We get a homotopy $G : N \times I \to \mathbb{R}^3 \times I$ between standard embeddings $N \to \mathbb{R}^3$ (Figure 7.2). This homotopy is obtained from the identity isotopy by Casson finger moves.

We conjecture that the nontrivial embedding f of Example 5.8(b) can be obtained from explicitly defined Borromean rings $S^n \sqcup S^n \sqcup S^{2m-2n-3} \subset \mathbb{R}^m$ [67], [125, Proposition 8.3] by 'wedging' $S^n \sqcup S^n$. We

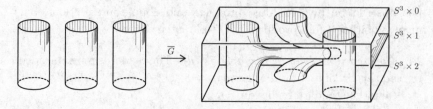

Fig. 7.2. Casson finger moves

also conjecture that by joining the two n-spheres of the above linking with a tube, we obtain a nontrivial embedding $S^n \sqcup S^{2m-2n-3} \to \mathbb{R}^m$ with trivial α-invariant (although this would be harder to prove: either we assume that $m - n \notin \{2, 4, 8\}$ and need to check that the linking coefficient of the obtained link is $[\iota_{m-n-1}, \iota_{m-n-1}] \neq 0$, or we need to apply the β-invariant (cf. [67, §3], [183]).

Example 7.1. There exists a 2-polyhedron N non-embeddable into $\mathbb{R}^4$ but for which there exists a PL nondegenerate almost-embedding $f : N \to \mathbb{R}^4$.

(The definition of a nondegenerate almost embedding was given before Lemma 4.4.)

Remark. The polyhedron N from Example 7.1 is even topologically non-embeddable into $\mathbb{R}^4$.

Before constructing Example 7.1 let us explain its meaning. By the definition of the van Kampen obstruction (see Section 4), Example 7.1 implies the Freedman-Krushkal-Teichner example, i.e. Theorem 4.2 for $n = 2$, as follows. Take a triangulation T of N from the definition of an almost embedding $f : N \to \mathbb{R}^4$. Then the Gauss map $\widetilde{f} : \widetilde{T} \to S^3$ is well-defined on $\widetilde{T}$. Since $\widetilde{T}$ is an equivariant deformation retract of $\widetilde{N}$, it follows that there exists an equivariant mapping $\widetilde{N} \to S^3$. So Example 7.1 implies non-embeddability in Example 5.10(c) for $m = 2n = 4$. However, it gives even more and shows the non-surjectivity not only of the Haefliger-Wu invariant, but also of the generalized Haefliger-Wu invariants (see Section 5).

Preliminary construction for Example 7.1.
Let Q be the 2-skeleton of the 6-simplex minus the interior of one 2-simplex from this 2-skeleton. Recall from Section 4.2 that Q contains two disjoint spheres with the following property:

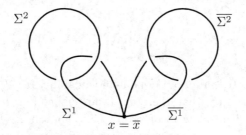

Fig. 7.3.

$$\text{for each embedding } Q \to \mathbb{R}^4, \text{ these spheres} \\ \text{link with an odd linking number.} \tag{$\sharp$}$$

(An alternative proof of this fact is presented in the proof of Lemma 7.2 below.)

Let us denote these spheres by Σ^2 and Σ^1. We write $\bar{Q}$ to denote a copy of the space Q (for a subset $A \subset Q$ its copy is denoted by $\bar{A} \subset \bar{Q}$).Embed $Q \sqcup \bar{Q}$ into $\mathbb{R}^4$ in the standard way, i.e. so that

(a) the copies Q and $\bar{Q}$ are far away from one another;
(b) both Σ^2 and $\bar{\Sigma}^2$ are unknotted.

Then Σ^2 and $\bar{\Sigma}^2$ are unlinked. Take any point $x \in \Sigma^1$; join the points x and $\bar{x}$ by an arc in $\mathbb{R}^4$, and pull small neighbourhoods in Q and $\bar{Q}$ of these points to each other along this arc. We obtain an embedding $Q \vee \bar{Q} \subset \mathbb{R}^4$ (Figure 7.3).

In Figure 7.3 each sphere Σ^1, $\bar{\Sigma}^1$, Σ^2 and $\bar{\Sigma}^2$ of dimensions 1, 1, 2 and 2 respectively, is shown as 1-dimensional. Consider the wedge $\Sigma^1 \vee \bar{\Sigma}^1$ with the base point $x = \bar{x}$. Then the inclusion $\Sigma^1 \vee \bar{\Sigma}^1 \subset \mathbb{R}^4 - (\Sigma^2 \sqcup \bar{\Sigma}^2)$ induces an isomorphism of the fundamental groups. Take generators a and $\bar{a}$ of the group $\pi_1(\Sigma^1 \vee \bar{\Sigma}^1)$ represented by the two (arbitrarily oriented) circles of the figure eight.

Sketch of the Freedman-Krushkal-Teichner construction. Here we sketch a construction of an example for Theorem 4.2 for $n = 2$. This construction is a bit simpler than that of Example 7.1, but it makes vanishing of the obstruction less clear, and it gives neither Example 7.1 nor the example 5.10(c) of non-embeddability.

Take a map $S^1 \to \Sigma^1 \vee \bar{\Sigma}^1$ representing the element $[a, \bar{a}] = a\bar{a}a^{-1}\bar{a}^{-1}$. Let N' be the mapping cone of the composition of this map with the inclusion $\Sigma^1 \vee \bar{\Sigma}^1 \subset Q \vee \bar{Q}$, i.e.

$$N' = B^2 \bigcup_{[a,\bar{a}]:\partial B^2 \to \Sigma^1 \vee \bar{\Sigma}^1} (Q \vee \bar{Q}).$$

Let us sketch the proof of the non-embeddability of N' into $\mathbb{R}^4$. Suppose to the contrary that there exists an embedding $h : N' \to \mathbb{R}^4$. The nontrivial element $[a, \bar{a}]$ of $\pi_1(\Sigma^1 \vee \bar{\Sigma}^1)$ goes under h to a loop in $\mathbb{R}^4 - h(\Sigma^2 \sqcup \bar{\Sigma}^2)$, which extends to hD^2 and hence is null-homotopic. This is a contradiction because h induces a *monomorphism*

$$\pi_1(\Sigma^1 \vee \bar{\Sigma}^1) \to \pi_1(\mathbb{R}^4 - h(\Sigma^2 \sqcup \bar{\Sigma}^2)).$$

If both $h\Sigma^2$ and $h\bar{\Sigma}^2$ are unknotted in $\mathbb{R}^4$, then the above assertion is clear. In general (i.e. when the spheres are knotted) the above assertion is proved using Stallings' theorem on the lower central series of groups [192, 49], see below.

We have $V(N') = 0$ because the van Kampen obstruction only detects the *homology* property that the loop $[a, \bar{a}]$ is null-homologous (for a detailed proof see [49]).

Since the van Kampen obstruction is a complete obstruction for the existence of an equivariant map $\widetilde{N'} \to S^3$ (for these dimensions), we obtain non-embeddability in Example 5.10(c) for $m = 2n = 4$. □

Construction of Example 7.1. Take an embedding $Q \sqcup \bar{Q} \subset \mathbb{R}^4$ with the properties (a) and (b) from the preliminary construction (p. 310). Take any points $x \in \Sigma^1$ and $y \in \Sigma^2$. Join points x to $\bar{x}$ and y to $\bar{y}$ by two arcs in $\mathbb{R}^4$. Pull small neighbourhoods in Q and $\bar{Q}$ of these points to each other along this arc (Figure 7.4). We obtain an embedding

$$K := Q \bigcup_{x=\bar{x},\, y=\bar{y}} \bar{Q} \subset \mathbb{R}^4.$$

Push a 2-dimensional finger from a small disc $D^2 \subset \Sigma^2$ near $y = \bar{y}$ intersecting $\bar{\Sigma}^2$ near $y = \bar{y}$. We get a new PL map $g : K \to \mathbb{R}^4$ which has transversal self-intersection points (Figure 7.4).

We can represent a disc neighbourhood B^4 of an arbitrary intersection point $c \in \mathbb{R}^4$ as the product $B^2 \times B^2$ of balls, where 0×0 corresponds to the intersection while $B^2 \times 0$ and $0 \times B^2$ correspond to the images of Σ^2 and $\bar{\Sigma}^2$ (we denote by 0 the centre of B^2). In a neighbourhood of the point c we have the *distinguished* or *characteristic* torus $\partial B^2 \times \partial B^2$, cf. [21, 103, 50]. In Figure 7.4 the 2-dimensional distinguished torus is

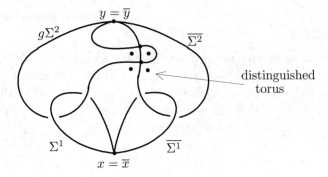

Fig. 7.4.

shown as 0-dimensional. By (b) we have

$$\pi_1(\mathbb{R}^4 - \Sigma^2 \vee \bar{\Sigma}^2) \cong \pi_1(S^1 \vee S^1).$$

Denote by a and $\bar{a}$ the elements of this group represented by homeomorphisms $S^1 \to z \vee S^1$ and $S^1 \to S^1 \vee z$ (for some point $z \in S^1$), respectively (with some orientations). With appropriate orientations the inclusions of $\partial B^2 \times z$ and $z \times \partial B^2$ into $\mathbb{R}^4 - \Sigma^2 \vee \bar{\Sigma}^2$ are homotopic to a and $\bar{a}$, respectively. Since the map

$$a\bar{a}a^{-1}\bar{a}^{-1} : S^1 \to S^1 \vee S^1 \cong (z \times \partial B^2) \vee (\partial B^2 \times z)$$

extends to a map $B^2 \to \partial B^2 \times \partial B^2$, it follows that $a\bar{a}a^{-1}\bar{a}^{-1}$ is null-homotopic in $\mathbb{R}^4 - g(\Sigma^2 \vee \bar{\Sigma}^2)$. Then there exists a PL map $r : B^2 \to \mathbb{R}^4 - g(\Sigma^2 \vee \bar{\Sigma}^2)$ such that $r|_{\partial B^2} : \partial B^2 \to \Sigma^1 \vee \bar{\Sigma}^1$ represents the commutator of the inclusions $\Sigma^1 \, '\subset \Sigma^1 \vee \bar{\Sigma}^1$ and $\bar{\Sigma}^1 \subset \Sigma^1 \vee \bar{\Sigma}^1$. Roughly speaking, $r(B^2)$ is a torus $0 \times \partial B^2 \times \partial B^2$. Set

$$N = B^2 \bigcup_{\partial B^2 = \partial D^2} (K - \mathring{D}^2) \cup r(B^2).$$

Analogously to the proof of the Freedman-Krushkal-Teichner example above, N does not embed into $\mathbb{R}^4$ (the details are analogous to the construction of Example 5.10(c) below).

We have $N \supset (K - \mathring{D}^2) \cup B^2 \cong K$. Define a map $f : N \to \mathbb{R}^4$ on $(K - \mathring{D}^2) \cup B^2$ as the composition of a homeomorphism with K and g, and on $r(B^2)$ as the identity. Then $\Sigma(f) \subset B^2 \cup \bar{D}^2$. By the construction of the balls D^2 and $\bar{D}^2$ it follows that the balls B^2 and $\bar{D}^2$ are contained in the interiors of some adjacent 2-simplices of some triangulation T

of N. Hence f is a nondegenerate almost embedding (whose image is $g(K) \cup r(B^2)$). $\square$

Construction of the non-embeddability example 5.10(c). The case $m = 3$ is proved in [58] using different ideas. The case $4 \leq m \leq n + 1$ can either be proved analogously to [58] or is covered by the case $4 \leq m \geq n + 2$. So we present the proof for $m \geq n + 2$. This is a higher-dimensional generalization of Example 7.1.

Let $l = m - n - 1 \geq 1$. Denote by $\Delta^k_{a_0 \ldots a_s}$ the k-skeleton of the s-simplex with vertices $a_0 \ldots a_s$.

(The definition of $\Delta^n_{012 \ldots m+2}$ makes sense even for $n = l + 1$, which case is outside the dimension range of Example 5.10(c). If $n = l + 1 = 1$, then $\Delta^n_{012 \ldots m+2}$ is one of the Kuratowski nonplanar graphs, namely K_5, and if $n = l + 1 > 1$ then $\Delta^n_{012 \ldots m+2}$ is an n-dimensional polyhedron non-embeddable in $\mathbb{R}^{2n}$.)

Set

$$Q = \Delta^n_{12 \ldots m+2} \cup \mathrm{Con}(\Delta^l_{12 \ldots m+2} - \mathrm{Int}\, \Delta^l_{12 \ldots l+1}, 0)$$

and set

$$K = Q \bigcup_{0 = \bar{0},\, m = \bar{m}} \bar{Q}\;, \quad \Sigma^l = \partial \Delta^{l+1}_{01 \ldots l+1} \quad \text{and} \quad \Sigma^n = \partial \Delta^{n+1}_{l+2 \ldots m+2}.$$

The polyhedron Q embeds in $\mathbb{R}^m$ (this was actually proved in the first two paragraphs of the proof of [171, Lemma 1.1]). Embed into $\mathbb{R}^m$ two copies of Q which are far apart. Since $m \geq n+2$, we can join two points of Σ^n and $\bar{\Sigma}^n$ by an arc and pull the points of the spheres together along this arc. Making the same construction for Σ^l and $\bar{\Sigma}^l$ we obtain an embedding $K \to \mathbb{R}^m$; so we assume that K is a subset of $\mathbb{R}^m$. We may assume that the wedge $\Sigma^n \vee \bar{\Sigma}^n$ is unknotted in $\mathbb{R}^m$. (If $m > n+2$, then this holds for any embedding $K \subset \mathbb{R}^m$ [114, Theorem 8]; if $m = n+2$, then for our embedding $Q \to \mathbb{R}^m$ the sphere Σ^n is unknotted in $\mathbb{R}^m$, and we can choose an embedding $K \to \mathbb{R}^m$ so that $\Sigma^n \vee \bar{\Sigma}^n$ is unknotted in $\mathbb{R}^m$.)

Take a triangulation T of K. Let $D^n \subset \Sigma^n$ and $\bar{D}^n \subset \bar{\Sigma}^n$ be PL discs, each one of them in the interior of an n-simplex of T containing the common point $m = \bar{m}$ of Σ^n and $\bar{\Sigma}^n$. Take points $a \in \mathring{D}^n$, $\bar{a} \in \mathring{\bar{D}}^n$ and a small arc $s \subset \mathbb{R}^m$ joining a to $\bar{a}$. By general position $s \cap K = \{a, \bar{a}\}$. Construct a new embedding $g : D^n \to \mathbb{R}^m$ obtained from the old one by pushing an n-dimensional finger from D^n along the arc s. Let $g|_{K - \mathring{D}^n}$ be the inclusion. We get a new PL map $g : K \to \mathbb{R}^m$ such that $g|_{K - \mathring{D}^n}$ is an embedding but $g(D^n) \cap g(\bar{D}^n) \neq \emptyset$ (Figure 7.4).

By general position $\dim(g(D^n) \cap \bar{D}^n) \leq 2n - m$ and $g(D^n)$ intersects $\bar{D}^n$ transversally. Denote by 0 the centre of B^k. We can represent a disc neighbourhood B^m of an arbitrary point c of this intersection as the product $B^{2n-m} \times B^{l+1} \times B^{l+1}$ of balls, where $B^{2n-m} \times 0 \times 0$ corresponds to the intersection, while

$$B^{2n-m} \times B^{l+1} \times 0 \quad \text{and} \quad B^{2n-m} \times 0 \times B^{l+1}$$

correspond to $g(D^n)$ and $\widetilde{D}^n$, respectively. In a neighbourhood of the point c we have the *distinguished* or *characteristic* torus $0 \times \partial B^{l+1} \times \partial B^{l+1}$.

Since $\Sigma^n \vee \bar{\Sigma}^n$ is unknotted in $\mathbb{R}^m \subset S^m$, it follows that $\pi_l(\mathbb{R}^m - \Sigma^n \vee \bar{\Sigma}^n) \cong \pi_l(S^l \vee S^l)$. Denote by α and $\bar{\alpha}$ the elements of this group represented by the inclusions of components of the wedge (with some orientations). Take a point $y \in \partial B^{l+1}$. With appropriate orientations the inclusions of

$$0 \times \partial B^{l+1} \times y \quad \text{and} \quad 0 \times y \times \partial B^{l+1} \quad \text{into} \quad \mathbb{R}^m - \Sigma^n \vee \bar{\Sigma}^n$$

are homotopic to α and $\bar{\alpha}$, respectively. Since the Whitehead product

$$[\alpha, \bar{\alpha}] : S^{2l-1} \to S^l \vee S^l \cong (0 \times y \times \partial B^{l+1}) \vee (0 \times \partial B^{l+1} \times y)$$

extends to a map $B^{2l} \to 0 \times \partial B^{l+1} \times \partial B^{l+1}$ [21, 103, 50], it follows that $[\alpha, \bar{\alpha}]$ is null-homotopic in $\mathbb{R}^m - \Sigma^n \vee \bar{\Sigma}^n$.

Denote the linking coefficient by $\mathrm{link}(\cdot, \cdot)$: let $p = \mathrm{link}(\Sigma^l, \Sigma^n)$ and $\bar{p} = \mathrm{link}(\bar{\Sigma}^l, \bar{\Sigma}^n)$. The inclusions of Σ^l and $\bar{\Sigma}^l$ into $\mathbb{R}^m - \Sigma^n \vee \bar{\Sigma}^n$ represent the elements $p\alpha$ and $\bar{p}\bar{\alpha}$ of the group $\pi_l(\mathbb{R}^m - \Sigma^n \vee \bar{\Sigma}^n)$, respectively. Since

$$[p\alpha, \bar{p}\bar{\alpha}] = p\bar{p}[\alpha, \bar{\alpha}] = 0 \in \pi_{2l-1}(\mathbb{R}^m - \Sigma^n \vee \bar{\Sigma}^n),$$

it follows that the Whitehead product of the (arbitrarily oriented) inclusions of Σ^l and $\bar{\Sigma}^l$ into $\mathbb{R}^m - \Sigma^n \vee \bar{\Sigma}^n$ is null-homotopic. Hence there exists a PL map $r : B^{2l} \to \mathbb{R}^m - \Sigma^n \vee \bar{\Sigma}^n$ whose restriction to ∂B^{2l} represents this Whitehead product. Set

$$N = B^n \bigcup_{\partial B^n = \partial D^n} (K - \mathring{D}^n) \cup r(B^{2l}).$$

Since $m \leq \frac{3n}{2} + 1$, it follows that $2l \leq n$ and hence $\dim N = n$. Analogously to the proof of Example 7.1, there exists an almost embedding $N \to \mathbb{R}^m$. So it remains to prove that N does not embed into $\mathbb{R}^m$.

Proof of the PL non-embeddability of N into $\mathbb{R}^m$. Suppose to the contrary that there is a PL embedding $h : N \to S^m$. Let

$$\Sigma_1^n = (\Sigma^n - \mathring{D}^n) \bigcup_{\partial B^n = \partial D^n} B^n \subset N.$$

The map $h \circ r|_{\partial B^{2l}}$ can be extended to map

$$h \circ r : B^{2l} \to \mathbb{R}^m - h(\Sigma_1^n \vee \bar{\Sigma}^n).$$

Hence $h \circ r|_{\partial B^{2l}}$ is homotopically trivial in $\mathbb{R}^m - h(\Sigma_1^n \vee \bar{\Sigma}^n)$. Now we shall show the contrary and get a contradiction. Let

$$q = \mathrm{link}(h\Sigma^l, h\Sigma_1^n) \quad \text{and} \quad \bar{q} = \mathrm{link}(h\bar{\Sigma}^l, h\bar{\Sigma}^n).$$

In the case $m > n + 2$ by [114, Theorem 8] (cf. the construction of N above) we have $S^m - h(\Sigma_1^n \vee \bar{\Sigma}^n) \simeq S^l \vee S^l$. Denote by β and $\bar{\beta}$ the elements of the group $\pi_l(\mathbb{R}^m - h(\Sigma_1^n \vee \bar{\Sigma}^n))$ represented by the homeomorphisms $S^l \to y \vee S^l$ and $S^l \to S^l \vee y$ (where $y \in S^l$), respectively (with chosen orientations). Hence the homotopy class of the map

$$h \circ r|_{\partial B^{2l}} : \partial B^{2l} \to \mathbb{R}^m - h(\Sigma_1^n \vee \bar{\Sigma}^n)$$

is

$$q\bar{q}[\beta, \bar{\beta}] \in \pi_{2l-1}(\mathbb{R}^m - h(\Sigma_1^n \vee \bar{\Sigma}^n)).$$

By the Hilton theorem (see [143, pp. 231, 257], or [85, p. 511]) the map $\varphi : \pi_{2l-1}(S^{2l-1}) \to \pi_{2l-1}(S^l \vee S^l)$ defined by $\varphi(\gamma) = [\beta, \bar{\beta}] \circ \gamma$ is injective (this can also be proved by using the homotopy exact sequence [85, V.3]). Hence $[\beta, \bar{\beta}]$ has infinite order. This implies that the element $q\bar{q}[\beta, \bar{\beta}]$ is nontrivial because both q and $\bar{q}$ are nonzero, in analogy with the property ($\sharp$) which was used in the preliminary construction on page 310 (see Lemma 7.2 below).

In the case $m = n + 2$ we have $l = 1$. Consider the compositions

$$\Sigma^1 \subset \Sigma^1 \vee \bar{\Sigma}^1 \to \mathbb{R}^{n+2} - h(\Sigma^n \vee \bar{\Sigma}^n)$$
$$\bar{\Sigma}^1 \subset \Sigma^1 \vee \bar{\Sigma}^1 \to \mathbb{R}^{n+2} - h(\Sigma^n \vee \bar{\Sigma}^n).$$

They are *homologous* to $q\beta$ and $\bar{q}\bar{\beta}$, respectively. The commutator of the homotopy classes of the above compositions is nonzero because the inclusion $\Sigma^1 \vee \bar{\Sigma}^1 \subset \mathbb{R}^{n+2} - (\Sigma^n \vee \bar{\Sigma}^n)$ induces a monomorphism of the fundamental groups. The latter is proved analogously to [49, proof of Lemmas 7 and 8], using the Stallings theorem [192] and the observation that by the linking lemma (Lemma 7.2 below)

$$\text{link}(\Sigma^n, \Sigma^1) \equiv \text{link}(\bar{\Sigma}^n, \bar{\Sigma}^1) \equiv 1 \mod 2$$
$$\text{link}(\Sigma^n, \bar{\Sigma}^1) = \text{link}(\bar{\Sigma}^n, \Sigma^1) = 0$$

as required.   □

Lemma 7.2 (The linking lemma). *For any PL embedding $K \subset \mathbb{R}^m$ of the above polyhedron K the pairs $(\Sigma^n, \bar{\Sigma}^l)$ and $(\bar{\Sigma}^n, \Sigma^l)$ are unlinked, and $\text{link}(\Sigma^n, \Sigma^l)$ is odd (cf. [171, Lemma 1.4], [49, Lemmas 6, 7 and 8]).*

Sketch of the proof. The unlinking part follows because Σ^l (resp. $\bar{\Sigma}^l$) bounds a disc $\Delta^{l+1}_{01...l+1}$ (resp. $\bar{\Delta}^{l+1}_{01...l+1}$) in $K - \bar{\Sigma}^n$ (resp. in $K - \Sigma^n$).

We illustrate the idea of the proof of the linking part by proving its particular case $m = 2n = 4l = 4$ (for which, however, there exists a simpler proof). Recall the formulation for this case:

Let $Q = \Delta^2_{01...6} - \mathring{\Delta}^2_{012}$. Let $\Sigma^1 = \partial\Delta^2_{012}$ and let Σ^2 be the union of 2-simplices disjoint from Δ^2_{012}. Then for each embedding $Q \to \mathbb{R}^4$ these spheres link with an odd linking number.

First we prove a simpler result: that $\text{link}(f\Sigma^2, f\Sigma^1) \neq 0$ for each embedding $f : Q \to \mathbb{R}^4$. (The higher-dimensional analogue of this simpler result is sufficient for the proof of the non-embeddability for $m \geq n+3$.) If $\text{link}(f\Sigma^2, f\Sigma^1) = 0$, then $f\Sigma^1$ spans a 2-disc outside $f\Sigma^2$. Hence we can construct an almost embedding $\widetilde{\Delta^2_{01...6}} \to \mathbb{R}^4$. Therefore there is an equivariant map $\widetilde{\Delta^2_{01...6}} \to S^3$.

Then by [215], any equivariant map $\widetilde{\Delta^2_{01...6}} \to S^4$ should induce a *trivial* homomorphism in $H^{eq}_4(\cdot, \mathbb{Z}_2)$. But there is an embedding $g : \Delta^2_{01...6} \to \mathbb{R}^5$ such that $\widetilde{g} : \widetilde{\Delta^2_{01...6}} \to S^4$ induces a *nontrivial* homomorphism in $H^{eq}_4(\cdot, \mathbb{Z}_2)$.

Indeed, $\Delta^2_{01...6} \cong \Delta^2_{01...5} \cup \text{Con}\,\Delta^1_{01...5}$, where $\Delta^5_{01...5}$ is a regular 5-simplex inscribed into the standard unit 4-sphere in $\mathbb{R}^5$ and the vertex of the cone is 0. This homeomorphism defines an embedding $g : \Delta^2_{01...6} \to \mathbb{R}^5$. The union of 4-cells in the simplicial deleted product $\widetilde{\Delta^2_{01...6}}$ is a $\mathbb{Z}_2$-equivariant 4-cycle.

Let p be a vertex of $\Delta^5_{01...5}$. The map $\widetilde{g}$ maps a small neighbourhood in $\widetilde{\Delta^2_{01...6}}$ of the point $(p, 0)$ homeomorphically onto a neighbourhood in S^4 of p. Hence the $\widetilde{g}$-image of the above 4-cycle is nontrivial, and so $\widetilde{g}$ induces a nontrivial homomorphism in $H^{eq}_4(\cdot, \mathbb{Z}_2)$.

In order to prove the full strength of the particular case $m = 2n = 4l = 4$, assume to the contrary that $\text{link}(f\Sigma^n, f\Sigma^l)$ is even. Let Δ' be a polyhedron obtained from Δ^2_{012} by removing the interiors of an even number $2r$ of disjoint 2-discs in $\mathring{\Delta}^2_{012}$ and running r pairwise disjoint tubes between the holes thus formed. Then $f\Sigma^1$ spans Δ' outside $f\Sigma^2$. Let

$$X = Q \bigcup_{\partial\Delta^2_{012} = \partial\Delta'} \Delta'.$$

Then there is an equivariant map $\widetilde{X} \to S^3$. Therefore by [215] any equivariant map $\widetilde{X} \to S^4$ should induce a trivial homomorphism in $H^{eq}_4(\cdot, \mathbb{Z}_2)$.

Let $p : \Delta \to \Delta^2_{01\ldots 6}$ be a map which is the identity on Q and such that $p(\mathring{\Delta}') = \mathring{\Delta}^2_{012}$. Then $\widetilde{p} : \widetilde{X} \to \widetilde{\Delta^2_{01\ldots 6}}$ induces an epimorphism in $H^{eq}_4(\cdot, \mathbb{Z}_2)$ [171, p. 278]. Then $\widetilde{f} \circ \widetilde{p}$ induces a nontrivial homomorphism in H^{eq}_4, which is a contradiction. $\square$

Sketch proof of the TOP non-embeddability of N into $\mathbb{R}^m$. For $m \geq n+ 3$ TOP non-embeddability follows from PL non-embeddability, by [17].

For $m = n+2$ we argue as follows. There are arbitrarily close approximations to the embedding h by PL almost embeddings $h' : N \to \mathbb{R}^m$ (for certain triangulations of N). By general position we may assume that $h'|_{\Sigma^1}$ and $h'|_{\bar{\Sigma}^1}$ are PL embeddings. The rest of the proof is analogous to the PL case (in which we need to replace h by h'), because we only used the Linking Lemma 7.2 but not the fact that h is an embedding. $\square$

Borromean rings and the Boy immersion

The following was proved in [4], cf. [48, 7]. Let $h : \mathbb{R}P^2 \to \mathbb{R}^3$ be the Boy immersion. Fix any orientation on the sphere S^2 and on the double points circle $\Delta(h)$. Take a small closed ball $D^3 \subset \mathbb{R}^3$ containing the triple point of h. Let $r : S^2 \to \mathbb{R}P^2$ be the standard double covering. Denote by $\pi_1 : \mathbb{R}^3 \times \mathbb{R} \to \mathbb{R}^3$ and $\pi_2 : \mathbb{R}^3 \times \mathbb{R} \to \mathbb{R}$ the projections. Take a general position smooth map $\bar{h} : (S^2 - h_1^{-1}\mathring{D}^3) \to \mathbb{R}^3 \times \mathbb{R}$ such that $\pi_1 \circ \bar{h} = h \circ r$ and for each points $x, y \in S^2 - (h \circ r)^{-1}D^3$ such that $hrx = hry$ and $\pi_2\bar{h}x > \pi_2\bar{h}y$ the following three vectors form a *positive* basis of $\mathbb{R}^3$ at the point $hrx = hry$: the vector along the orientation of Δ, the normal vector to the small sheet of $h(\mathbb{R}P^2)$ containing x and the normal vector to the small sheet of $h(\mathbb{R}P^2)$ containing y. Then $\bar{h}|_{(hor)^{-1}\partial D^3} \to \partial D^3 \times \mathbb{R}$ forms the Borromean rings linking (after the identification $\partial D^3 \times \mathbb{R} \cong \mathbb{R}^3$).

8 The disjunction method

In this section we illustrate the disjunction method by sketching some ideas of the proofs of surjectivity in the Haefliger-Weber theorem (Theorem 5.5) for the PL case, and of Theorem 5.6. The first of these results can be restated as follows.

Theorem 8.1 (The Weber theorem [207]). *Suppose that N is an n-polyhedron, $2m \geq 3n + 3$ and $\Phi : \widetilde{N} \to S^{m-1}$ is an equivariant map. Then there is a PL embedding $f : N \to \mathbb{R}^m$ such that $\widetilde{f} \simeq_{eq} \Phi$ on $\widetilde{N}$.*

Simplices of any triangulation T are assumed to be linearly ordered with respect to increasing dimension. The lexicographical ordering on $T \times T$ is used.

8.1 Proof of the Weber theorem for $m = 2n + 1$

We present the proof for $m = 3$ and $n = 1$ (the general case $m = 2n + 1$ is proved analogously). Take a general position map $f : N \to \mathbb{R}^3$. Then it is an embedding, so we only need to modify it in order to obtain the property $\widetilde{f} \simeq_{eq} \Phi$. This property does not follow by general position. We obtain it by applying the *van Kampen finger moves*, i.e. by winding edges of the graph N around images of other edges.

Note that van Kampen invented his finger moves for the proof of Lemma 4.4; here we present a *generalization* of the van Kampen finger moves.

Proposition 8.2. *Let T be a triangulation of a 1-polyhedron N (i.e. T is a graph representing this 1-polyhedron). For each pair of edges $\sigma, \tau \in T$ such that $\sigma \leq \tau$ there exists a PL embedding $f : N \to \mathbb{R}^3$ such that*

$$\widetilde{f} \simeq_{eq} \Phi \quad on \quad J_{\sigma\tau} := \bigcup_{\sigma \times \tau > \alpha \times \beta \in \widetilde{T}} \alpha \times \beta.$$

Theorem 8.1 for $m = 2n + 1 = 3$ follows from Proposition 8.2 by taking σ and τ to be the last simplices of T.

Proof of Proposition 8.2, 1st step: construction of balls. By induction on $\sigma \times \tau$. If both σ and τ are the first edge of T, then $\dim J_{\sigma\tau} = 1$, hence Proposition 8.2 is true by general position. Now suppose as inductive hypothesis that $\widetilde{f} \simeq_{eq} \Phi$ on $J_{\sigma\tau}$ for an embedding $f : N \to \mathbb{R}^3$. We need to prove that for $\sigma \cap \tau = \emptyset$ there exists a map

$$f^+ : N \to \mathbb{R}^3 \quad such\ that \quad \widetilde{f^+} \simeq_{eq} \Phi \quad on \quad J_{\sigma\tau} \cup \sigma \times \tau \cup \tau \times \sigma.$$

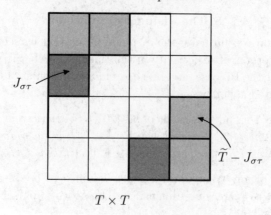

$T \times T$

Fig. 8.1.

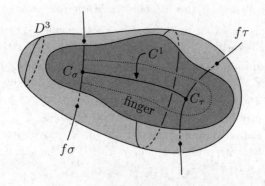

Fig. 8.2.

Take points $C_\sigma \in \mathring{\sigma}$ and $C_\tau \in \mathring{\tau}$ (Figure 8.2). Join their images by an arc $C^1 \subset \mathbb{R}^3$ such that $C^1 \cap fN = \{fC_\sigma, fC_\tau\}$. Let D^3 be a small ball neighbourhood of C^1 in $\mathbb{R}^3$ such that $f^{-1}D^3$ is a disjoint union of arcs $D_\sigma \subset \mathring{\sigma}$ and $D_\tau \subset \mathring{\tau}$ containing points C_σ and C_τ, respectively.

We may assume that fD_σ is unknotted in D^3. Hence a homotopy equivalence $h : D^3 - fD_\sigma \to S^1$ is constructed, analogously to the homotopy equivalence $S^m - fS^q \to S^{m-q-1}$ from (3.1) (which was used in defining the linking coefficients). □

Proof of Proposition 8.2, 2nd step: the van Kampen finger move. In order to construct such an f^+ we shall wind the arc $f|_{D_\tau}$ around fD_σ in $D^3 - fD_\sigma$ (Figure 8.3).

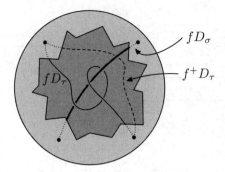

Fig. 8.3.

Take any embedding $f^+ : D_\sigma \sqcup D_\tau \to D^3$ such that $f^+ = f$ on $D_\sigma \sqcup \partial D_\tau$. Let D_+ be a copy of D_τ; identify S^1 with $D_\tau \underset{\partial D_\tau = \partial D_+}{\bigcup} D_+$.

Identify S^1 with $D^3 - fD_\sigma$ by the homotopy equivalence h given earlier in (3.1). Define a map

$$h_{ff^+} : S^1 = D_\tau \underset{\partial D_\tau = \partial D_+}{\bigcup} D_+ \to D^3 - fD_\sigma = S^1$$

by

$$h_{ff^+}(x) = \begin{cases} h(f(x)) & x \in D_\tau \\ h(f^+(x)) & x \in D_+. \end{cases}$$

Since $f^+ = f$ on $D_\sigma \sqcup \partial D_\tau$, it follows that there is a homotopy $f_t : D_\sigma \sqcup D_\tau \to D^3$ from f to f^+ fixed on $D_\sigma \sqcup \partial D_\tau$. Denote by $\widetilde{f}$, Φ and $\widetilde{f}^+$ the restrictions of these maps to $D_\sigma \times D_\tau$, and by $\widetilde{f}_t$ the restriction of this map to $\partial(D_\sigma \times D_\tau)$. Define a map $H_{\widetilde{f}\widetilde{f}^+} : \partial(D_\sigma \times D_\tau \times I) \to S^2$ by

$$H_{\widetilde{f}\widetilde{f}^+}|_{D_\sigma \times D_\tau \times 0} = \widetilde{f}, \quad H_{\widetilde{f}\widetilde{f}^+}|_{D_\sigma \times D_\tau \times 1} = \widetilde{f}^+, \quad H_{\widetilde{f}\widetilde{f}^+}|_{\partial(D_\sigma \times D_\tau) \times I} = \widetilde{f}_t.$$

By [207, Lemma 1] we have

$$[H_{\widetilde{f}\widetilde{f}^+}] = \Sigma[h_{ff^+}] \in \pi_2(S^2).$$

By the equivariant analogue of the Borsuk homotopy extension theorem, there is an equivariant extension $\Psi : \widetilde{K} \to S^2$ of $\widetilde{f}|_{J_{\sigma\tau} \cup (\sigma \times \tau - \mathring{D}_\sigma \times \mathring{D}_\tau)}$ such that $\Psi \simeq_{eq} \Phi$. We may assume that $\Psi = \Phi$, so that $\Phi = \widetilde{f}$ on $\partial(D_\sigma \times D_\tau)$. Therefore $H_{\Phi\widetilde{f}^+}$ can be defined analogously to above; we

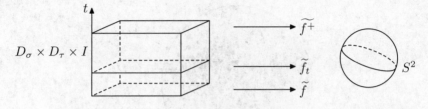

$$D_\sigma \times D_\tau \times I$$

Fig. 8.4.

can also define $H_{\Phi \widetilde{f}}$ similarly, using the constant homotopy between Φ and $\widetilde{f}$ on $\partial(D_\sigma \times D_\tau)$. Then

$$[H_{\Phi \widetilde{f}+}] = [H_{\Phi \widetilde{f}}] + [H_{\widetilde{f}\widetilde{f}+}] = [H_{\Phi \widetilde{f}}] + \Sigma[h_{ff+}] \in \pi_2(S^2).$$

For *every* element $\beta \in \pi_1(S^1)$ there is an embedding

$$f^+ : D_\tau \to D^3 - fD_\sigma$$

such that $[h_{ff+}] = \beta$. Therefore, by the suspension theorem there exists a map $f^+ : D_\tau \to D^3 - fD_\sigma$ such that $[H_{\Phi \widetilde{f}+}] = 0$. Extend this f^+ to all N by f. Then f^+ is as required. $\square$

8.2 Proof of the Weber theorem for $m = 2n \geq 6$

Let us introduce some natural and useful definitions.

Definition 8.3. Fix a triangulation T of an n-polyhedron N. Then a map $f : N \to \mathbb{R}^m$ is an embedding if and only if the following conditions hold:

- f is (T)-*nondegenerate*, i.e. $f|_\alpha$ is an embedding for each $\alpha \in T$;
- f is a (T)-*almost embedding*, i.e. $f\alpha \cap f\beta = \emptyset$ for each $\alpha \times \beta \subset \widetilde{T}$;
- f is a T-*immersion*, i.e. $f\alpha \cap f\beta = f(\alpha \cap \beta)$ for each $\alpha, \beta \in T$ such that $\alpha \cap \beta \neq \emptyset$.

Plan of the proof of the Weber theorem 8.1 for $m = 2n \geq 6$. Take a triangulation T of N and a general position map $f : N \to \mathbb{R}^m$ that is linear on simplices of T. Hence f is nondegenerate. By general position, the properties of almost embedding and T-immersion hold unless $\dim \alpha = \dim \beta = n$.

Step 1. Wind n-simplices of T around $(n-1)$-simplices (analogously to the case $m = 2n + 1$) and thus modify f to obtain additionally the condition $f \simeq_{eq} \Phi$ on the $(2n-1)$-skeleton of $\widetilde{T}$ (see Figure 8.5). Such

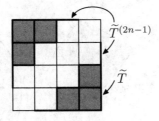

Fig. 8.5.

a *van Kampen finger move* is a higher-dimensional generalization of the fifth Reidemeister move (Figure 4.2.V).

Step 2. Now it is possible to remove intersections of disjoint n-simplices of T and thus modify f so as to make it additionally an almost embedding. This construction is a higher-dimensional generalization of the first and second Reidemeister moves (Figures 4.2.I and 4.2.II), which are called the *Penrose-Whitehead-Zeeman trick* and the *Whitney trick* respectively. The details for Step 2 are given as Proposition 8.4 below.

Step 1 and Step 2 together are analogues of Lemma 4.4.

Step 3. Wind n-simplices of T around n-simplices (analogously to the case $m = 2n + 1$) and thus modify f to obtain additionally the condition $f \simeq_{eq} \Phi$ on $\widetilde{T}$. These are the van Kampen finger moves in other dimensions.

Step 4. Remove unnecessary intersections of n-simplices of T which have a common face, and thus modify f so as to make it additionally a T-immersion (and hence an embedding). This construction is a higher-dimensional generalization of the fourth Reidemeister move (Figure 4.2.IV), an analogue of the Freedman-Krushkal-Teichner lemma (Lemma 4.5), and is called the *Freedman-Krushkal-Teichner trick*. ☐

Proposition 8.4 (cf. Proposition 8.2). *Let T be a triangulation of an n-polyhedron N. For each pair of n-simplices $\sigma, \tau \in T$ such that $\sigma < \tau$ there exists a nondegenerate PL map $f : N \to \mathbb{R}^{2n}$ such that*

$$f\alpha \cap f\beta = \emptyset \quad if \quad \sigma \times \tau > \alpha \times \beta \in \widetilde{T}.$$

Proof. As inductive hypothesis, assume that we have such f for a pair $\sigma \times \tau \subset \widetilde{T}$. We need to prove that there exists a map $f^+ : N \to \mathbb{R}^{2n}$ such that

$$f\alpha \cap f\beta = \emptyset \quad if \quad \sigma \times \tau \geq \alpha \times \beta \in \widetilde{T}.$$

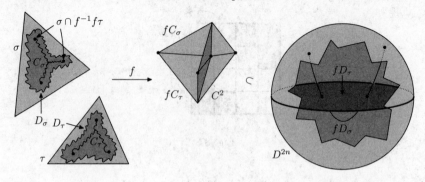

Fig. 8.6.

By the inductive hypothesis $f\sigma \cap f\tau = f\mathring{\sigma} \cap f\mathring{\tau}$ is a finite set of points. Let $C_\sigma \subset \mathring{\sigma}$ be an arc containing the points of $\sigma \cap f^{-1}\tau$ (Figure 8.6). Let $C_\tau \subset \mathring{\tau}$ be an arc containing the points of $\tau \cap f^{-1}\sigma$ 'in the same order' as C_σ. Let $C^2 \subset \mathbb{R}^{2n}$ be a union of discs such that $C^2 \cap f\sigma = fC_\sigma$ and $C^2 \cap f\tau = fC_\tau$.

Let D^{2n} be a small neighbourhood of C^2 in $\mathbb{R}^{2n}$ such that $f^{-1}D^{2n}$ is disjoint union of n-balls $D_\sigma \subset \mathring{\sigma}$ and $D_\tau \subset \mathring{\tau}$, which are small neighbourhoods of the arcs C_σ and C_τ in N. We have the following:

- $f|_{D_\sigma}$ and $f|_{D_\tau}$ are proper embeddings into D^{2n};
- $f\sigma \cap f\tau \subset \mathring{D}^{2n}$;
- $D_\sigma = \sigma \cap f^{-1}D^{2n}$ and $D_\tau = \tau \cap f^{-1}D^{2n}$; and
- $f\partial D_\sigma \cap fD_\tau = f\partial D_\tau \cap fD_\sigma = \emptyset$.

Since $n \geq 3$, it follows that a homotopy equivalence $h : D^{2n} - fD_\sigma \to S^{n-1}$ can be constructed in analogous fashion to the homotopy equivalence (3.1). The *coefficient of the intersection* of fD_σ and fD_τ is the homotopy class

$$I(fD_\sigma, fD_\tau) = [\partial D_\tau \xrightarrow{f|_{\partial D_\tau}} D^{2n} - fD_\sigma \xrightarrow{h} S^{n-1}] \in \pi_{n-1}(S^{n-1}).$$

We have

$$\pm\Sigma^n I(fD_\sigma, fD_\tau) = [\widetilde{f}|_{\partial(D_\sigma \times D_\tau)}] = [\Phi|_{\partial(D_\sigma \times D_\tau)}] = 0 \in \pi_{2n-1}(S^{2n-1}).$$

Here the first equality is [207, Proposition 1], the second equality holds because $\widetilde{f} \simeq \Phi$ on $\partial(D_\sigma \times D_\tau)$ by the inductive hypothesis, and the third equality holds since Φ is defined over $\widetilde{T} \supset D_\sigma \times D_\tau$.

By the Freudenthal suspension theorem $I(fD_\sigma, fD_\tau) = 0$. Hence the embedding $f|_{\partial D_\tau}$ extends to a *map* $f^+ : D_\tau \to D^{2n} - fD_\sigma$. Using the

Penrose-Whitehead-Zeeman trick we modify f' to an *embedding*. Extend f^+ over the entire N by f. By general position C^2 (and hence D^{2n} and $f^+ D_\tau$) is disjoint with $f(N - \text{st}\,\sigma - \text{st}\,\tau)$. Therefore f^+ is as required. $\square$

8.3 Definition of a regular neighbourhood

The notions of *collapse* and *regular neighbourhood* are used in our proofs and are generally important in topology.

A polyhedron Y is said to be obtained from a polyhedron K by an *elementary collapse* if $K = Y \cup B^n$ and $Y \cap B^n = B^{n-1}$, where B^{n-1} is a face of the ball B^n. This elementary collapse is said to be made *from* $\text{Cl}(\partial B^n - B^{n-1})$ *along* B^n *to* B^{n-1}. A polyhedron K *collapses to* Y (denoted by $K \searrow Y$) if there exists a sequence of elementary collapses $K = K_0 \searrow K_1 \searrow \cdots \searrow K_n = Y$. A polyhedron K is *collapsible* if it collapses to a point.

Clearly, the ball B^n is collapsible, since it is collapsible to its face B^{n-1}, thence to B^{n-2}, and so on by induction. Moreover, a cone cK on a compact polyhedron K is collapsible (to its vertex). Indeed, note that for each simplex $A \subset K$, the cone cA collapses from A to $c(\partial A)$, hence cK collapses to a point inductively via simplices of decreasing dimension.

A collapsing $K \searrow Y$ generates a deformation retraction $r : K \to Y$, given by deformations of each ball B^n to its face B^{n-1}. Consider a homotopy H_t between the identity map $K \to K$ and the deformation retraction $r : K \to Y$, given by the collapse $K \searrow Y$. The *trace* of a subpolyhedron S of K under the collapse $K \searrow Y$ is the union of $H_t(S)$ over $t \in [0, 1]$ (this in fact depends on the homotopy, not only on the collapse).

Suppose that K is a subpolyhedron of a PL manifold M. A neighbourhood N of K in M is called *regular* if N is a compact bounded manifold and $N \searrow K$. The same polyhedron can have distinct regular neighbourhoods. However, the regular neighbourhood is unique up to homeomorphism and even up to isotopy fixed on K [161]. The regular neighbourhood of a collapsible polyhedron is a ball [161]. (The converse statement, i.e. that $K \subset B^n$ and $K \searrow *$ imply $B^n \searrow K$, is true only in codimension ≥ 3 [90].)

Let us fix the following convention. The notation $R_M(K)$ means 'a sufficiently small regular neighbourhood of K in M', when it first appears, and 'the regular neighbourhood of K in M', after the first appearance.

8.4 Proof of the Weber theorem for the general case
$$2m \geq 3n + 3$$

The proof of the Weber theorem (Theorem 8.1) for the general case consists of two steps:

(i) the construction of a nondegenerate almost embedding (an analogue of Lemma 4.4, the details for this step are given as Proposition 8.5 below); and

(ii) the construction of an embedding from a nondegenerate almost embedding (a generalization of the Freedman-Krushkal-Teichner lemma 4.5, the details are given in [179, 153]).

Proposition 8.5. *Suppose that N is an n-polyhedron with a triangulation T, $2m \geq 3n + 3$ and $\Phi : \widetilde{N} \to S^{m-1}$ is an equivariant map. Then for each $\sigma \times \tau \in \widetilde{T}$ such that $\sigma \leq \tau$ there exists a nondegenerate PL map $f : N \to \mathbb{R}^m$ such that*

$$f\alpha \cap f\beta = \emptyset \qquad \text{for each } \alpha \times \beta < \sigma \times \tau \qquad (*)$$

and

$$\widetilde{f} \simeq_{eq} \Phi \quad on \quad J_{\sigma\tau} := \bigcup \{\alpha \times \beta \cup \beta \times \alpha \subset \widetilde{T} \mid \alpha \times \beta < \sigma \times \tau\}. \ (**)$$

Theorem 8.1 follows from Proposition 8.5 by taking σ and τ to be the last simplex of T and then applying a generalization of the Freedman-Krushkal-Teichner lemma [179]. Theorem 8.1 can also be proved by first constructing the immersion and then modifying it to an embedding [181].

Proof of Proposition 8.5. Take a general position map $f : N \to \mathbb{R}^m$ that is linear on the simplices of the triangulation T. The map f is already nondegenerate. By the induction hypothesis on $\sigma \times \tau$ we may assume that f is nondegenerate and the properties $(*)$ and $(**)$ hold. Suppose that $p+q \geq m-1$ (otherwise the inductive step holds by general position).

The first part of the proof (a generalization of Proposition 8.4 and the Whitney trick) is to obtain the property $f\sigma \cap f\tau = \emptyset$. The second part of the proof (a generalization of Proposition 8.2 and the van Kampen finger moves) is getting the property $\widetilde{f} \simeq_{eq} \Phi$ on $J_{\sigma\tau} \cup \sigma \times \tau \cup \tau \times \sigma$. $\square$

Constructions of balls D_σ, D_τ and D^m

The first step in the proof of Proposition 8.5 generalizes the construction of the arcs l_1, l_2 and the disc D in the Whitney trick, or the construction of D_σ, D_τ and D^{2n} in the case $m = 2n$ above.

Let $\Sigma = f\sigma \cap f\tau$. By $(*)$ we have

$$f\sigma \cap f\partial\tau = f\partial\sigma \cap f\tau = \emptyset.$$

Hence $\Sigma = f\mathring{\sigma} \cap f\mathring{\tau}$. By general position, $\dim \Sigma \leq p + q - m$.

Let $C_\sigma \subset \mathring{\sigma}$ be the trace of the polyhedron $\sigma \cap f^{-1}\tau$ under some collapse $\sigma \searrow$ (a point in $\mathring{\tau}$). Define analogously $C_\tau \subset \mathring{\tau}$. The polyhedra C_σ, C_τ are generalizations of the arcs l_1, l_2 from the Whitney trick, and of C_σ, C_τ above. They are collapsible and satisfy

$$\Sigma \subset fC_\sigma \cap fC_\tau \quad \text{and} \quad \dim C_\sigma, \dim C_\tau \leq p + q - m + 1.$$

Consider a collapse from some PL m-ball J^m in $\mathbb{R}^m$, containing Σ in its interior, to a point in $\mathring{J}^m$. Let C be the trail of $C_\sigma \cup C_\tau$ under this collapse. The polyhedron C is a generalization of the disc C from the Whitney trick. It is collapsible and contains $C_\sigma \cup C_\tau$, and $\dim C \leq p + q - m + 2$. Hence by general position, $C \cap f\sigma = C_\sigma$ and $C \cap f\tau = C_\tau$.

Take the regular neighbourhoods of polyhedra C_σ, C_τ and C in some sufficiently fine (agreeing) triangulations of σ, τ and $\mathbb{R}^m$, respectively. They are PL balls $D_\sigma^p \subset \mathring{\sigma}$, $D_\tau^q \subset \mathring{\tau}$ and $D^m \subset \mathbb{R}^m$ such that

(a) $f|_{D_\sigma}$ and $f|_{D_\tau}$ are proper embeddings into D^m;
(b) $f\sigma \cap f\tau \subset \mathring{D}^m$;
(c) $D_\sigma = \sigma \cap f^{-1}D^m$ and $D_\tau = \tau \cap f^{-1}D^m$;
(d) $D^m \cap fP = \emptyset$, where $P = N - \operatorname{st}\sigma - \operatorname{st}\tau$.

Only the last property needs a proof. By $(*)$ we have $C_\sigma \cap P = \emptyset$. By general position $\dim(fP \cap f\tau) \leq n + q - m$, hence

$$\dim(fP \cap f\tau) + \dim C_\tau < q$$

and so $C_\tau \cap P = \emptyset$. Therefore $C \cap fP = \emptyset$, which implies (d).

A generalization of the Whitney trick

Take PL balls D^m, D_σ and D_τ as above. Since f is nondegenerate and $(*)$ holds, it follows that $f\partial D_\sigma \cap fD_\tau = f\partial D_\tau \cap fD_\sigma = \emptyset$. Since $m - p \geq 3$, it follows that a homotopy equivalence $h : D^m - fD_\sigma \to S^{m-p-1}$ may be constructed analogously to the homotopy equivalence (3.1). The *coefficient of the intersection* of $f|_{\partial D_\sigma}$ and $f|_{\partial D_\tau}$ is the homotopy class

$$I(f|_{D_\tau}, f|_{D_\tau}) := [\partial D_\tau \xrightarrow{f|_{\partial D_\tau}} D^m - fD_\sigma \xrightarrow{h} S^{m-p-1}] \in \pi_{q-1}(S^{m-p-1}).$$

We have

$$\Sigma^p I(f|_{D_\sigma}, f|_{D_\tau}) = (-1)^{m-p} [\tilde{f}|_{\partial(D_\sigma \times D_\tau)}] = [\Phi|_{\partial(D_\sigma \times D_\tau)}] = 0.$$

Here the first equality holds by [207, Proposition 1]. The second equality holds since $\tilde{f} \simeq \Phi$ on $\partial(D_\sigma \times D_\tau)$ by the inductive hypothesis. The third equality holds since Φ is defined over $\tilde{T} \supset D_\sigma \times D_\tau$. Since $2p + q \leq 2m - 3$ we have $q - 1 \leq 2(m - p - 1) - 2$. So by the Freudenthal suspension theorem, the homomorphism Σ^p is a monomorphism. Hence the embedding $f|_{\partial D_\tau}$ extends to a *map* $f' : D_\tau \to D^m - fD_\sigma$. Since $2q - m + 1 \leq m - p - 2$, by the Penrose-Whitehead-Zeeman-Irwin embedding theorem (Theorem 2.9) it follows that f' is homotopic rel ∂D_τ to an *embedding* $f^+ : D_\tau \to D^m - fD_\sigma$. Here we again use the inequality $p + 2q \leq 2m - 3$. Since $m - q \geq 3$, by the relative version of Theorem 2.7(a) [218, Corollary 1 to Theorem 9] it follows that there is an ambient isotopy $h_t : D^m \to D^m$ rel ∂D^m carrying $f|_{D_\tau}$ to f^+. Extend f^+ over N by the formula

$$f^+(x) = \begin{cases} h_1(f(x)) & \text{if } f(x) \in D^m \text{ and } x \in \gamma \text{ for some } \gamma \supset \sigma^q \\ f(x) & \text{otherwise.} \end{cases}$$

It is easy to check that f^+ is a nondegenerate PL map satisfying the properties $(*)$ and $(**)$ and such that $f^+\sigma \cap f^+\tau = \emptyset$.

A generalization of the van Kampen finger moves

We begin with the analogous construction of PL balls. By general position we can take points $C_\sigma \in \overset{\circ}{\sigma}$ and $C_\tau \in \overset{\circ}{\tau}$ such that the restrictions of f to some small neighbourhoods of C_σ and C_τ are embeddings. Since $x, y \leq m - 2$, we can join points fC_σ and fC_τ by an arc $C \subset \mathbb{R}^m$ such that $C \cap fN = \{fC_\sigma, fC_\tau\}$. Let $D^m = R_{\mathbb{R}^m}(C)$. Then $f^{-1}D^m$ is the disjoint union of PL discs $D_\sigma \subset \overset{\circ}{\sigma}$ and $D_\tau \subset \overset{\circ}{\tau}$, which are regular neighbourhoods in N of C_σ and C_τ, respectively.

By the Borsuk homotopy extension theorem, there is an extension $\Psi : \tilde{T} \to S^{m-1}$ of the map $\tilde{f}|_{J_{\sigma\tau} \cup (\sigma \times \tau - \mathring{D}_\sigma \times \mathring{D}_\tau)}$ such that $\Psi \simeq \Phi$. So $\Psi = \tilde{f}$ on $\partial(D_\sigma \times D_\tau)$. We may assume $\Psi = \Phi$. By [207, Lemma 1], for each map $f' : D_\sigma \sqcup D_\tau \to D^m$ such that $f' = f$ on $D_\sigma \sqcup \partial D_\tau$ and

$f'D_\sigma \cap f'D_\tau = \emptyset$ and each homotopy f_t rel $D_\sigma \sqcup \partial D_\tau$ from f to f', we have

$$[H_{\Phi \widetilde{f_t} \widetilde{f'}}] = [H_{\Phi \widetilde{f}}] + [H_{\widetilde{f} \widetilde{f_t} \widetilde{f'}}] = [H_{\Phi \widetilde{f}}] + (-1)^{m-p} \Sigma^p [h_{ff'}] \in \pi_{p+q}(S^{m-1}).$$

Here Φ, $\widetilde{f}$ and $\widetilde{f}^+$ denote the restrictions of these maps to $D_\sigma \times D_\tau$; $\widetilde{f_t}$ denotes the restriction of this map to $\partial(D_\sigma \times D_\tau)$, and the maps H, $h_{ff'}$ are defined as in the second step of the proof of Proposition 8.2. Since $2p + q \le 2m - 3$, we have $q \le 2(m - p - 1) - 1$. So by the Freudenthal suspension theorem Σ^p is an epimorphism. Since for *every* element $\beta \in \pi_q(S^{m-p-1})$ there is a map (not necessarily an embedding) $f' : D_\tau \to D^m - fD_\sigma$ such that

$$[h_{ff'}] = \beta \quad \text{and} \quad f' = f \quad \text{on} \quad D_\sigma \sqcup \partial D_\tau,$$

it follows that we can take f' so that $[H_{\Phi \widetilde{f_t} \widetilde{f'}}] = 0$.

The rest of the proof is the same as in the generalization of the Whitney trick.

8.5 Generalization of the Weber theorem

We illustrate some of the ideas of the proof of Theorem 5.6 by proving the following weaker result for $d \in \{0, 1\}$.

Theorem 8.6. *Suppose that N is a d-connected closed PL n-manifold, $2m \ge 3n + 2 - d$, $m \ge n + 3$ and $\Phi : \widetilde{N} \to S^{m-1}$ is an equivariant map. Then there is a PL embedding $f : N \to \mathbb{R}^m$ [178].*

We first recall some classical results and their generalizations, which are required to prove Theorem 8.6.

Lemma 8.7 (The engulfing lemma). *Suppose that N is a $(2k + 2 - n)$-connected closed n-manifold and $K \subset N$ is a k-polyhedron such that $n - k \ge 3$ and the inclusion $K \subset N$ is null-homotopic. Then K can be engulfed in N, i.e. is contained in an n-ball $B \subset N$ [141, 218].*

Theorem 8.8. *Suppose that N is a closed homologically $(3n - 2m + 2)$-connected PL n-manifold, $m - n \ge 3$ and $g : N \to \mathbb{R}^m$ is a map such that $\Sigma(g)$ is contained in some PL n-ball $B \subset N$. Then there is an embedding $f : N \to \mathbb{R}^m$ such that $f = g$ on $N - \overset{\circ}{B}$.*

Proof. The theorem is essentially proved in [79], cf. [178, Theorem 2.1.2].

Let $M = \mathbb{R}^m - \operatorname{Int} R(g(N - \mathring{B}), g\partial B)$. Since N is homologically $(3n - 2m + 2)$-connected, we have by Alexander duality

$$H_i(M) \cong H^{m-1-i}(\mathbb{R}^m - M) \cong H^{m-1-i}(N - \mathring{B}) \cong H_{n-m+1+i}(N) = 0$$

for $i \le 2n - m + 1$. Since $m - n \ge 3$, it follows that M is simply connected. Therefore by the Hurewicz isomorphism theorem we have that M is $(2n - m + 1)$-connected. Hence by the Penrose-Whitehead-Zeeman-Irwin embedding theorem (Theorem 2.9) the embedding $g : \partial B \to \partial M$ extends to an embedding $f : B \to M$. Extending f by g outside B we complete the proof. $\qquad\square$

Theorem 8.9. *Let N be an n-polyhedron with triangulation T. If $m - n \ge 3$ and there exists an equivariant map $\Phi : \widetilde{N} \to S^{m-1}$, then there exists a general position nondegenerate PL map $f : N \to \mathbb{R}^m$ such that*

$$f\sigma \cap f\tau = f(\sigma \cap \tau) \quad \text{if} \quad p = \dim \sigma \le \dim \tau = q \quad \text{and} \quad p + q + n \le 2m - 3.$$

Sketch of the proof. This is analogous to the proof of Theorem 8.1. Recall that Theorem 8.1 is proved by induction on $\sigma \times \tau \in \widetilde{T}$. If $\dim \sigma = p$ and $\dim \tau = q$, then we need the following dimensional restrictions:

- $p + 2q \le 2m - 3$ to apply the Freudenthal suspension theorem (twice) and the Penrose-Whitehead-Zeeman trick;
- $p + q + n \le 2m - 3$ to get the property $D^m \cap f(N - \operatorname{st}\sigma - \operatorname{st}\tau) = \emptyset$.

$$\square$$

A reduction of Theorem 8.6. It suffices to prove the following.

Claim. *For some fine triangulation T of N and a map $f : N \to \mathbb{R}^m$ as given by Theorem 8.9, the inclusion $\Sigma(f) \subset N$ is homotopic to a map to some d-dimensional subpolyhedron of N.*

Indeed, since N is d-connected, the claim implies that the inclusion $\Sigma(f) \subset N$ is null-homotopic. Since N is $(2(2n - m) - n + 2) = d$-connected, by Lemma 8.7 it follows that $\Sigma(f)$ is contained in some PL n-ball in N. Then N embeds into $\mathbb{R}^m$ by Theorem 8.8. $\qquad\square$

Note that if a map f given by Theorem 8.9 is a PL immersion (i.e. a local embedding), then $\Sigma(f)$ does not intersect the $(2m - 2n - 3)$-skeleton of T. Hence it retracts to the $(n - 1 - (2m - 2n - 3)) = d$-skeleton of a triangulation, dual to T, i.e. the claim holds.

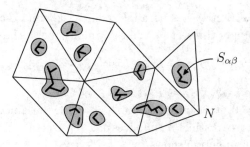

Fig. 8.7.

Proof of the claim for $d = 0$. We may assume that $2m = 3n + 2$. Hence $f\sigma \cap f\tau = f(\sigma \cap \tau)$ unless $\dim \sigma = \dim \tau = n$. For n-simplices α and β let

$$S_{\alpha\beta} = \begin{cases} \alpha \cap f^{-1} f\beta & \alpha \cap \beta = \emptyset \\ f^{-1} \operatorname{Cl}[(f\alpha \cap f\beta) - f(\alpha \cap \beta)] & \alpha \cap \beta \neq \emptyset \end{cases}$$

(see Figure 8.7).

Then

$$\Sigma(f) = \bigcup_{\alpha \neq \beta} S_{\alpha\beta} \quad \text{and} \quad S_{\alpha\beta} \cap S_{\gamma\delta} = \emptyset$$

when $\alpha\beta \neq \gamma\delta$. Here $\alpha\beta$ is the ordered pair (α, β) when $\alpha \cap \beta = \emptyset$ and the non-ordered pair $\{\alpha, \beta\}$ when $\alpha \cap \beta \neq \emptyset$. Therefore the contractibility of α (and of $\alpha \cup \beta$ for $\alpha \cap \beta \neq \emptyset$) yields the claimed homotopy. $\square$

Proof of the claim for $d = 1$. We may assume that $2m = 3n + 1$. Define $\alpha\beta$ and $S_{\alpha\beta}$ as in the case $d = 0$. We denote such pairs $\alpha\beta$ by Latin letters i, j, k, l. First we prove that:

(a) $S_i \cap S_j \cap S_k = \emptyset$ for distinct $i, j, k = 1, \ldots, s$;
(b) for each $i = 1, \ldots, s$ there is a contractible polyhedron $A_i \subset N$ which contains S_i.

If $S_i \cap S_j \neq \emptyset$, then there is a contractible polyhedron $A_{ij} \subset N$, containing $A_i \cup A_j$.

Indeed, we may assume that a triangulation T of N is such that, for each $x \in N$, the star $\operatorname{st}^2 x = \operatorname{st} \operatorname{st} x$ is contractible. By Theorem 8.9 $S_{\alpha\beta} \neq \emptyset$ only when

$$\dim \alpha = \dim \beta = n \quad \text{or} \quad \{\dim \alpha, \dim \beta\} = \{n, n - 1\}.$$

By general position, f has no triple points. Therefore each nonempty intersection of any three of $S_1, \ldots, S_s$ can only be of the form

$$S_{\alpha_1\beta} \cap S_{\alpha_2\beta} \cap S_{\alpha_3\beta} = S_{\alpha\beta} \quad (\text{or } S_{\beta\alpha_1} \cap S_{\alpha_2\beta} \cap S_{\beta\alpha_3} = S_{\beta\alpha})$$

for some $\alpha_1^n, \alpha_2^n, \alpha_3^n, \beta^n, \alpha \in T$, $\alpha = \alpha_1 \cap \alpha_2 \cap \alpha_3$.

Since $S_{\alpha\beta} \neq \emptyset$, it follows that $\dim \alpha = n - 1$; but since N is a closed manifold, no three distinct n-simplices of T intersect in an $(n-1)$-simplex of T. This contradiction shows that (a) is true.

For part (b), let

$$A_{\alpha\beta} = \begin{cases} \alpha & \alpha \cap \beta = \emptyset, \\ \alpha \cup \beta & \alpha \cap \beta \neq \emptyset. \end{cases}$$

If $S_i \cap S_j \neq \emptyset$, then take a point $a_{ij} \in S_i \cap S_j$ and let $A_{ij} = \mathrm{st}^2 a_{ij}$. From the definition of S_i and A_i it follows that $S \subset A_i$ and $A_i \cup A_j \subset A_{ij}$. By the choice of T, A_i and A_{ij} are contractible.

Now we construct a homotopy of $\Sigma(f)$ onto its 'reduced' nerve. From (a) it follows that the sets $S_i \cap S_j$ are disjoint for distinct non-ordered pairs $i, j = 1, \ldots, s$. Take disjoint regular neighbourhoods U_{ij} of $S_i \cap S_j$ in $\bigcup_{i=1}^s S_i$. Since A_i is contractible, it follows that there is a homotopy

$$F_i : \mathrm{Cl}\left(S_i - \bigcup_{j \neq i} U_{ij}\right) \times I \to A_i$$

between the inclusion and a constant map to some point $a_i \in A_i$.

Suppose that $S_i \cap S_j \neq \emptyset$. Since A_{ij} is contractible, it follows that there is an arc $l_{ij} \subset A_{ij}$ joining a_i and a_j. Also we can extend homotopies F_i and F_j over U_{ij} to a homotopy $F_{ij} U_{ij} \times I \to N$ between the inclusion and a map of U_{ij} to l_{ij} (we can do this first for vertices and then for edges). Since all U_{ij} are disjoint and all F_{ij} are extensions of F_i and F_j, then all the constructed homotopies define a homotopy $F : \left(\bigcup_{i=1}^s S_i\right) \times I \to N$ between the inclusion and a map onto the following subgraph of N:

$$\left(\bigcup_{i=1}^s a_i\right) \cup \left(\bigcup\{l_{ij} | 1 \leq i < j \leq s \text{ and } S_i \cap S_j \neq \emptyset\}\right).$$

This completes the proof. $\qquad\square$

It follows from our proof that Theorem 8.6 for $2m \geq 3n+1$ is true even if there only exists an equivariant map to S^{m-1} from the $(\left[\frac{4m}{3}\right] - 2)$-skeleton of $\widetilde{T}$.

Remark. It would be interesting to know if Theorem 8.6 remains true when N has singularities of dimension at most $m-n-2$. This is not clear, contrary to what is written in [178] (for example, in the suspension of a homology sphere the complement to vertices is not simply-connected, so we cannot apply the engulfing lemma).

Bibliography

[1] M. Adachi. *Embeddings and immersions*, volume 124 of *Translations of Mathematical Monographs*. Amer. Math. Soc., Providence, RI, 1993. Translated from the Japanese original by Kiki Hudson.

[2] P. M. Akhmetiev. Mapping an n-sphere in a $2n$-Euclidean space: its realization. *Trudy Mat. Inst. im. Steklova*, 212:37–45, 1996. Translation in *Proc. Math. Steklov Inst.*, 212:32–39, 1996.

[3] P. M. Akhmetiev. On isotopic and discrete realization of mappings from n-dimensional sphere to Euclidean space. *Mat. Sbornik*, 187(7):3–34, 1996. Translation in *Sb. Math.*, 187(7):951–980, 1996.

[4] P. M. Akhmetiev. Prem-mappings, triple points of orientable surface and the Rohlin signature theorem. *Mat. Zametki*, 59(6):803–810, 1996. Translation in *Math. Notes*, 59(6):581–585, 1996.

[5] P. M. Akhmetiev. Embeddings of compacta, stable homotopy groups of spheres and singularity theory. *Uspekhi Mat. Nauk*, 55(3):3–62, 2000. Translation in *Russian Math. Surveys*, 55(3):405–462, 2000.

[6] P. M. Akhmetiev, D. Repovš, and A. B. Skopenkov. Embedding products of low–dimensional manifolds in $\mathbb{R}^m$. *Topol. Appl.*, 113:7–12, 2001.

[7] P. M. Akhmetiev, D. Repovš, and A. B. Skopenkov. Obstructions to approximating maps of n-surfaces to $\mathbb{R}^{2n}$ by embeddings. *Topol. Appl.*, 123:3–14, 2002.

[8] E. Akin. Manifold phenomena in the theory of polyhedra. *Trans. Amer. Math. Soc.*, 143:413–473, 1969.

[9] J. W. Alexander. On the subdivision of 3-space by a polyhedron. *Proc. Nat. Acad. Sci. USA*, 10:6–8, 1924.

[10] D. R. Bausum. Embeddings and immersions of manifolds in Euclidean space. *Trans. Amer. Math. Soc.*, 213:263–303, 1975.

[11] J. C. Becker and H. H. Glover. Note on the embedding of manifolds in Euclidean space. *Proc. Amer. Math. Soc.*, 27:405–410, 1971.

[12] M. Bestvina, M. Kapovich, and B. Kleiner. Van Kampen's embedding obstruction for discrete groups. *Invent. Math.*, 150(2):219–235, 2002.

[13] J. Boéchat. Plongements de variétés différentiables orientées de dimension $4k$ dans r^{6k+1}. *Comment. Math. Helv.*, 46(2):141–161, 1971.

[14] J. Boéchat and A. Haefliger. Plongements différentiables des variétés orientées de dimension 4 dans $\mathbb{R}^7$. In *Essays on Topology and Related Topics (Mémoires dédiés à Georges de Rham)*, pages 156–166. Springer, 1970.

[15] W. Browder. Embedding smooth manifolds. In *Proc. Internat. Congr. Math. (Moscow, 1966)*, pages 712–719. Izdat. "Mir", Moscow, 1968.

[16] R. L. Brown. Immersions and embeddings up to cobordism. *Canad. J. Math*, 23:1102–1115, 1971.

[17] J. L. Bryant. Approximating embeddings of polyhedra in codimension 3. *Trans. Amer. Math. Soc.*, 170:85–95, 1972.

[18] R. L. Bryant and W. Mio. Embeddings of homology manifolds in codimension ≥ 3. *Topology*, 38(4):811–821, 1999.

[19] R. L. Bryant and W. Mio. Embeddings in generalized manifolds. *Trans. Amer. Math. Soc.*, 352(3):1131–1137, 2000.

[20] S. E. Cappell and J. L. Shaneson. Embeddings and immersions of four-dimensional manifolds in $\mathbb{R}^6$. In J. C. Cantrell, editor, *Geometric topology (Proc. Georgia Topology Conf., Athens, Ga., 1977)*, pages 301–303. Academic Press, New York, 1979.

[21] A. Casson. Three lectures on new infinite constructions in 4-dimensional manifolds. In L. Guillou and A. Marin, editors, *À la recherche de la topologie perdue*, volume 62 of *Progr. Math.*, pages 201–244. Birkhäuser Boston, Boston, MA, 1986.

[22] A. Cavicchioli, D. Repovš, and A. B. Skopenkov. Open problems on graphs arising from geometric topology. *Topol. Appl.*, 84:207–226, 1998.

[23] M. Cencelj and D. Repovš. On embeddings of tori in Euclidean spaces. *Acta Math. Sin. (Engl. Ser.)*, 21(2):435–438, 2005.

[24] M. Cencelj, D. Repovš, and A. B. Skopenkov. On the Browder-Levin-Novikov embedding theorems. *Trudy Math. Inst. Ross. Akad. Nauk*, 247:280–290, 2004. Translation in *Proc. Steklov Math. Inst.*, 247(4):259–268, 2004.

[25] M. Cencelj, D. Repovš, and M. Skopenkov. Knotted tori and the β-invariant. Preprint.

[26] A. V. Chernavskii. Piecewise linear approximations of embeddings of cells and spheres. *Mat. Sb.*, 80(122):339–364, 1969. Translation in *Math. USSR Sb.*, 9:21–344, 1969.

[27] S. Claytor. Topological immersions of Peanian continua in a spherical surface. *Ann. of Math. (2)*, 35(4):809–835, 1934.

[28] S. Claytor. Peanian continua not imbeddable in a spherical surface. *Ann. of Math. (2)*, 38(3):631–646, 1937.

[29] R. L. Cohen. The immersion conjecture for differentiable manifolds. *Ann. of Math. (2)*, 122:237–328, 1985.

[30] P. E. Conner and E. E. Floyd. Fixed point free involutions and equivariant maps. *Bull. Amer. Math. Soc.*, 66:416–441, 1960.

[31] F. X. Connolly and B. Williams. Embedding up to homotopy and geometric suspension of manifolds. *Quart. J. Math. Oxford Ser. (2)*, 29(116):385–401, 1978.

[32] J. H. Conway and C. M. A. Gordon. Knots and links in spatial graphs. *J. Graph Theory*, 7:445–453, 1983.

[33] G. Cooke. Embedding certain complexes up to homotopy in Euclidean space. *Ann. of Math. (2)*, 90:144–156, 1969.

[34] R. J. Daverman. *Decompositions of manifolds*. Academic Press, New York, 1986.

[35] R. de Sapio. Embedding π-manifolds. *Ann. of Math. (2)*, 82:213–224, 1965.

[36] C. T. J. Dodson and P. E. Parker. *A user's guide to algebraic topology*, volume 387 of *Mathematics and its Applications*. Kluwer, Doldrecht-Boston-London, 1997.

[37] S. K. Donaldson. The orientation of Yang-Mills moduli spaces and 4-manifold topology. *J. Diff. Geom.*, 26:397–428, 1987.

[38] A. N. Dranišnikov, D. Repovš, and E. V. Ščepin. On intersections of compacta of complementary dimensions in Euclidean space. *Topol. Appl,* 38:237–253, 1991.

[39] A. N. Dranišnikov, D. Repovš, and E. V. Ščepin. On intersections of compacta in Euclidean space: the metastable case. *Tsukuba J. Math.,* 17:549–564, 1993.

[40] R. B. Edwards. The equivalence of close piecewise linear embeddings. *General Topology and its Applications,* 5:147–180, 1975.

[41] F. Fang. Embedding four-manifolds in $\mathbb{R}^7$. *Topology,* 33(3):447–454, 1994.

[42] F. Fang. Orientable 4-manifolds topologically embed into $\mathbb{R}^7$. *Topology,* 41:927–930, 2002.

[43] S. Finashin, H. Kreck, and O. Ya. Viro. Exotic knottings of surfaces in the 4-sphere. *Bull. Amer. Math. Soc. (N.S.),* 17(2):287–290, 1987.

[44] S. M. Finashin, H. Kreck, and O. Ya. Viro. Non-diffeomorphic but homeomorphic knottings of surfaces in the 4-sphere. In *Topology and geometry—Rohlin Seminar,* volume 1346 of *Lecture Notes in Math.,* pages 157–198. Springer, 1988.

[45] A. Flores. Über n-dimensionale Komplexe die im E^{2n+1} absolute Selbstverschlungen sind. *Ergeb. Math. Koll.,* 6:4–7, 1934.

[46] A. T. Fomenko, D. B. Fuchs, and V. L. Gutenmacher. *Homotopic topology.* Akadémiai Kiadó, Budapest, 1986. Translated from the Russian by K. Mályusz.

[47] A. T. Fomenko and D. B. Fuks. *Kurs gomotopicheskoi topologii.* "Nauka", Moscow, 1989. With an English summary.

[48] G. K. Francis. *A topological picturebook.* Springer-Verlag, Berlin, 1987.

[49] M. H. Freedman, V. S. Krushkal, and P. Teichner. Van Kampen's embedding obstruction is incomplete for 2-complexes in $\mathbb{R}^4$. *Math. Res. Letters,* 1:167–176, 1994.

[50] M. H. Freedman and F. S. Quinn. *Topology of 4-manifolds,* volume 39 of *Princeton Mathematical Series.* Princeton Univ. Press, Princeton, 1990.

[51] D. B. Fuchs and A. S. Schwarz. Cyclic powers of polyhedra and the imbedding problem. *Dokl. Akad. Nauk SSSR,* 125:285–288, 1959. In Russian.

[52] M. Galecki. On embeddability of CW-complexes in Euclidean space. Preprint, 1992.

[53] D. Gillman, S. V. Matveev, and D. Rolfsen. Collapsing and reconstruction of manifolds. In *Geometric topology (Haifa, 1992),* volume 164 of *Contemp. Math.,* pages 35–39. Amer. Math. Soc., Providence, RI, 1994.

[54] D. Gillman and D. Rolfsen. Three-manifolds embed in small 3-complexes. *Internat. J. Math.,* 3(2):179–183, 1992.

[55] S. Gitler. Immersion and embedding of manifolds. *Proc. Symp. Pura Appl. Math.,* 22:87–96, 1971.

[56] H. Gluck. Unknotting S^1 in S^4. *Bull. Amer. Math. Soc.,* 69(1):91–94, 1963.

[57] H. Gluck. Geometric characterization of differentiable manifolds in Euclidean space. II. *Michigan Math. J.,* 15(1):33–50, 1968.

[58] D. Gonçalves and A. Skopenkov. Embeddings of homology equivalent manifolds with boundary. *Topology Appl.,* 153(12):2026–2034, 2006.

[59] T. Goodwillie and M. Weiss. Embeddings from the point of view of immersion theory. II. *Geometry and Topology,* 3:103–118, 1999.

[60] C. M. A. Gordon. Embeddings of PL-manifolds with boundary. *Proc. Camb. Phil. Soc.*, 72:21–25, 1972.

[61] M. Gromov. *Partial differential relations*. Ergebnisse der Mathematik und ihrer Grenzgebiete (3). Springer-Verlag, Berlin-New York, 1986.

[62] N. Habegger. Obstruction to embedding disks. II. A proof of a conjecture by Hudson. *Topol. Appl.*, 17(2):123–130, 1984.

[63] N. Habegger. Knots and links in codimension greater than 2. *Topology*, 25(3):253–260, 1986.

[64] N. Habegger and U. Kaiser. Link homotopy in 2–metastable range. *Topology*, 37(1):75–94, 1998.

[65] D. D. J. Hacon. Embeddings of S^p in $S^1 \times S^q$ in the metastable range. *Topology*, 7:1–10, 1968.

[66] A. Haefliger. Plongements différentiables de variétés dans variétés. *Comment. Math. Helv.*, 36:47–82, 1961.

[67] A. Haefliger. Differentiable links. *Topology*, 1:241–244, 1962.

[68] A. Haefliger. Knotted $(4k - 1)$-spheres in $6k$-space. *Ann. of Math. (2)*, 75:452–466, 1962.

[69] A. Haefliger. Plongements de variétés dans le domaine stable. *Seminare Bourbaki*, 245, 1962/63.

[70] A. Haefliger. Plongements différentiables dans le domaine stable. *Comment. Math. Helv.*, 36:155–176, 1962/63.

[71] A. Haefliger. Differentiable embeddings of S^n in S^{n+q} for $q > 2$. *Ann. of Math. (2)*, 83:402–436, 1966.

[72] A. Haefliger. Lissage des immersions. II. Preprint, 1966.

[73] A. Haefliger. Enlacements de sphères en codimension supérieure à 2. *Comment. Math. Helv.*, 41:51–72, 1966/67.

[74] A. Haefliger. Lissage des immersions. I. *Topology*, 6:221–240, 1967.

[75] A. Haefliger. Knotted spheres and related geometric topics. In *Proc. Int. Congr. Math. (Moscow, 1966)*, pages 437–445. Izdat. "Mir", Moscow, 1968.

[76] A. Haefliger and M. W. Hirsch. Immersions in the stable range. *Ann. of Math. (2)*, 75(2):231–241, 1962.

[77] A. Haefliger and M. W. Hirsch. On existence and classification of differential embeddings. *Topology*, 27:129–135, 1963.

[78] M. W. Hirsch. The imbedding of bounding manifolds in Euclidean space. *Ann. of Math. (2)*, 74:494–497, 1961.

[79] M. W. Hirsch. On embedding 4-manifolds in R^7. *Proc. Camb. Phil. Soc.*, 61:657–658, 1965.

[80] M. W. Hirsch. Embeddings and compressions of polyhedra and smooth manifolds. *Topology*, 4(4):361–369, 1966.

[81] K. Horvatič. Embedding manifolds with low-dimensional spine. *Glasnik Mat.*, 4(24):1:101–116, 1969.

[82] K. Horvatič. On embedding polyhedra and manifolds. *Trans. Amer. Math. Soc.*, 157:417–436, 1971.

[83] W. C. Hsiang, J. Levine, and R. H. Sczarba. On the normal bundle of a homotopy sphere embedded in Euclidean space. *Topology*, 3:173–181, 1965.

[84] W. C. Hsiang and R. H. Szarba. On the embeddability and non-embeddability of sphere bundles over spheres. *Ann. of Math. (2)*, 80(2):397–402, 1964.

[85] S.-T. Hu. *Homotopy theory*. Academic Press, New York, 1959.

[86] S. T. Hu. Isotopy invariants of topological spaces. *Proc. Royal Soc.*, A255:331–366, 1960.

[87] J. F. P. Hudson. Knotted tori. *Topology*, 2:11–22, 1963.

[88] J. F. P. Hudson. Extending piecewise linear isotopies. *Proc. London Math. Soc. (3)*, 16:651–668, 1966.

[89] J. F. P. Hudson. Piecewise linear embeddings. *Ann. of Math. (2)*, 85(1):1–31, 1967.

[90] J. F. P. Hudson. *Piecewise linear topology*. University of Chicago Lecture Notes. W. A. Benjamin, Inc., New York, Amsterdam, 1969.

[91] J. F. P. Hudson. Concordance, isotopy and diffeotopy. *Ann. of Math. (2)*, 91(3):425–448, 1970.

[92] J. F. P. Hudson. Obstructions to embedding disks. In *Topology of Manifolds (Proc. Inst., Univ. of Georgia, Athens, Ga., 1969)*, pages 407–415. Markham, Chicago, Ill., 1970.

[93] J. F. P. Hudson. Embeddings of bounded manifolds. *Proc. Camb. Phil. Soc.*, 72:11–20, 1972.

[94] J. F. P. Hudson and W. B. R. Lickorish. Extending piecewise linear concordances. *Quart. J. Math. (2)*, 22:1–12, 1971.

[95] J. F. P. Hudson and E. C. Zeeman. On regular neighbourhoods. *Proc. Lond. Math. Soc. (3)*, 14:719–745, 1964. Cf. Correction to "On regular neighbourhoods", *ibid.*, 21:513–524, 1970.

[96] M. C. Irwin. Embeddings of polyhedral manifolds. *Ann. of Math. (2)*, 82:1–14, 1965.

[97] C. Kearton. Obstructions to embeddings and isotopy in the metastable range. *Math. Ann.*, 243:103–113, 1979.

[98] J. Keesling and D. C. Wilson. Embedding T^n-like continua in Euclidean space. *Topol. Appl.*, 21:241–249, 1985.

[99] A. Kervaire and J. W. Milnor. On 2-spheres in 4-manifolds. *Proc. Nat. Acad. Sci. USA*, 47:1651–1657, 1961.

[100] M. A. Kervaire. An interpretation of G. Whitehead's generalization of H. Hopf's invariant. *Ann. of Math. (2)*, 62:345–362, 1959.

[101] M. A. Kervaire. Smooth homology spheres and their fundamental groups. *Trans. Amer. Math. Soc.*, 144:67–72, 1969.

[102] R. C. Kirby. 4-manifold problems. In R. Kirby, editor, *Four-manifold theory (Durham, N.H., 1982)*, volume 35 of *Contemp. Math.*, pages 513–528. Amer. Math. Soc., Providence, RI, 1984.

[103] R. C. Kirby. *The topology of 4-manifolds*, volume 1374 of *Lecture Notes in Mathematics*. Springer-Verlag, Berlin, 1989.

[104] U. Koschorke. Link maps and the geometry of their invariants. *Manuscripta Math.*, 61(4):383–415, 1988.

[105] U. Koschorke. On link maps and their homotopy classification. *Math. Ann.*, 286(4):753–782, 1990.

[106] A. Kosiński. On Alexander's theorem and knotted spheres. In *Topology of 3-manifolds and related topics (Proc. The Univ. of Georgia Institute, 1961)*, pages 55–57. Prentice-Hall, Englewood Cliffs, NJ, 1962.

[107] M. Kreck and A. B. Skopenkov. Classification of smooth embeddings of 4-manifolds in 7-space. Submitted, cf. arXiv math.GT/0512594.

[108] V. S. Krushkal. Embedding obstructions and 4-dimensional thickenings of 2-complexes. *Proc. Amer. Math. Soc.*, 128(12):3683–3691, 2000.

[109] K. Kuratowski. Sur le problème des courbes gauche en topologie. *Fund. Math.*, 15:271–283, 1930.

[110] V. Kurlin. Basic embeddings into products of graphs. *Topol. Appl.*, 102:113–137, 2000.

[111] M. Lackenby. The Whitney trick. *Topol. Appl.*, 71:115–118, 1996.

[112] J. Lannes. La conjecture des immersions. *Astérisque*, 92/93:331–346, 1982.

[113] J. Levine. A classification of differentiable knots. *Ann. of Math. (2)*, 82:15–50, 1965.

[114] W. B. R. Lickorish. The piecewise linear unknotting of cones. *Topology*, 4:67–91, 1965.

[115] W. B. R. Lickorish and L. C. Siebenmann. Regular neighborhoods and the stable range. *Trans. Amer. Math. Soc.*, 139:207–230, 1969. .

[116] A. Liulevicius. Immersions up to cobordism. *Illinois J. Math*, 19:149–164, 1975.

[117] S. MacLane and V. W. Adkisson. Extensions of homeomorphisms on the sphere. In *Lectures in Topology*, pages 223–236. University of Michigan Press, Ann Arbor, Mich., 1941.

[118] M. Mahowald and R. D. Thompson. The *EHP* sequence and periodic homotopy. In *Handbook of algebraic topology*, pages 397–423. North-Holland, Amsterdam, 1995.

[119] J. Malešič, D. Repovš, and A. B. Skopenkov. On incompleteness of the deleted product obstruction for embeddings. *Bol. Soc. Mat. Mexicana (3)*, 9:165–170, 2003.

[120] R. Mandelbaum. Four-dimensional topology: an introduction. *Bull. Amer. Math. Soc. (N.S.)*, 2(1):1–159, 1980.

[121] S. Mardešić and J. Segal. ε-mappings and generalized manifolds. *Michigan Math. J.*, 14:171–182, 1967.

[122] W. Massey and D. Rolfsen. Homotopy classification of higher dimensional links. *Indiana Univ. Math. J.*, 34:375–391, 1986.

[123] W. S. Massey. On the Stiefel-Whitney classes of a manifold. *Amer. J. Math*, 82:92–102, 1960.

[124] W. S. Massey. On the Stiefel-Whitney classes of a manifold. II. *Proc. Amer. Math. Soc.*, 13:938–942, 1962.

[125] W. S. Massey. Homotopy classification of 3-component links of codimension greater than 2. *Topol. Appl.*, 34:269–300, 1990.

[126] M. C. McCord. Embedding P-like compacta in manifolds. *Canad. J. Math.*, 19:321–332, 1967.

[127] S. Melikhov. On maps with unstable singularities. *Topol. Appl.*, 120:105–156, 2002.

[128] S. Melikhov. Sphere eversions and realization of mappings. *Trudy Math. Inst. Ross. Akad. Nauk*, 247:1–21, 159–181, 2004. Cf. arXiv `math.GT/0305158`.

[129] R. J. Milgram. On the Haefliger knot groups. *Bull. Amer. Math. Soc.*, 78(5):861–865, 1972.

[130] R. J. Milgram and E. Rees. On the normal bundle to an embedding. *Topology*, 10:299–308, 1971.

[131] J. Milnor. Link groups. *Ann. of Math. (2)*, 59:177–195, 1954.

[132] J. W. Milnor and M. A. Kervaire. Bernoulli numbers, homotopy groups, and a theorem of Rohlin. In *Proc. Internat. Congress Math. 1958*, pages 454–458. Cambridge Univ. Press, New York, 1960.

[133] J. W. Milnor and J. D. Stasheff. *Characteristic classes*. Number 76 in Annals of Math. Studies. Princeton Univ. Press, Princeton, NJ, 1974.

[134] P. Minc. Embedding simplicial arcs into the plane. *Topol. Proc.*, 22:305–340, 1997.

[135] J. Minkus. On embeddings of highly connected manifolds. *Trans. Amer. Math. Soc.*, 115:525–540, 1965.

[136] J. Mukai and A. B. Skopenkov. A direct summand in a homotopy group of the mod 2 Moore space. *Kyushu J. Math.*, 58(1):203–209, 2004.

[137] S. Negami. Ramsey-type theorems for spatial graphs and good drawings. *J. Combin. Theory Ser. B*, 72(1):53–62, 1998.

[138] L. Neuwirth. An algorithm for the construction of 3-manifolds from 2-complexes. *Proc. Camb. Phil. Soc.*, 64:603–613, 1968.

[139] S. P. Novikov. Imbedding of simply-connected manifolds in Euclidean space. *Dokl. Akad. Nauk SSSR*, 138:775–778, 1961. In Russian.

[140] G. F. Paechter. On the groups $\pi_r(V_{mn})$. I. *Quart. J. Math. Oxford Ser. (2)*, 7(28):249–265, 1956. See also: part II, *ibid.* 9(33):8–27, 1958; part III, *ibid.* 10(37):17–37, 1959; part IV, *ibid.* 10(40):241–260, 1959; part V, *ibid.* 11(41):1–16, 1960 .

[141] R. Penrose, J. H. C. Whitehead, and E. C. Zeeman. Embeddings of manifolds in a Euclidean space. *Ann. of Math. (2)*, 73:613–623, 1961.

[142] L. S. Pontryagin. Characteristic cycles of smooth manifolds. *Dokl. Akad. Nauk SSSR*, 35(2):35–39, 1942. In Russian.

[143] M. M. Postnikov. *Homotopy theory of CW-complexes.* Nauka, Moscow, 1985. In Russian.

[144] V. Prasolov and M. Skopenkov. Ramsey link theory. *Mat. Prosveschenie*, 9, 2005. In Russian.

[145] V. V. Prasolov. *Elements of homology theory.* 2005. In Russian; English translation in preparation.

[146] V. V. Prasolov. *Elements of combinatorial and differential topology*, volume 74 of *Graduate Studies in Mathematics*. Amer. Math. Soc., Providence, RI, 2006. Translated from the Russian original by O. Sipacheva.

[147] T. M. Price. Equivalence of embeddings of k-complexes in E^n for $n \geq 2k + 1$. *Michigan Math. J.*, 13:65–69, 1966.

[148] E. Rees. Some embeddings of Lie groups in Euclidean spaces. *Mathematika*, 18:152–156, 1971.

[149] E. Rees. Problems concerning embeddings of manifolds. *Advances in Math.*, 19(1):72–79, 1990.

[150] D. Repovš, N. B. Brodsky, and A. B. Skopenkov. A classification of 3-thickenings of 2-polyhedra. *Topol. Appl.*, 94:307–314, 1999.

[151] D. Repovš and A. B. Skopenkov. Embeddability and isotopy of polyhedra in Euclidean spaces. *Trudy Math. Inst. Ross. Akad. Nauk*, 212:173–188, 1996. Translation in *Proc. Steklov Inst. Math.* 212:163–178.

[152] D. Repovš and A. B. Skopenkov. A deleted product criterion for approximability of a map by embeddings. *Topol. Appl.*, 87:1–19, 1998.

[153] D. Repovš and A. B. Skopenkov. New results on embeddings of polyhedra and manifolds into Euclidean spaces. *Uspekhi Mat. Nauk.*, 54(6):61–109, 1999. Translation in *Russian Math. Surveys*, 54(6):1149–1196, 1999.

[154] D. Repovš and A. B. Skopenkov. Obstruction theory for beginners. *Mat. Prosveschenie*, 4, 2000. In Russian.

[155] D. Repovš and A. B. Skopenkov. On contractible n-dimensional compacta, non-embeddable into $\mathbb{R}^{2n}$. *Proc. Amer. Math. Soc.*, 129:627–628, 2001.

[156] D. Repovš and A. B. Skopenkov. On projected embeddings and desuspension of the α-invariant. *Topol. Appl.*, 124:69–75, 2002.

[157] D. Repovš, A. B. Skopenkov, and E. V. Ščepin. On embeddability of $X \times I$ into Euclidean space. *Houston J. Math.*, 21:199–204, 1995.

[158] N. Robertson, P. D. Seymour, and R. Thomas. Sachs' linkless embedding conjecture. *J. Combin. Theory, Ser. B*, 64(2):185–227, 1995.

[159] V. A. Rohlin. The embedding of non-orientable three-dimensional manifolds in the five-dimensional Euclidean space. *Dokl. Akad. Nauk SSSR*, 160:549–551, 1965. In Russian.

[160] C. Rourke and B. Sanderson. The compression theorem. I, II. *Geom. Topol. (electronic)*, 5:399–429, 431–440, 2001.

[161] C. P. Rourke and B. J. Sanderson. *Introduction to piecewise-linear topology*. Ergebnisse der Mathematik und ihrer Grenzgebiete, Band 69. Springer-Verlag, Berlin-Heidelberg-New York, 1972. .

[162] D. Ruberman. Imbedding four-manifolds and slicing links. *Math. Proc. Camb. Phil. Soc*, 91:107–110, 1982.

[163] T. B. Rushing. *Topological embeddings*. Academic Press, New York, 1973.

[164] H. Sachs. On spatial representation of finite graphs. In *Finite and infinite sets, Vol. I, II (Eger, 1981)*, volume 37 of *Colloq. Math. Soc. János Bolyai*, pages 649–662. North-Holland, Amsterdam, 1984.

[165] K. S. Sarkaria. Kuratowski complexes. *Topology*, 30:67–76, 1991.

[166] K. S. Sarkaria. A one-dimensional Whitney trick and Kuratowski's graph planarity criterion. *Israel J. Math.*, 73:79–89, 1991.

[167] M. Scharlemann. Isotopy and cobordism of homology spheres in spheres. *J. London Math. Soc. (2)*, 16(3):559–567, 1977.

[168] E. V. Schepin. Soft mappings of manifolds. *Uspekhi Mat. Nauk*, 39(5):209–224, 1984. Translation in *Russian Math. Surveys*, 39(5):251–270.

[169] E. V. Schepin and M. A. Shtanko. A spectral criterion for embeddability of compacta in Euclidean space. In *Proc. Leningrad Int. Topol. Conf.*, pages 135–142. 1983. In Russian.

[170] J. Segal, A. B. Skopenkov, and D. Spież. Embeddings of polyhedra in $\mathbb{R}^m$ and the deleted product obstruction. *Topol. Appl.*, 85:225–234, 1998.

[171] J. Segal and S. Spież. Quasi embeddings and embeddings of polyhedra in $\mathbb{R}^m$. *Topol. Appl.*, 45:275–282, 1992.

[172] A. Shapiro. Obstructions to the embedding of a complex in a Euclidean space. I. the first obstruction. *Ann. of Math. (2)*, 66:256–269, 1957.

[173] K. Sieklucki. Realization of mappings. *Fund. Math.*, 65:325–343, 1969.

[174] A. B. Skopenkov. Algebraic topology from geometrical point of view. MCCME, Moscow, to appear (in Russian). See also http://dfgm.math.msu.su/files/skopenkov/obstruct2.ps.

[175] A. B. Skopenkov. Classification of smooth embeddings of 3-manifolds in the 6-space. Submitted, cf. arXiv math/0603429.

[176] A. B. Skopenkov. A new invariant and parametric connected sum of embeddings. Submitted to *Fund. Math.*, cf. arXiv math/0509621.

[177] A. B. Skopenkov. A description of continua basically embeddable in $\mathbb{R}^2$. *Topol. Appl.*, 65:29–48, 1995.

[178] A. B. Skopenkov. On the deleted product criterion for embeddability of manifolds in $\mathbb{R}^m$. *Comment. Math. Helv.*, 72:543–555, 1997.

[179] A. B. Skopenkov. On the deleted product criterion for embeddability in $\mathbb{R}^m$. *Proc. Amer. Math. Soc.*, 126(8):2467–2476, 1998.

[180] A. B. Skopenkov. On the generalized Massey-Rolfsen invariant for link maps. *Fund. Math.*, 165:1–15, 2000.

[181] A. B. Skopenkov. On the Haefliger-Hirsch-Wu invariants for embeddings and immersions. *Comment. Math. Helv.*, 77:78–124, 2002.

[182] A. B. Skopenkov. On the Kuratowski graph planarity criterion. *Mat. Prosveschenie*, 9:116–128, 2005. Cf. *ibid.*, 10:276–277, 2006.

[183] A. B. Skopenkov. Classification of embeddings below the metastable dimension. Preprint, cf. arXiv `math.GT/0607422`, 2006.

[184] M. Skopenkov. Embedding products of graphs into Euclidean spaces. *Fund. Math.*, 179:191–198, 2003.

[185] M. Skopenkov. On approximability by embeddings of cycles in the plane. *Topol. Appl.*, 134:1–22, 2003.

[186] M. Skopenkov. A formula for the group of links in the 2-metastable dimension. *Proc. Amer. Math. Soc.*, to appear. See also arXiv `math.GT/0610320`, 2006.

[187] S. Smale. On the structure of manifolds. *Amer. J. Math.*, 84:387–399, 1962.

[188] S. Spież. Imbeddings in $\mathbb{R}^{2m}$ of m-dimensional compacta with $\dim(X \times X) < 2m$. *Fund. Math.*, 134:105–115, 1990.

[189] S. Spież and H. Toruńczyk. Moving compacta in $\mathbb{R}^m$ apart. *Topol. Appl.*, 41:193–204, 1991.

[190] J. Stallings. On topologically unknotted spheres. *Ann. of Math. (2)*, 77:490–503, 1963.

[191] J. Stallings. The embedding of homotopy type into manifolds, 1965. Mimeographed notes, Princeton University.

[192] J. Stallings. Homology and central series of groups. *J. of Algebra*, 2:170–181, 1965.

[193] Y. Sternfeld. Hilbert's 13th problem and dimension. In *Geometric aspects of functional analysis (1987–88)*, volume 1376 of *Lecture Notes in Math.*, pages 1–49. Springer, Berlin, 1989.

[194] K. Taniyama. Homology classification of spatial embeddings of a graph. *Topol. Appl.*, 65:205–228, 1995.

[195] R. S. Tindell. Knotting tori in hyperplanes. In *Conference on the Topology of Manifolds (Michigan State Univ., 1967)*, pages 147–153. Prindle, Weber & Schmidt, Boston, Mass., 1968.

[196] H. Toda. *Composition methods in homotopy groups of spheres.* Number 49 in Annals of Math. Studies. Princeton Univ. Press, Princeton, NJ, 1962.

[197] E. R. van Kampen. Komplexe in euklidische Räumen. *Abh. Math. Sem. Hamburg*, 9:72–28, 152–153, 1932.

[198] V. A. Vassiliev. *Complements of discriminants of smooth maps: Topology and applications.* Amer. Math. Soc., Providence, RI, 1992. Translated from the Russian by B. Goldfarb.

[199] J. Vrabec. Knotting a k-connected closed PL m-manifold in $\mathbb{R}^{2m-k}$. *Trans. Amer. Math. Soc.*, 233:137–165, 1977.

[200] J. Vrabec. Deforming a PL submanifold of Euclidean space into a hyperplane. *Trans. Amer. Math. Soc.*, 312(1):155–178, 1989.

[201] C. T. C. Wall. Differential topology. IV. Theory of handle decompositions, 1964. Mimeographed notes.

[202] C. T. C. Wall. All 3-manifolds imbed in 5-space. *Bull. Amer. Math. Soc.*, 71:490–503, 1965.

[203] C. T. C. Wall. Unknotting tori in codimension one and spheres in codimension two. *Proc. Camb. Phil. Soc.*, 61:659–664, 1965.

[204] C. T. C. Wall. Classification problems in differential topology. IV. Thickenings. *Topology*, 5:73–94, 1966.

[205] C. T. C. Wall. Classification problems in differential topology. V. On certain 6-manifolds. *Invent. Math.*, 1:355–374, 1966. Cf. *corrigendum, ibid.* 2 (1966).

[206] C. T. C. Wall. *Surgery on compact manifolds*. Academic Press, London, 1970.

[207] C. Weber. Plongements de polyèdres dans le domain metastable. *Comment. Math. Helv.*, 42:1–27, 1967.

[208] C. Weber. Deux remarques sur les plongements d'un AR dans un éspace euclidien. *Bull. Acad. Polon. Sci. Ser. Sci. Math. Astronom. Phys*, 16:851–855, 1968.

[209] M. Weiss. Second and third layers in the calculus of embeddings. Preprint.

[210] H. Whitney. Differentiable manifolds in Euclidean space. *Proc. Nat. Acad. Sci. USA*, 21(7):462–464, 1935.

[211] H. Whitney. The self-intersections of a smooth n-manifolds in $2n$-space. *Ann. of Math (2)*, 45:220–246, 1944.

[212] W.-T. Wu. On the realization of complexes in an Euclidean space. I, II, III. *Sci Sinica*, 7:251–297, 1958. See also *ibid.*,7:365–387, 1958; 8:133–150, 1959.

[213] W.-T. Wu. On the isotopy of a finite complex in a Euclidean space. II. *Sci. Record (N.S.)*, 3:342–351, 1959.

[214] W.-T. Wu. *A theory of embedding, immersion and isotopy of polytopes in a Euclidean space*. Science Press, Peking, 1965.

[215] C.-T. Yang. On theorems of Borsuk-Ulam, Kakutani-Yamabe-Yujobô and Dyson. I. *Ann. of Math. (2)*, 60:262–282, 1954.

[216] E. C. Zeeman. Unknotting spheres. *Ann. of Math. (2)*, 72:350–360, 1960.

[217] E. C. Zeeman. Isotopies and knots in manifolds. In *Topology of 3-manifolds and related topics (Proc. The Univ. of Georgia Institute, 1961)*, pages 187–193. Prentice-Hall, Englewood Cliffs, NJ, 1962.

[218] E. C. Zeeman. Unknotting combinatorial balls. *Ann. of Math. (2)*, 78:501–526, 1963.

[219] L. Zhongmou. Every 3-manifold with boundary embeds in Triod × Triod × I. *Proc. Amer. Math. Soc.*, 122(2):575–579, 1994.

[220] A. V. Zhubr. A classification of simply-connected spin 6-manifolds. *Izvestiya Akad. Nauk. SSSR*, 39(4):839–859, 1975. In Russian.

[221] A. V. Zhubr. Classification of simply-connected topological 6-manifolds. In *Topology and geometry—Rohlin Seminar*, volume 1346 of *Lecture Notes in Math.*, pages 325–339. Springer, Berlin, 1989.

Department of Differential Geometry
Faculty of Mechanics and Mathematics
Moscow State University
Moscow 119992,
Russia
skopenko@mccme.ru

and

Independent University of Moscow
B. Vlasyevskiy, 11
Moscow 119002
Russia

On Maxwellian and Boltzmann distributions

Vladimir V. Ten

Introduction

The foundations of the establishment of thermal equilibrium constitute
an old problem, first investigated by Boltzmann in [1]. Although Boltzmann's kinetic equations assume a probabilistic nature of the system, the
starting point was a deterministic and conservative mechanical system
consisting of a finite number of elastically colliding balls. This model is
called the *Boltzmann–Gibbs gas*.

In this article the normal distribution of velocities is put on a firm
foundation for conservative mechanical systems with a large number of
degrees of freedom, without any additional assumptions of a statistical
or random kind. At the end, the Boltzmann distribution for density in
configuration space is also justified. So, here some of Boltzmann's ideas
(see [2]) are proved in a rigorous way.

Summary of the article

In Section 1 it is shown that at most points of an n-sphere, the difference
between the joint density of the Cartesian coordinates and the density
of a normal distribution vanishes as n tends to infinity. Our approach is
elementary but requires some technical lemmas whose proofs are deferred
to Section 2.

In Section 3 the whole energy level is considered – the product of a
sphere and a compact configuration space. Then, using the individual
ergodic theorem and Lemma 3.1, deviations of the distribution of velocities from normal for individual solutions with different initial conditions
are investigated.

Finally, it is shown that for systems with sufficiently many degrees

343

of freedom, for most initial conditions the deviation from the normal distribution is small at almost every moment of time.

Sections 1–3 comprise a revised version of the author's paper [4].

1 Convergence to the normal distribution

Consider a conservative mechanical system consisting of a huge number of identical simple systems: the standard model of an ideal gas provides an example. The phase space of such a system is the product of n simple phase spaces. For simplicity we shall assume that simple mechanical systems are one-dimensional, but the analysis works for more dimensions once the notation is sufficiently extended.

Consider the *space of velocities* $\mathbb{R}^n \{v_1, \ldots, v_n\}$ – the tangent space over a point of configuration space. Let us set the total mass and average square of velocities to be equal to one:

$$\sum_{k=1}^{n} \frac{1}{n} = 1, \qquad \sum_{k=1}^{n} \frac{v_k^2}{n} = 1.$$

The general case may be reduced to this by the simple coordinate change $u_k = \sqrt{\frac{M}{2E}} \, v_k$. The case of different masses of particles also can be considered using a proper coordinate change. Here the total mass M and energy E are assumed to be constants (i.e. do not depend on n).

The energy level in the tangent space over a configuration of n particles is an $(n-1)$-dimensional sphere $\mathbf{S}_n$ with centre at the origin and radius equal to $\sqrt{n}$. Introduce the natural uniform density on this sphere. The density of the magnitude of the coordinate v_1 on the sphere is proportional to the $(n-2)$-volume of the intersection of $\mathbf{S}_n$ and with the hyperplane $v_1 = \text{const}$. The result is an $(n-2)$-sphere with radius $\sqrt{n - v_1^2}$, whose volume is proportional to

$$\left(1 - \frac{v_1^2}{n}\right) \left(\sqrt{n - v_1^2}\right)^{n-2} = \sqrt{n} \left(\sqrt{n - v_1^2}\right)^{n-3} = \sqrt{n}\,(n - v_1^2)^{\frac{n-3}{2}}$$

(the first multiplier is due to the cosine of the angle of projection to the v_1-axis). After normalization the previous expression has the form

$$\frac{1}{K_n} \left(1 - \frac{v_1^2}{n}\right)^{\frac{n-3}{2}}. \tag{1.1}$$

Lemma 1.1. *The following inequality is valid:*

$$\left| e^z - \left(1 + \frac{z}{n}\right)^{n-a} \right| < \frac{C(a)}{n}, \qquad -n \le z \le 0.$$

Lemma 1.2. *For any $p \in \mathbb{N}$ there exists a constant $C_p(a)$ such that*

$$|z|^p \left| e^z - \left(1 + \frac{z}{n}\right)^{n-a} \right| < \frac{C_p(a)}{n}, \quad -n \le z \le 0.$$

Lemma 1.3. *There is an absolute constant C, such that*

$$\left| \frac{1}{K_n} - \frac{1}{\sqrt{2\pi}} \right| < \frac{C}{n}.$$

The proofs of these lemmas will be given in Section 2.

How can we define 'similarity' of two densities (in a rigorous sense) when one is a continuous function and the other is a sum of delta-functions? In our case the latter corresponds to the set of n velocities of our dynamical system at a fixed moment of time, and the density is

$$\frac{1}{n} \delta(v - v_1) + \cdots + \frac{1}{n} \delta(v - v_n).$$

In this paper the difference between their Fourier transforms is suggested as a measure of their 'similarity'. It is justified by the similarity between the density of a discrete set of points, represented as a finite histogram, and the normal density, in the case of small difference of their Fourier transform.

The Fourier transform of the density of n points $v_1, \ldots, v_n$ is

$$\Phi_n(s) = \int_{-\infty}^{\infty} e^{isv} \frac{1}{n} \sum_{k=1}^{n} \delta(v - v_k) dv = \frac{1}{n} \sum_{k=1}^{n} e^{isv_k}$$

and the Fourier transform for the standard Gaussian (normal) density $\frac{1}{\sqrt{2\pi}} e^{-\frac{1}{2}v^2}$ is

$$\Phi_e(s) = \int_{-\infty}^{\infty} e^{isv} \frac{1}{\sqrt{2\pi}} e^{-\frac{1}{2}v^2} dv = e^{-\frac{1}{2}s^2}.$$

We shall show that for the majority (in measure) of points on the energy level $\mathbf{S}_n$, the deviation $|\Phi_n(s) - \Phi_e(s)|$ vanishes uniformly in s as n goes to infinity.

Consider the sphere $\mathbf{S}_n$ with normalized constant density as a probability space: the random variables are the measurable functions on $\mathbf{S}_n$. Denote the measure of a subset as $\mathbb{P}(M)$, $M \subset \mathbf{S}_n$. This makes it possible to use the language and apparatus of probability theory. For example, the expected value (EV) is the mean value of a function.

Let us define the function $\Psi(v_1)$ as given by (1.1) on the interval $[-\sqrt{n}, \sqrt{n}]$ and equal to zero for the rest of the line. Thus Ψ is the density obtained by projecting the uniform density on $\mathbf{S}_n$ to the v_1-axis.

Define functions $\xi_k = \Re(e^{isv_k}) = \cos(sv_k)$. In terms of probability theory they are random variables, though they have nothing to do with randomness in our context; the use of probabilistic terms is for convenience and easy recognition of metric results of the theory.

We have

$$\mathbf{M}\xi_1 = \int_{\mathbf{S}_n} \cos(sv_1)\, d\omega$$

$$= \int_{-\infty}^{\infty} \cos(sv_1)\, \Psi(v_1)\, dv_1$$

$$= \frac{1}{K_n} \int_{-\sqrt{n}}^{\sqrt{n}} \cos(sv_1) \left(1 - \frac{v_1^2}{n}\right)^{\frac{n-3}{2}} dv_1,$$

The EV on measure with normal distribution

$$\frac{1}{(2\pi)^{\frac{n}{2}}}\, e^{-\frac{1}{2}(v_1^2 + \cdots + v_n^2)}$$

on the probability space $\mathbb{R}^n\{v_1, \ldots, v_n\}$ will be denoted by $\mathbb{E}$. For a random variable which is constant in every coordinate hyperplane (say $v_1 = \text{const}$)

$$\mathbb{E}\xi_1 = \frac{1}{(2\pi)^{\frac{n}{2}}} \int_{-\infty}^{\infty} \cdots \int_{-\infty}^{\infty} \cos(sv_1)\, e^{-\frac{1}{2}(v_1^2 + \cdots + v_n^2)}\, dv_1 \ldots dv_n$$

$$= \frac{1}{\sqrt{2\pi}} \int_{-\infty}^{\infty} \cos(sv_1) e^{-\frac{1}{2}v_1^2}\, dv_1.$$

Define the random variable

$$\eta_n = \frac{1}{n} \sum_{k=1}^{n} \xi_k = \frac{1}{n} \sum_{k=1}^{n} \cos(sv_k)$$

(this is just the real part of the Fourier transform $\Phi_n(s)$). Clearly

$$\mathbf{M}\eta_n = \mathbf{M}\xi_1 = \cdots = \mathbf{M}\xi_n \quad \text{and} \quad \mathbb{E}\eta_n = \mathbb{E}\xi_1 = \cdots = \mathbb{E}\xi_n.$$

We can now state the main lemma of this section.

Lemma 1.4. *The following estimates hold:*

(i) $|\mathbf{M}\xi_1 - \mathbb{E}\xi_1| \le \frac{5}{n\sqrt{2\pi}} C_1(\frac{3}{2})$,

(ii) $|\mathbf{M}\xi_1^2 - \mathbb{E}\xi_1^2| \le \frac{5}{n\sqrt{2\pi}} C_1(\frac{3}{2})$,

(iii) $|\mathbf{M}\xi_1\xi_2 - \mathbb{E}\xi_1\xi_2| \le \frac{5}{n\sqrt{2\pi}} C_1(\frac{3}{2})$,

(iv) $|(\mathbf{M}\xi_1)^2 - (\mathbb{E}\xi_1)^2| \le \frac{10}{n\sqrt{2\pi}} C_1(\frac{3}{2})$.

Let $\mathbf{D}$ denote the variance of a 'random variable' on the probability space $\mathbf{S}_n$. Then

$$
\begin{aligned}
\mathbf{D}\eta_n &= \mathbf{M}(\eta_n - M\eta_n)^2 \\
&= \mathbf{M}(\eta_n^2 - 2\eta_n \mathbf{M}\eta_n + (\mathbf{M}\eta_n)^2) \\
&= \frac{1}{n^2} \mathbf{M}\left(\sum_{k=1}^{n} \xi_k^2 + 2\sum_{i<j\le n} \xi_i\xi_j \right) - (\mathbf{M}\eta_n)^2 \\
&= \frac{1}{n^2}\left(n\mathbf{M}\xi_1^2 + 2\frac{n(n-1)}{2}\mathbf{M}\xi_1\xi_2 \right) - (\mathbf{M}\xi_1)^2 \\
&= \frac{1}{n}\mathbf{M}\xi_1^2 + \frac{n-1}{n}\mathbf{M}\xi_1\xi_2 - (\mathbf{M}\xi_1)^2 \\
&= \frac{1}{n}\left(\mathbf{M}\xi_1^2 - \mathbf{M}\xi_1\xi_2\right) + \mathbf{M}\xi_1\xi_2 - (\mathbf{M}\xi_1)^2
\end{aligned}
$$

Since $\mathbb{E}\xi_1\xi_2 = \mathbb{E}\xi_1 = 0$, we can estimate this using Lemma 1.4 to get

$$
\begin{aligned}
\mathbf{D}\eta_n &= \mathbf{M}(\eta_n - M\eta_n)^2 \\
&= \frac{1}{n}\left(\mathbf{M}\xi_1^2 - \mathbf{M}\xi_1\xi_2\right) + \mathbf{M}\xi_1\xi_2 - (\mathbf{M}\xi_1)^2 - \left(\mathbb{E}\xi_1\xi_2 - (\mathbb{E}\xi_1)^2\right) \\
&\le \frac{2}{n} + \frac{15}{n\sqrt{2\pi}}C_1(\tfrac{3}{2}) \\
&\le \frac{16}{n\sqrt{2\pi}}C_1(\tfrac{3}{2}).
\end{aligned}
$$

$$(1.2)$$

Remark. Note that for functions on a measure space which are independent (in the sense of probability theory) the result is very similar, and so we could say that our functions (one-dimensional projections of uniform measure on n-sphere) are 'almost independent'. Moreover, the 'dependency' is of order n^{-1}.

Lemma 1.5 (Chebyshev's inequality).

$$
\mathbb{P}(|\eta - \mathbf{M}\eta| \ge \varepsilon) \le \frac{\mathbf{D}\eta}{\varepsilon^2}.
$$

Combining Chebyshev's inequality with the variance estimate (1.2) gives

$$\mathbb{P}\left(|\eta_n - \mathbf{M}\xi_1| \geq \frac{\varepsilon}{2}\right) \leq \frac{64}{\varepsilon^2 n \sqrt{2\pi}} C_1(\tfrac{3}{2}).$$

By Lemma 1.4 we have $|\mathbf{M}\xi_1 - \mathbb{E}\xi_1| \leq \frac{5}{n\sqrt{2\pi}} C_1(\tfrac{3}{2})$ and

$$\mathbb{P}\left(|\eta_n - \mathbb{E}\xi_1| \geq \frac{\varepsilon}{2} + \frac{5}{n\sqrt{2\pi}} C_1(\tfrac{3}{2})\right) \leq \frac{64}{\varepsilon^2 n \sqrt{2\pi}} C_1(\tfrac{3}{2})$$

so that

$$\mathbb{P}\left(|\eta_n - \mathbb{E}\xi_1| \geq \varepsilon\right) \leq \frac{64}{\varepsilon^2 n \sqrt{2\pi}} C(\tfrac{3}{2}) \quad \text{for } n > \frac{10}{\varepsilon\sqrt{2\pi}} C_1(\tfrac{3}{2}).$$

For $\xi_k = \Im(e^{isv_k}) = \sin(sv_k)$ the calculations are analogous. Hence

$$\mathbb{P}\left(\left|\frac{\cos(sv_1) + \cdots + \cos(sv_n)}{n} - e^{-\frac{1}{2}s^2}\right| \geq \varepsilon\right) \leq \frac{64 C_1(\tfrac{3}{2})}{\varepsilon^2 n \sqrt{2\pi}},$$

$$\mathbb{P}\left(\left|\frac{\sin(sv_1) + \cdots + \sin(sv_n)}{n}\right| \geq \varepsilon\right) \leq \frac{64 C_1(\tfrac{3}{2})}{\varepsilon^2 n \sqrt{2\pi}},$$

because $\mathbb{E}\cos(sv_1) = e^{-\frac{1}{2}s^2}$ and $\mathbb{E}\sin(sv_1) = 0$.

Thus

$$\mathbb{P}\left(\left|\Phi_n(s) - e^{-\frac{1}{2}s^2}\right| \geq \varepsilon\right) \leq \frac{256\, C_1(\tfrac{3}{2})}{\varepsilon^2 n \sqrt{2\pi}} \tag{1.3}$$

In words: for every $\varepsilon > 0$, the set of points on $\mathbf{S}_n$ where the Fourier transform deviates from normal by more than ε can be made arbitrarily small (in measure) by taking sufficiently large n.

Remark. For higher-dimensional simple systems one uses the multi-dimensional Fourier transform. For instance, in three dimensions we have

$$\iiint e^{i(s_1v_1 + s_2v_2 + s_3v_3)} \sum_{k=1}^{n} \frac{1}{n} \delta(v_1 - v_{1,k}, v_2 - v_{2,k}, v_3 - v_{3,k})\, dv_1 dv_2 dv_3,$$

The integral is similarly reduced to a finite sum. For fixed (s_1, s_2, s_3) we may introduce new orthogonal coordinates

$$w_k = s_1 v_{1,k} + s_2 v_{2,k} + s_3 v_{3,k}$$

and hence the problem is reduced to estimates obtained for one-dimensional simple systems.

Note that using this technique we may also prove convergence of the 'finite histogram' of this discrete density to the corresponding 'finite histogram' of the normal distribution. Indeed, let

$$\Theta_{ab}(x) = \begin{cases} \frac{1}{b-a} & \text{if } x \in [a,b], \\ 0 & \text{otherwise.} \end{cases}$$

For sake of convenience we *redefine* our variables ξ_k by setting

$$\xi_k = \Theta_{ab}(v_k) \quad k = 1, \ldots, n,$$

and let

$$\eta_n = \frac{\xi_1 + \cdots + \xi_n}{n}.$$

As before it can be shown that

$$|\mathbf{M}\xi_1 - \mathbb{E}\xi_1| \leq \frac{1}{n} + \frac{C}{n} = \frac{C+1}{n},$$

$$|\mathbf{M}\xi_1\xi_2 - \mathbb{E}\xi_1\xi_2| \leq \frac{1}{n} + \frac{C_2}{n} = \frac{C_2+1}{n},$$

and

$$\mathbf{D}\eta_n = \frac{1}{n}\left(\mathbf{M}\xi_1^2 - \mathbb{E}\xi_1\xi_2\right) + \mathbf{M}\xi_1\xi_2 - \mathbb{E}\xi_1\xi_2 - (\mathbf{M}\xi_1)^2 + (\mathbb{E}\xi_1)^2$$

$$\leq \frac{2}{n(b-a)} + \frac{C_3}{n} = \frac{1}{n}\left(\frac{2}{b-a} + C_3\right).$$

Therefore

$$\mathbb{P}\left(\left|\frac{1}{n}\sum_{k=1}^{n}\Theta_{ab}(v_k) - \frac{1}{b-a}\int_a^b e^{-\frac{1}{2}v^2}\, dv\right| \geq \varepsilon\right) \leq \frac{4}{\varepsilon^2 n}\left(\frac{2}{b-a} + C_3\right).$$

This estimate may be known but we were unable to find a reference in the literature.

2 Proofs of the lemmas

Proof of Lemma 1.1. It is known that

$$\lim_{n\to\infty}\left(1 + \frac{z}{n}\right)^{n-a} = e^z$$

for every fixed z. For convenience let $a = 0$.

Consider the difference

$$\left(1 + \frac{z}{n+1}\right)^{n+1} - \left(1 + \frac{z}{n}\right)^n.$$

and with a change of coordinates $x = 1 + z/n$ let

$$a_n(x) = \left(\frac{n}{n+1}x + \frac{1}{n+1}\right)^{n+1} - x^n$$

(as z varies from $-n$ to 0, x varies from 0 to 1).

Suppose the derivative of $a_n(x)$ vanishes at some point x_0:

$$0 = a'_n(x_0) = n\left(\frac{n}{n+1}x_0 + \frac{1}{n+1}\right)^n - nx_0^{n-1}.$$

Then

$$a_n(x_0) = \left[\left(\frac{n}{n+1}x_0 + \frac{1}{n+1}\right) - x_0\right]x_0^{n-1} = \frac{(1-x_0)x_0^{n-1}}{n+1}.$$

We may assume that the function $a_n(x)$ reaches its extremal value at $x = x_0$ (for at the ends of the segment $[0,1]$ it takes the values $(n+1)^{-(n+1)}$ and 0). Now, the maximum value of $\varphi(s) = (1-s)s^{n-1}$ on $[0,1]$ is attained at the unique solution $s_0 \in (0,1)$ of

$$0 = \varphi'(s) = (n-1)s^{n-2} - ns^{n-1}.$$

The root s_0 of the equation is equal to $\left(1 - \frac{1}{n}\right)$, and

$$\varphi(s_0) = \left(1 - \left(1 - \frac{1}{n}\right)\right)\left(1 - \frac{1}{n}\right)^{n-1} \le \frac{2e^{-1}}{n} \le \frac{1}{n}.$$

Hence

$$a_n(x) \le a_n(x_0) = \frac{\varphi(x_0)}{n+1} \le \frac{\varphi(s_0)}{n+1} \le \frac{1}{n(n+1)} \le \frac{1}{n^2}, \quad x \in [0,1].$$

Thus the terms of the series

$$\sum \left(1 + \frac{z}{n+1}\right)^{n+1} - \left(1 + \frac{z}{n}\right)^n$$

allow a uniform estimate $1/n^2$ for z on the negative semi-axis. Therefore the absolute accuracy of the series may be estimated as $1/n$.

For nonzero a the calculations are more tedious, but in qualitative terms they repeat almost verbatim those for the zero case. In general the estimate has the form $2e^{a-1}/n$, for example $C(\frac{3}{2}) = 2\sqrt{e}$. $\qquad\square$

Proof of Lemma 1.2. The structure of the proof is very similar to that of Lemma 1.1. Let $b_n(x) = n^p(1-x)^p a_n(x)$; the expression carries the same sense as $a_n(x)$ from previous proof. Again let x_0 be such that

$$0 = b'_n(x_0)$$

$$= -p\left(\left(\frac{nx_0 + 1}{n+1}\right)^{n+1} - x_0^n\right) + n(1 - x_0)\left(\left(\frac{nx_0 + 1}{n+1}\right)^n - x_0^{n-1}\right)$$

Then

$$\left(\frac{nx_0 + 1}{n+1}\right)^n = \frac{(n+1)((p+n)x_0 - n)}{(n(p+n+1)x_0 + p - n^2 - n)} x_0^{n-1}.$$

Suppose the function $b_n(x)$ attains its maximum at $x = x_0$ (otherwise we immediately get the estimate by values at 0 and 1). Introduce the notation

$$\varphi(s) = n^p(1-s)^p s^{n-1}\left(\frac{ns+1}{n+1}\frac{(n+1)((p+n)s-n)}{(n(p+n+1)s+p-n^2-n)} - s\right);$$

thus $\varphi(x_0) = b_n(x_0)$. Note that the roots of the equation $\varphi'(s) = 0$ lie to the right of the point $s_0 = 1 - \frac{p+1}{n}$.

Consider two possibilities: if x_0 is to the right of s_0 then the estimate is easily deduced using Lemma 1.3, since $n^p(1 - x_0)^p$ is bounded by $(p+1)^p$; if x_0 is to the left of s_0 then it is sufficient to estimate $\varphi(s_0)$, since $\varphi'(s)$ does not change its sign on the interval $[0, s_0]$ and $\varphi(0) = 0$.

We observe that

$$\lim_{n\to\infty} \varphi(s_0)n^2 = (p+1)^{p+2}e^{-(p+1)}.$$

Hence, for sufficiently large n

$$b_n(x) \le b_n(x_0) \le \max\left(\varphi(x_0), \frac{(p+1)^p C}{n^2}\right)$$

$$< \frac{\max\{2(p+1)^{p+2}e^{-(p+1)}, (p+1)^p C\}}{n^2}.$$

Therefore

$$|z|^p\left|e^{-z} - \left(1 + \frac{z}{n}\right)^n\right| < \frac{\max\{2(p+1)^{p+2}e^{-(p+1)}, (p+1)^p C\}}{n} \le \frac{C_p}{n}$$

for $z \le 0$. $\qquad\square$

Proof of Lemma 1.3. We have

$$K_n = \int_{-\sqrt{n}}^{\sqrt{n}} (1 - v^2/n)^{\frac{n-3}{2}} \, dv$$

$$= \sqrt{n} \int_{-\pi/2}^{\pi/2} \cos^{n-2} u \, du = 2\sqrt{n} \, \frac{n-3}{n-2} \frac{n-5}{n-4} \cdots \frac{3}{4} \frac{1}{2} \frac{\pi}{2}.$$

By Wallis' formula $\lim_{n \to \infty} K_n = \sqrt{2\pi}$. Now

$$K_n = \sqrt{1 + \frac{2}{n-2}} \left(1 - \frac{1}{n-2}\right) K_{n-2}$$

and so

$$K_n - K_{n-2} = \left(\sqrt{1 + \frac{2}{n-2}} \left(1 - \frac{1}{n-2}\right) - 1\right) K_{n-2}.$$

Since K_n converges to $\sqrt{2\pi}$, for sufficiently large n we have $K_{n-2} < 2\sqrt{\pi}$. Hence

$$|K_n - K_{n-2}| < 2\sqrt{\pi} \left| \sqrt{1 + \frac{2}{n-2}} \left(1 - \frac{1}{n-2}\right) - 1 \right| \leq \frac{4\sqrt{\pi}}{n^2}.$$

Therefore

$$|K_n - \sqrt{2\pi}| < \frac{8\sqrt{\pi}}{n},$$

and this is equivalent to the statement of the lemma. □

Proof of Lemma 1.4. By Lemma 1.2

$$\left| e^{-\frac{1}{2} v_1^2} - \left(1 - \frac{v_1^2}{n}\right)^{\frac{n-3}{2}} \right| = \left| e^{-\frac{1}{2} v_1^2} - \left(1 - \frac{v_1^2}{2} \frac{2}{n}\right)^{\frac{n}{2} - \frac{3}{2}} \right| \leq \frac{2C(\frac{3}{2})}{n v_1^2}.$$

$$(2.1)$$

Using the definitions of 'expected values' $\mathbf{M}$ and $\mathbb{E}$ we have

$$|\mathbf{M}\xi_1 - \mathbb{E}\xi_1| \leq \frac{1}{\sqrt{2\pi}} \frac{4C_1(\frac{3}{2})}{n} + \frac{C}{n} + o\left(\frac{1}{n}\right) \leq \frac{5C_1(\frac{3}{2})}{n\sqrt{2\pi}},$$

using the estimate (2.1) for the interval $[-\sqrt{n}, \sqrt{n}]$ and Lemma 1.3. This proves the first inequality (i). Since $|\xi_1| \leq 1$ the second inequality (ii) is proved similarly.

For (iii) it is sufficient to note that

$$\cos(sv_1)\cos(sv_2) = \cos(s(v_1 + v_2)) + \sin(sv_1)\sin(sv_2)$$

and the 'expected value' of the second term equals zero since sin is an odd function. The calculcation of $\mathbf{M}\xi_1\xi_2$ is thus reduced to a one-dimensional integration in $v_1 + v_2$.

Finally, using (i)

$$|(\mathbf{M}\xi_1)^2 - (\mathbb{E}\xi_1)^2| = |\mathbf{M}\xi_1 - \mathbb{E}\xi_1||\mathbf{M}\xi_1 + \mathbb{E}\xi_1| \le \frac{10C_1(\frac{3}{2})}{n\sqrt{2\pi}}.$$

and (iv) is proved. □

Proof of Lemma 1.5 (Chebyshev's inequality). This is standard but we sketch the proof for the reader's convenience.

Introduce a new random variable $\nu = (\eta - \mathbf{M}\eta)^2$: then $\varepsilon\mathbb{P}(\nu \ge \varepsilon) \le \mathbf{M}\nu$, since $\mathbf{M}\nu$ is the integral of ν over a probability space and $\varepsilon\mathbb{P}(\nu \ge \varepsilon)$ does not exceed the integral of ν over the region where $\nu \ge \varepsilon$. Change ε to ε^2 and note that $\mathbf{M}\nu = \mathbf{D}\eta$. □

3 Deviations on individual solutions

Lemma 3.1. *Let (Ω, μ) be a probability measure space. Let $f : \Omega \to \mathbb{R}$ be such that $f(\omega) \ge 0$ and $\int_\Omega f(\omega)\,d\mu(\omega) = D$. Then*

$$\mu(f \ge \sqrt{D}) = \int_{f(\omega) \ge \sqrt{D}} d\mu(\omega) \le \sqrt{D}.$$

Proof of Lemma 3.1. As in the proof of Lemma 1.5, it can be shown that

$$\int_{f(\omega) \ge \varepsilon} d\mu(\omega) \le \frac{1}{\varepsilon} \int_\Omega f(\omega)d\mu(\omega)$$

(this is *Markov's inequality*). Taking $\varepsilon = \sqrt{D}$ gives

$$\int_{f(\omega) \ge \sqrt{D}} d\mu(\omega) \le \frac{D}{\sqrt{D}} = \sqrt{D}.$$

and the lemma is proved. □

The canonical invariant measure in the phase space of a mechanical system, restricted to the energy level $\mathbf{U}_n = \mathbf{K}_n \times \mathbf{S}_n$ (the *microcanonical distribution*), is proportional to uniform measure on the velocity sphere over every point of configuration space $\mathbf{K}_n$ (the coefficient depends only on the configuration coordinates). Therefore, the measure of any subset like $\mathbf{K}_n \times \mathbf{V}$ (where $\mathbf{V} \subset \mathbf{S}_n$) is equal to $\mathbb{P}(\mathbf{V})$ on the sphere $\mathbf{S}_n$ after normalizing the measure.

We denote by $d\mu$ the microcanonical distribution on $\mathbf{U}_n$. For $\omega \in \mathbf{U}_n$ let $f_n(\omega)$ be the absolute value of the difference between the Fourier transform and the normal distribution at the point ω. For any positive ε

$$\int\limits_{\mathbf{U}_n} f_n(\omega)d\mu(\omega) = \int\limits_{f_n(\omega)\leq\varepsilon} f_n(\omega)d\mu(\omega) + \int\limits_{f_n(\omega)\geq\varepsilon} f_n(\omega)d\mu(\omega) \leq \varepsilon + \frac{C}{\varepsilon^2 n}$$

provided $\mu(f_n(\omega) \geq \varepsilon)$ is bounded by the earlier estimate (1.3) and $f_n(\omega)$ is bounded by 1. The expression $\varepsilon + C(\varepsilon^2 n)^{-1}$ takes its minimal value at

$$\varepsilon = \frac{(2C)^{\frac{1}{3}}}{n^{\frac{1}{3}}}$$

and inserting this ε into the previous expression we obtain

$$\int\limits_{\mathbf{U}_n} f_n(\omega)d\mu(\omega) \leq 3\left(\frac{C}{4}\right)^{\frac{1}{3}}\frac{1}{n^{\frac{1}{3}}}.$$

Thus as n grows the mean value (over phase space) of the variation $f_n(\omega)$ tends to zero.

Let

$$g_n(\omega) = \lim_{T\to\infty}\frac{1}{T}\int\limits_0^T f_n(g^t\omega)dt,$$

where $g^t\omega$ is the result of the phase flow of our mechanical system with initial condition ω. By the individual ergodic theorem† of Birkhoff-Khinchin

$$\int\limits_{\mathbf{U}_n} g_n(\omega)d\mu(\omega) = \int\limits_{\mathbf{U}_n} f_n(\omega)d\mu(\omega)$$

and so

$$\int\limits_{\mathbf{U}_n} g_n(\omega)d\mu(\omega) \leq 3\left(\frac{C}{4}\right)^{\frac{1}{3}}\frac{1}{n^{\frac{1}{3}}}.$$

† The theorem holds for every dynamical system with invariant measure. The word 'ergodic' in its name refers to the nature of its corollary, not its condition.

Introduce the notation $A = 3 \left(\frac{C}{4} \right)^{\frac{1}{3}}$. By Lemma 3.1

$$\int\limits_{g_n(\omega) \geq \sqrt{A} \, n^{-\frac{1}{6}}} d\mu(\omega) \leq \sqrt{A} \, n^{-\frac{1}{6}}. \tag{3.1}$$

Therefore

$$\mu \big(g_n(\omega) \leq \sqrt{A} \, n^{-\frac{1}{6}} \big) > 1 - \sqrt{A} \, n^{-\frac{1}{6}}. \tag{3.2}$$

Define

$$g_n(\omega, T) = \frac{1}{T} \int\limits_0^T f_n(g^t \omega) dt.$$

Using (3.1) and the definition of $g_n(\omega)$ we get the inequality

$$g_n(\omega, T) \leq 2\sqrt{A} \, n^{-\frac{1}{6}}, \qquad \omega \in \big\{ g_n(\omega) \leq \sqrt{A} \, n^{-\frac{1}{6}} \big\}.$$

for sufficiently large T. Consider the segment of time $[0, T]$ with measure dt/T as Ω; then using Lemma 3.1 once again we have

$$\frac{1}{T} \int\limits_{f_n(g^t(\omega)) \geq B \, n^{-\frac{1}{12}}} dt \leq B \, n^{-\frac{1}{12}} \qquad \text{whenever } g_n(\omega) \leq \sqrt{A} \, n^{-\frac{1}{6}},$$

where $B = \sqrt{2} A^{\frac{1}{4}}$.

In other words, for all initial conditions except for a certain subset with measure of order $n^{-\frac{1}{6}}$ (see (3.2)), the variation from the normal distribution has the order of $n^{-\frac{1}{12}}$ for all moments of time except for segments with total relative measure less than $n^{-\frac{1}{12}}$.

Let us formulate the results achieved above as theorems.

Theorem 3.2. *For every $\varepsilon > 0$*

$$\mu \left(f_n \geq \varepsilon \right) \leq \frac{C}{n\varepsilon^2}.$$

Theorem 3.3. *For every $n \in \mathbb{N}$ there exists a set $\mathbf{B}_n \subset \mathbf{U}_n$, such that if $\omega \notin \mathbf{B}_n$ then*

$$f_n(g^t \omega) \leq C_1 n^{-\frac{1}{12}}$$

for every point of time $t \notin b_n^\omega \subset \mathbb{R}$. Here

$$\mu(\mathbf{B}_n) \leq C_2 n^{-\frac{1}{6}} \quad \text{and} \quad m(b_n^\omega) \leq C_3 n^{-\frac{1}{12}}$$

where the 'measure' m on the real axis is defined by

$$m(b) := \limsup_{T \to \infty} \frac{1}{T} \int_{b \cap [0,T]} dt.$$

Note that we did not use ergodicity or any other 'chaotic' assumptions. If the system is ergodic then the estimates may be improved as follows.

Theorem 3.4. *If the energy level is ergodic then for almost every ω the deviation satisfies*

$$f_n(g^t\omega) \le C'_1 n^{-\frac{1}{3}}$$

for every point of time $t \notin b_n^\omega \subset \mathbb{R}$, where

$$m(b_n^\omega) \le C'_3 n^{-\frac{1}{3}}.$$

We shall not give the proof in this article.

Remark. All estimates are essentially the same if, in the definition of $f_n(\omega)$, we work with a finite histogram instead of the Fourier transform: see the remarks at the end of Section 1.

4 Improvements of estimates

Using more advanced techniques, the estimate of Theorem 3.2 can be improved from the order of $(\varepsilon^2 n)^{-1}$ up to $e^{-\frac{\varepsilon^2 n}{\ln n}}$.

Theorem 4.1. *For every $\varepsilon > 0$*

$$\mu\left(f_n \ge \varepsilon\right) \le e^{-C\frac{\varepsilon^2 n}{\ln n}},$$

where C is a positive constant.

In the proof we use a technique developed in the theory of large deviations. For more details see the book [3] of A. Dembo and O. Zeitouni.

Proof of Theorem 4.1. Let

$$S_n = \sum_{k=1}^{n} (\xi_k - \mathbf{M}\xi_1),$$

where $|\xi_k - \mathbf{M}\xi_1| \le 1$.

Using the theory of large deviations one can show that

$$\nu\left(\left|\frac{S_n}{n}\right| \ge \varepsilon\right) \le e^{-\frac{\varepsilon^2 n}{2}},$$

where ν denotes measure with standard normal density in $\mathbb{R}^n$. Taking $\varepsilon = \sqrt{kn^{-1}\ln n}$ this gives

$$\nu\left(|S_n| \geq \sqrt{kn\ln n}\right) \leq n^{-\frac{k}{2}}. \tag{4.1}$$

Let us estimate the mean value of $|S_n|^k$ over the sphere $\mathbf{S}_n$. It is easy to show that

$$(2\pi)^{-\frac{n}{2}} \int_{\mathbb{R}^n} |S_n|^k e^{-\frac{x^2}{2}}\, dx \leq n.$$

Passing to polar coordinates

$$(2\pi)^{-\frac{n}{2}} \int_{\mathbb{R}^n} |S_n|^k e^{-\frac{x^2}{2}}\, dx = \int_0^\infty \frac{2r^{n-1}}{2^{\frac{n}{2}}\Gamma\left(\frac{n}{2}\right)} e^{-\frac{r^2}{2}} \left(\int_{\mathbf{S}_n(r)} |S_n|^k d\mu(\omega)\right) dr.$$

We introduce the notation

$$\Theta(r) = \frac{2r^{n-1}}{2^{\frac{n}{2}}\Gamma\left(\frac{n}{2}\right)} e^{-\frac{r^2}{2}},$$

$$G(r) = \int_{\mathbf{S}_n(r)} |S_n|^k d\mu(\omega),$$

where $\mathbf{S}_n(r)$ is the sphere of radius r ($\mathbf{S}_n(\sqrt{n}) = \mathbf{S}_n$). Then our earlier estimate may be written as

$$\int_0^\infty \Theta(r)G(r)dr \leq n.$$

By direct calculations we get

$$\lim_{n\to\infty} \Theta(\sqrt{n} + \delta) = \frac{e^{-\delta^2}}{\sqrt{\pi}}.$$

so that most of the measure with normal density is concentrated in a small neighbourhood of $\mathbf{S}_n$. We have

$$\int_{\sqrt{n}-1}^{\sqrt{n}+1} \Theta(r)G(r)dr \leq \int_0^\infty \Theta(r)G(r)dr \leq n.$$

Since $|\nabla S_n|^k \leq kn^{k-1}$ and $\Theta(r)$ are separated from zero on the interval $[\sqrt{n} - 1, \sqrt{n} + 1]$, the value $G(\sqrt{n})$ may be estimated using (4.1), and we have

$$G(\sqrt{n}) \leq C_1(kn\ln n)^{\frac{k}{2}}. \tag{4.2}$$

Clearly for any positive λ

$$\mu\left(\frac{S_n}{n} \geq \varepsilon\right) = \mu\left(e^{\lambda S_n} \geq e^{\lambda \varepsilon n}\right) \leq e^{(\Lambda(\lambda) - \varepsilon \lambda)n}$$

whenever

$$\Lambda(\lambda) \geq \frac{1}{n} \ln \mathbf{M}\left(e^{\lambda S_n}\right) .$$

Let us show that we can take Λ to be $C_2 \lambda^2 (\ln n)^{-1}$. Indeed,

$$\mathbf{M}e^{\lambda S_n} = \mathbf{M}\left(1 + \lambda S_n + \frac{1}{2}\lambda^2 S_n^2 + \cdots\right).$$

Using the inequality $\mathbf{M}(S_n^2) \leq C_3 n$ and (4.2) we have

$$\mathbf{M}e^{\lambda S_n} \leq 1 + \frac{1}{2}\lambda^2 C_3 n + C_1 \sum_{k=3}^{\infty} \frac{(\lambda^2 k n \ln n)^{\frac{k}{2}}}{k!}$$

$$\leq 1 + \frac{1}{2}\lambda^2 C_3 n + C_4 \sum_{k=3}^{\infty} \frac{(\lambda^2 n \ln n)^{\frac{k}{2}}}{\Gamma\left(\frac{k}{2}+1\right)} \leq C_5 e^{\lambda^2 n \ln n} .$$

Thus we may take $\Lambda(\lambda) = C_6 \lambda^2 \ln n$, as claimed.

The function $\varepsilon \lambda - C \lambda^2 \ln n$ is maximized at $\lambda_\varepsilon = \varepsilon(2C \ln n)^{-1}$ and the maximum value has the form $C_2 \varepsilon^2 (\ln n)^{-1}$. Hence

$$\mu\left(\frac{S_n}{n} \geq \varepsilon\right) \leq e^{-(\varepsilon \lambda - \Lambda(\lambda))n} \leq e^{-C_2 \frac{\varepsilon^2 n}{\ln n}}.$$

Therefore

$$\mu(f_n \geq \varepsilon) = \mu\left(\frac{S_n}{n} \geq \varepsilon\right) + \mu\left(-\frac{S_n}{n} \geq \varepsilon\right)$$

$$\leq e^{-C_2 \frac{\varepsilon^2 n}{\ln n}} + e^{-C_2 \frac{\varepsilon^2 n}{\ln n}}$$

and the theorem is proved. $\qquad\square$

The new estimate will help us to improve those from Theorems 3.3 and 3.4. The mean value of the deviation in phase space is

$$\int_{\mathbf{U}_n} f_n(\omega) d\mu(\omega) = \int_{f_n(\omega) \leq \frac{\ln n}{\sqrt{n}}} f_n(\omega) d\mu(\omega) + \int_{f_n(\omega) \geq \frac{\ln n}{\sqrt{n}}} f_n(\omega) d\mu(\omega)$$

$$\leq \frac{\ln n}{\sqrt{n}} + C e^{-\frac{(\ln n)^2 n}{n \ln n}}$$

$$= \frac{\ln n}{\sqrt{n}} + \frac{C}{n} \leq 2\frac{\ln n}{\sqrt{n}}.$$

(Note that in the i.i.d. case this deviation is of order $n^{-\frac{1}{2}}$.)

Theorem 4.2. *For every $n \in \mathbb{N}$ there exists a set $\mathbf{B}_n \subset \mathbf{U}_n$, such that if $\omega \notin \mathbf{B}_n$ then*

$$f_n(g^t \omega) \leq C_1 (\ln n)^{\frac{1}{4}} n^{-\frac{1}{8}}$$

for every moment of time $t \notin b_n^\omega \subset \mathbb{R}$, where

$$\mu(\mathbf{B}_n) \leq C_2 (\ln n)^{\frac{1}{2}} n^{-\frac{1}{4}}, \qquad m(b_n^\omega) \leq C_3 (\ln n)^{\frac{1}{4}} n^{-\frac{1}{8}}.$$

Again, this can be improved in the ergodic case:

Theorem 4.3. *For almost every ω, the deviation*

$$f_n(g^t \omega) \leq C_1' n^{-\frac{1}{2}} \ln n$$

for every moment of time $t \notin b_n^\omega \subset \mathbb{R}$, where

$$m(b_n^\omega) \leq C_3' n^{-\frac{1}{2}} \ln n.$$

We omit the proof.

5 Density of particles on configuration space

It is well-known that the density of a gas under potential forces is given by

$$D(x) = \exp\left(\frac{M}{2E} V(x_1)\right).$$

Consider the projection onto configuration space $\mathbf{K}_n$ of the microcanonical distribution on the energy level $\mathbf{U}_n$ of a system of n particles. This projected density is proportional to the 'area' of a sphere with radius $\sqrt{T_n(x)}$, which equals

$$\left(\frac{M}{2n}(v_1^2 + \cdots + v_n^2)\right)^{\frac{n-1}{2}} = (T_n(x))^{\frac{n-1}{2}}$$

where $T_n(x) = E - V_n(x)$ is the kinetic energy at the point $x \in K^n$.

Let us assume the potential energy is defined by the sum

$$V_n(x_1, \ldots, x_n) = \frac{M}{n} \sum_{k=1}^{n} V(x_k).$$

Then the density projected from phase space to configuration space is proportional to

$$\rho_n(x_1, \ldots, x_n) = \left(1 - \frac{M}{En} \sum_{k=1}^{n} V(x_k)\right)^{\frac{n-1}{2}}.$$

According to Lemma 1.1 the quantity converges to $D(x)$ at every point.

The projection to one instance of elementary configuration space K is equal to

$$\rho_0(x_1) = \int_{K^{n-1}} \rho_n(x_1, \ldots, x_n)\, dx_2 \ldots dx_n$$

Therefore

$$\rho_0(x_1) - D(x_1)$$
$$= \int_{K^{n-1}} \rho_n(x_1, \ldots, x_n) - \exp\left(\frac{M}{2E} \sum_{k=1}^{n} V(x_k)\right) dx_2 \ldots dx_n.$$

For simplicity we assume that $\int_K dx = 1$ and the potential function V is bounded in absolute magnitude by some $\Delta > 0$. Then

$$|\rho_0(x_1) - D(x_1)|$$
$$\leq \int_{K^{n-1}} \left|\rho_n(x_1, \ldots, x_n) - \exp\left(\frac{M}{2E} \sum_{k=1}^{n} V(x_k)\right)\right| dx_2 \ldots dx_n \leq \Delta \frac{C}{n}.$$

To obtain theorems which are analogous to those from the previous section, it is sufficient to prove lemmas analogous to Lemmas 1.3 and 1.4. This can be done using the calculations stated above.

Acknowledgements

The author thanks Valery V. Kozlov, Dmitry V. Treschev and Robert S. MacKay for useful discussions and remarks. The work has been supported by the London Mathematical Society, by the RFFI (grants 02–01–00400, 02–01–01059, 99-01-01096, 99-01-00953, 00-15-96146) and by INTAS 00–221.

Bibliography

[1] L. Boltzmann. Weitere Studien über das Wärmegleichgewicht unter Gasmolekülen. *Sitzungsber. Kais. Akad. Wiss. Wien Math. Naturwiss.*, 66:275–370, 1872.

[2] L. Boltzmann. Zu Hrn. Zermelo's Abhandlung über die mechanische Erklärung irreversibler vorgänge. *Ann. Phys.*, 60:392–398, 1897.

[3] A. Dembo and O. Zeitouni. *Large deviations techniques and applications.* Jones and Bartlett Publishers, Boston, MA, 1993.

[4] V. V. Ten. On normal distribution in velocities. *Regul. Chaotic Dyn.*, 7(1):11–20, 2002. **DOI:** 10.1070/RD2002v007n01ABEH000191.

NC KazCosmos

263 Rozybakieva St.

Almaty 480000

Kazakhstan

vladimir.ten@gmail.com